Universitext

Vladimir I. Arnold

Ordinary Differential Equations

Translated from the Russian
by Roger Cooke

With 272 Figures

Springer

Author:
Vladimir I. Arnold
Russian Academy of Sciences
Steklov Mathematical Institute
ul. Gubkina 8
117966 Moscow, Russia
e-mail: arnold@genesis.mi.ras.ru

and

CEREMADE
Université de Paris-Dauphine
Place du Maréchal de Lattre de Tassigny
75775 Paris Cedex 16, France
e-mail: arnold@ceremade.dauphine.fr

Translator:
Roger Cooke
Department of Mathematics
University of Vermont
Burlington, VT 05405
USA

Title of the original Russian edition: *Obyknovennye differentsial'nye uravneniya*, 3rd edition, Publisher Nauka, Moscow 1984

The first printing of this book appeared as V. I. Arnol'd *Ordinary Differential Equations*, Springer Textbook 1992, ISBN 3-540-54813-0

Library of Congress Control Number: 2006927373

Mathematics Subject Classification (1991): 34-01

1st ed. 1992. 2nd printing 2006
ISBN-10 3-540-34563-9 Springer-Verlag Berlin Heidelberg New York
ISBN-13 978-3-540-34563-3 Springer-Verlag Berlin Heidelberg New York

Springer is a part of Springer Science+Business Media
springer.com

Printed in Germany

Cover design: Erich Kirchner, Heidelberg
Typeset by the author using a Springer LaTeX macro package
Production: LE-TeX Jelonek, Schmidt & Vöckler GbR, Leipzig

Printed on acid-free paper 46/3100YL - 5 4 3 2 1 0

Preface to the Third Edition

The first two chapters of this book have been thoroughly revised and significantly expanded. Sections have been added on elementary methods of integration (on homogeneous and inhomogeneous first-order linear equations and on homogeneous and quasi-homogeneous equations), on first-order linear and quasi-linear partial differential equations, on equations not solved for the derivative, and on Sturm's theorems on the zeros of second-order linear equations. Thus the new edition contains all the questions of the current syllabus in the theory of ordinary differential equations.

In discussing special devices for integration the author has tried throughout to lay bare the geometric essence of the methods being studied and to show how these methods work in applications, especially in mechanics. Thus to solve an inhomogeneous linear equation we introduce the delta-function and calculate the retarded Green's function; quasi-homogeneous equations lead to the theory of similarity and the law of universal gravitation, while the theorem on differentiability of the solution with respect to the initial conditions leads to the study of the relative motion of celestial bodies in neighboring orbits.

The author has permitted himself to include some historical digressions in this preface. Differential equations were invented by Newton (1642–1727). Newton considered this invention of his so important that he encoded it as an anagram whose meaning in modern terms can be freely translated as follows: "The laws of nature are expressed by differential equations."

One of Newton's fundamental analytic achievements was the expansion of all functions in power series (the meaning of a second, long anagram of Newton's to the effect that to solve any equation one should substitute the series into the equation and equate coefficients of like powers). Of particular importance here was the discovery of Newton's binomial formula (not with integer exponents, of course, for which the formula was known, for example, to Viète (1540–1603), but – what is particularly important – with fractional and negative exponents). Newton expanded all the elementary functions in "Taylor series" (rational functions, radicals, trigonometric, exponential, and logarithmic functions). This, together with a table of primitives compiled by Newton (which entered the modern textbooks of analysis almost unaltered), enabled him, in his words, to compare the areas of any figures "in half of a quarter of an hour."

Newton pointed out that the coefficients of his series were proportional to the successive derivatives of the function, but did not dwell on this, since he correctly considered that it was more convenient to carry out all the computations in analysis not by repeated differentiation, but by computing the first terms of a series. For Newton the connection between the coefficients of a series and the derivatives was more a means of computing derivatives than a means of constructing the series.

On of Newton's most important achievements is his theory of the solar system expounded in the *Mathematical Principles of Natural Philosophy* (the *Principia*) without using mathematical analysis. It is usually assumed that Newton discovered the law of universal gravitation using his analysis. In fact Newton deserves the credit only for proving that the orbits are ellipses (1680) in a gravitational field subject to the inverse-square law; the actual law of gravitation was shown to Newton by Hooke (1635–1703) (cf. § 8) and seems to have been guessed by several other scholars.

Modern physics begins with Newton's *Principia*. The completion of the formation of analysis as an independent scientific discipline is connected with the name of Leibniz (1646–1716). Another of Leibniz' grand achievements is the broad publicizing of analysis (his first publication is an article in 1684) and the development of its algorithms[1] to complete automatization: he thus discovered a method of teaching how to use analysis (and teaching analysis itself) to people who do not understand it at all – a development that has to be resisted even today.

Among the enormous number of eighteenth-century works on differential equations the works of Euler (1707–1783) and Lagrange (1736–1813) stand out. In these works the theory of small oscillations is first developed, and consequently also the theory of linear systems of differential equations; along the way the fundamental concepts of linear algebra arose (eigenvalues and eigenvectors in the n-dimensional case). The characteristic equation of a linear operator was long called the *secular equation*, since it was from just such an equation that the secular perturbations (long-term, i.e., slow in comparison with the annual motion) of planetary orbits were determined in accordance with Lagrange's theory of small oscillations. After Newton, Laplace and Lagrange and later Gauss (1777–1855) develop also the methods of perturbation theory.

When the unsolvability of algebraic equations in radicals was proved, Liouville (1809–1882) constructed an analogous theory for differential equations, establishing the impossibility of solving a variety of equations (including such classical ones as second-order linear equations) in elementary functions and quadratures. Later S. Lie (1842–1899), analyzing the problem of integration equations in quadratures, discovered the need for a detailed investigation of groups of diffeomorphisms (afterwards known as Lie groups) – thus from the

[1] Incidentally the concept of a matrix, the notation a_{ij}, the beginnings of the theory of determinants and systems of linear equations, and one of the first computing machines, are due to Leibniz.

theory of differential equations arose one of the most fruitful areas of modern mathematics, whose subsequent development was closely connected with completely different questions (Lie algebras had been studied even earlier by Poisson (1781–1840), and especially by Jacobi (1804–1851)).

A new epoch in the development of the theory of differential equations begins with the works of Poincaré (1854–1912), the "qualitative theory of differential equations," created by him, taken together with the theory of functions of a complex variable, lead to the foundation of modern topology. The qualitative theory of differential equations, or, as it is more frequently known nowadays, the theory of dynamical systems, is now the most actively developing area of the theory of differential equations, having the most important applications in physical science. Beginning with the classical works of A. M. Lyapunov (1857–1918) on the theory of stability of motion, Russian mathematicians have taken a large part in the development of this area (we mention the works of A. A. Andronov (1901–1952) on bifurcation theory, A. A. Andronov and L. S. Pontryagin on structural stability, N. M .Krylov (1879–1955) and N. N. Bogolyubov on the theory of averaging, A. N. Kolmogorov on the theory of perturbations of conditionally-periodic motions). A study of the modern achievements, of course, goes beyond the scope of the present book (one can become acquainted with some of them, for example, from the author's books, *Geometrical Methods in the Theory of Ordinary Differential Equations*, Springer-Verlag, New York, 1983; *Mathematical Methods of Classical Mechanics*, Springer-Verlag, New York, 1978; and *Catastrophe Theory*, Springer-Verlag, New York, 1984).

The author is grateful to all the readers of earlier editions, who sent their comments, which the author has tried to take account of in revising the book, and also to D. V. Anosov, whose numerous comments promoted the improvement of the present edition.

V. I. Arnol'd

From the Preface to the First Edition

In selecting the material for this book the author attempted to limit the material to the bare minimum. The heart of the course is occupied by two circles of ideas: the theorem on the rectification of a vector field (equivalent to the usual existence, uniqueness, and differentiability theorems) and the theory of one-parameter groups of linear transformations (i.e., the theory of autonomous linear systems).

The applications of ordinary differential equations in mechanics are studied in more detail than usual. The equation of the pendulum makes its appearance at an early stage,; thereafter efficiency of the concepts introduced is always verified through this example. Thus the law of conservation of energy appears in the section on first integrals, the "small parameter method" is derived from the theorem on differentiation with respect to a parameter, and the theory of linear equations with periodic coefficients leads naturally to the study of the swing ("parametric resonance").

The exposition of many topics in the course differs widely from the traditional exposition. The author has striven throughout to make plain the geometric, qualitative side of the phenomena being studied. In accordance with this principle there are many figures in the book, but not a single complicated formula. On the other hand a whole series of fundamental concepts appears, concepts that remain in the shadows in the traditional coordinate presentation (the phase space and phase flows, smooth manifolds and bundles, vector fields and one-parameter diffeomorphism groups). The course could have been significantly shortened if these concepts had been assumed to be known. Unfortunately at present these topics are not included in courses of analysis or geometry. For that reason the author was obliged to expound them in some detail, assuming no previous knowledge on the part of the reader beyond the standard elementary courses of analysis and linear algebra.

The present book is based on a year-long course of lectures that the author gave to second-year mathematics majors at Moscow University during the years 1968–1970.

In preparing these lectures for publication the author received a great deal of help from R. I. Bogdanov. The author is grateful to him and all the students and colleagues who communicated their comments on the mimeographed text of the lectures (MGU, 1969). The author is also grateful to the reviewers D. V. Anosov and S. G. Krein for their attentive review of the manuscript.

1971 *V. Arnol'd*

Frequently used notation

$\boldsymbol{R}$ – the set (group, field) of real numbers.
$\boldsymbol{C}$ – the set (group, field) of complex numbers.
$\boldsymbol{Z}$ – the set (group, ring) of integers.
$x \in X \subset Y$ – x is an element of the subset X of the set Y.
$X \cap Y$, $X \cup Y$ – the intersection and union of the sets X and Y.
$f : X \to Y$ – f is a mapping of the set X into the set Y.
$x \mapsto y$ – the mapping takes the point x to the point y.
$f \circ g$ – the composite of the mappings (g being applied first).
$\exists$, $\forall$ – there exists, for all.
$*$ – a problem or theorem that is not obligatory (more difficult).
$\boldsymbol{R}^n$ – a vector space of dimension n over the field $\boldsymbol{R}$.

Other structures may be considered in ths set $\boldsymbol{R}^n$ (for example, an affine structure, a Euclidean structure, or the direct product of n lines). Usually this will be noted specifically ("the affine space $\boldsymbol{R}^n$", "the Euclidean space $\boldsymbol{R}^n$", "the coordinate space $\boldsymbol{R}^n$", and so forth).

Elements of a vector space are called *vectors.* Vectors are usually denoted by bold face letters ($\boldsymbol{v}$, $\boldsymbol{\xi}$, and so forth). Vectors of the coordinate space $\boldsymbol{R}^n$ are identified with n-tuples of numbers. We shall write, for example, $\boldsymbol{v} = (v_1, \ldots, v_n) = v_1 \boldsymbol{e}_1 + \cdots + v_n \boldsymbol{e}_n$; the set of n vectors $\boldsymbol{e}_i$ is called a *coordinate basis* in $\boldsymbol{R}^n$.

We shall often encounter functions a real variable t called *time.* The derivative with respect to t is called *velocity* and is usually denoted by a dot over the letter: $\dot{x} = dx/dt$.

Contents

Chapter 1. Basic Concepts

§ 1. Phase Spaces

The theory of ordinary differential equations is one of the basic tools of mathematical science. This theory makes it possible to study all evolutionary processes that possess the properties of *determinacy, finite-dimensionality*, and *differentiability*. Before giving precise mathematical definitions, let us consider several examples.

1. Examples of Evolutionary Processes

A process is called *deterministic* if its entire future course and its entire past are uniquely determined by its state at the present time. The set of all states of the process is called the *phase space*.

Thus, for example, classical mechanics considers the motion of systems whose future and past are uniquely determined by the initial positions and initial velocities of all points of the system. The phase space of a mechanical system is the set whose elements are the sets of positions and velocities of all points of the given system.

The motion of particles in quantum mechanics is not described by a deterministic process. The propagation of heat is a semideterministic process: the future is determined by the present, but the past is not.

A process is called *finite-dimensional* if its phase space is finite-dimensional, i.e., if the number of parameters needed to describe its states is finite. Thus, for example, the Newtonian mechanics of systems consisting of a finite number of material points or rigid bodies belongs to this class. The dimension of the phase space of a system of n material points is $6n$, and that of a system of n rigid bodies is $12n$. The motions of a fluid studied in fluid mechanics, the vibrating processes of a string and a membrane, the propagation of waves in optics and acoustics are examples of processes that cannot be described using a finite-dimensional phase space.

A process is called *differentiable* if its phase space has the structure of a differentiable manifold, and the change of state with time is described by differentiable functions.

Thus, for example, the coordinates and velocities of the points of a mechanical system vary differentiably with time.

The motions studied in impact theory do not possess the property of differentiability.

Thus the motion of a system in classical mechanics can be described using ordinary differential equations, while quantum mechanics, the theory of heat transfer, fluid mechanics, elasticity theory, optics, acoustics, and impact theory require other methods.

We now give two more examples of deterministic finite-dimensional and differentiable processes: the process of radioactive decay and the process of multiplication of bacteria in the presence of a sufficient quantity of nutrients. In both cases the phase space is one-dimensional: the state of the process is determined by the quantity of the substance or bacteria. In both cases the process is described by an ordinary differential equation.

We remark that the form of the differential equation of the process, and also the very fact of determinacy, finite-dimensionality, and differentiability of a given process can be established only by experiment, and consequently only with limited accuracy. In what follows we shall not emphasize this circumstance every time, and we shall talk about real processes as if they coincided exactly with our idealized mathematical models.

2. Phase Spaces

A precise formulation of the general principles discussed above requires rather abstract concepts: *phase space* and *phase flow*. To become familiar with these concepts, let us consider an example where the mere introduction of the phase space makes it possible to solve a difficult problem.

Problem 1. (N. N. Konstantinov). Two nonintersecting roads lead from city A to city B (Fig. 1). It is known that two cars traveling from A to B over different roads and joined by a cord of a certain length less than $2l$ were able to travel from A to B without breaking the cord. Is it possible for two circular wagons of radius l whose centers move over these roads toward each other to pass without touching?

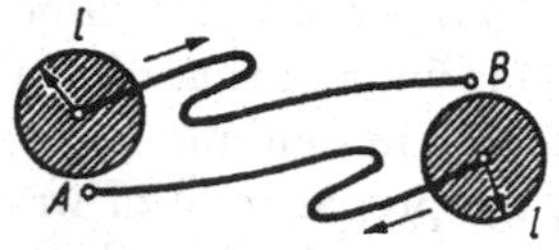

Fig. 1. The initial position of the wagons

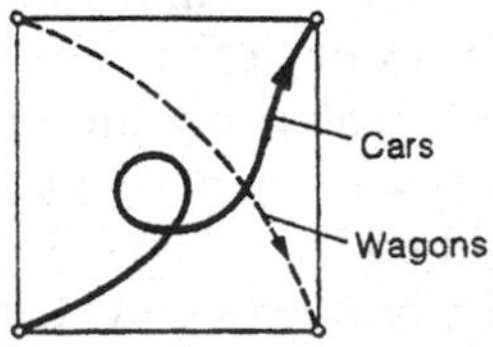

Fig. 2. The phase space of a pair of vehicles

Solution. Consider the square (Fig. 2)

$$M = \{x_1, x_2 : 0 \le x_i \le 1\}.$$

The position of two vehicles (one on the first road, the other on the second) can be indicated by a point of the square M: it suffices to denote by x_i the portion of the distance from A to B on road i contained between A and the vehicle on that road.

The set of all possible positions of the vehicles corresponds to all possible points of the square M. This square is called the ***phase space*** and its points are the ***phase points***. Thus each phase point corresponds to a definite position of the pair of vehicles, and every motion of the vehicles is depicted by the motion of a phase point in the phase space.

For example, the initial position of the cars (in city A) corresponds to the lower left corner of the square ($x_1 = x_2 = 0$), and the motion of the cars from A to B is depicted by a curve leading to the opposite corner.

In exactly the same way the initial position of the wagons corresponds to the lower right corner of the square ($x_1 = 1, x_2 = 0$), and the motion of the wagons is depicted by a curve leading to the opposite corner of the square.

But any two curves in the square joining different pairs of opposite vertices must intersect. Therefore no matter how the wagons move there will come a time at which the pair of wagons occupies a position at which the pair of cars was located at some instant. At that instant the distance between the centers of the wagons will be less than $2l$. Hence they will not be able to pass.

In the example under consideration differential equations played no part but the course of the reasoning is close to that which we shall study below: the description of the states of the process as the points of a suitable phase space often turns out to be extremely useful.

For example, in classical mechanics the state of a process consisting of the motion of a system of n material points is described by the values of the coordinates and velocities of all the material points. Consequently the phase space of such a system has dimension $6n$ (three coordinates and three components of velocity for each material point). The phase space of a system of three points (the sun, Jupiter, and Saturn) is 18-dimensional. The phase space of a system of n rigid bodies has dimension $12n$ (why?).

The motion of the entire system is described by the motion of a point over a curve in the phase space. The velocity of the motion of the phase point over this curve is defined by the point itself. Thus at each point of the phase space a vector is given – it is called the ***phase velocity vector***. The set of all phase velocity vectors forms the phase velocity vector field in the phase space. This vector field defines the differential equation of the process (the dependence of the velocity of the motion of a phase point on its position).

The fundamental problem of the theory of differential equations is to determine or study the motion of the system using the phase velocity vector field. This involves, for example, questions about the form of phase curves (the trajectories of the motion of the phase point): do the phase curves, say, of a given vector field in phase space go to infinity or remain in a bounded region?

In general form this problem does not yield to the methods of modern mathematics and is apparently unsolvable in a certain sense (this applies in particular to the three-body problem mentioned above). In the simplest special cases, with which we shall begin, the problem can be solved explicitly using the operation of integration. Computers make it possible to find approximately the solutions of differential equations on a finite interval of time, but do not answer the qualitative questions about the global behavior of phase curves. In what follows, along with methods for explicitly solving special differential equations, we shall also present some methods for studying them qualitatively.

The concept of a phase space reduces the study of evolutionary processes to geometric problems about curves defined by vector fields. We shall begin our study of differential equations with the following geometric problem.

3. The Integral Curves of a Direction Field

Let us assume that at each point of a certain region of the plane a straight line passing through this point has been chosen. In this case we say that a *direction field* has been defined in the region (Fig. 3).

Remark. Two smooth curves passing through the same point define the same direction at that point if they are tangent. Thus the straight lines in the definition of a direction field can be replaced by arbitrary smooth curves: only the tangent to the curve at the point matters. Figure 3 depicts only a small piece of the line near each point.

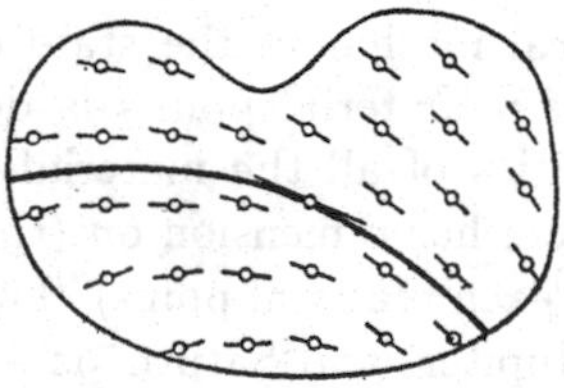

Fig. 3. A direction field and one of its integral curves

Remark. Here and below all the objects encountered (functions, mappings,...) are assumed smooth, i.e., continuously differentiable the necessary number of times, unless otherwise stated. The direction field is called *continuous* (resp. smooth) if the lines of the field depend continuously (resp. smoothly) on the point of attachment.

Remark. A direction field (of lines) in n-dimensional space (and also on any smooth manifold) is defined similarly.

Definition. A line that, at each of its points, is tangent to a vector field is called an *integral curve* of the direction field.

The name "integral curve" is motivated by the fact that in certain cases these curves can be found using the operation of integration.

Example. Assume that a continuous direction field on a plane maps into itself under all translations along a certain line and contains no directions parallel to that line (Fig. 4).

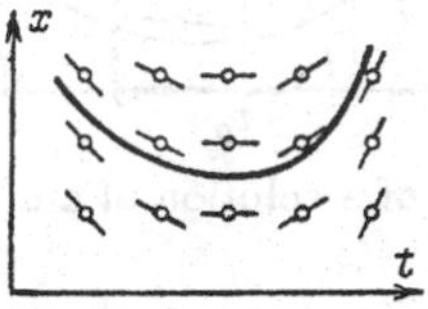

Fig. 4. A field invariant with respect to vertical translations

Theorem. *The problem of finding the integral curves of a field of this type is precisely the problem of integrating a given continuous function.*

Proof. We choose a system of coordinates in which the given line is the vertical ordinate axis, and the axis of abscissas is horizontal. An integral curve of a field without vertical directions is the graph of a function. The derivative of this function is the slope of the graph. The graph is an integral curve if and only if this slope is equal to the slope of the line of the given field. But this last slope is a known function of the abscissa (since the field maps into itself under translations along the axis of ordinates). Consequently a function whose graph is an integral curve has a known function as derivative and hence is a primitive of it, which was to be proved. □

Let us denote the abscissa by the letter t and the ordinate by the letter x. The slope of a line of the field is the known function $v(t)$, and an integral curve is the graph of an unknown function φ. The curve $x = \varphi(t)$ is an integral curve if and only if $\frac{d\varphi}{dt} \equiv v(t)$. By Barrow's theorem[1] $\varphi = \int v\,dt + C$.

In the general case the problem of finding integral curves does not reduce to the operation of integration: even for very simply defined direction fields in the plane the equations of the integral curves cannot be represented by finite combinations of elementary functions and integrals.[2]

4. A Differential Equation and its Solutions

The geometric problem of finding integral curves is written analytically as the problem of finding the solutions of a differential equation. Assume that a field

---

[1] Isaac Barrow (1630–1677), Newton's teacher, who devoted a book to the mutually inverse relation between the tangent and area problems.

[2] Example: Such is the case for a field in which the tangent of the angle between the line attached at the point (t, x) and the x-axis is equal to $x^2 - t$ (Liouville).

in the (t, x)-plane contains no vertical direction (is never parallel to the x-axis (Fig. 5)). Then the slope $v(t, x)$ of the field line attached at the point (t, x) is finite and the integral curves are the graphs of functions $x = \varphi(t)$.

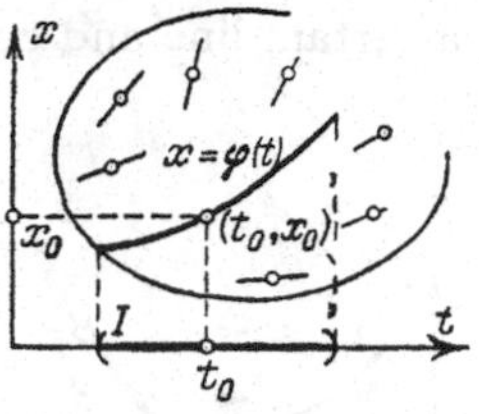

Fig. 5. The graph of a solution of a differential equation

We shall assume that the domain of definition of the function φ is an interval I of the t-axis. The following result is obvious.

Theorem. *A necessary and sufficient condition for the graph of a function φ to be an integral curve is that the following relation hold for all t in I:*

$$\frac{d\varphi}{dt} = v(t, \varphi(t)). \tag{1}$$

Definition. The function φ is called a ***solution*** of the differential equation

$$\dot{x} = v(t, x) \tag{2}$$

if it satisfies relation (1) (i.e., if "the equation becomes an identity when the function φ is substituted for x in the equation").

Definition. The solution φ satisfies the *initial condition* (t_0, x_0) if $\varphi(t_0) = x_0$.

Thus a solution is a function defined on the interval whose graph is an integral curve; the solution satisfies the initial condition (t_0, x_0) if the integral curve passes through the given point (Fig. 5).

Example. The solution of the simplest equation $\dot{x} = v(t)$ with initial condition (t_0, x_0) is given by *Barrow's formula*:

$$\varphi(t) = x_0 + \int_{t_0}^{t} v(\tau)\, d\tau.$$

Every differential equation (2) determines a direction field in the plane: the line attached at the point (t, x) has slope $v(t, x)$. This field is called the *direction field of v* or the *direction field of Eq.* (2) for short.

5. The Evolutionary Equation with a One-dimensional Phase Space

Consider the equation

$$\dot{x} = v(x), \quad x \in \mathbf{R}.$$

This equation describes an evolutionary process with a one-dimensional phase space. The right-hand side defines a *phase velocity vector field*: a vector $v(x)$ is attached at the point x (Fig. 6, left-hand side). Such an equation, whose right-hand side is independent of t, is called *autonomous*. The rate of evolution of an autonomous system, i.e., a system not interacting with other systems, is determined entirely by the state of the system: the laws of nature are time-independent.

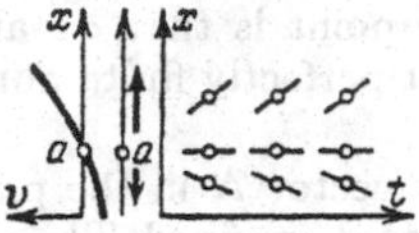

Fig. 6. The vector field and the direction field for the equation $\dot{x} = v(x)$

The points where v vanishes are called *equilibrium positions* (also *stationary points* or *singular points*) of the vector field. If a is an equilibrium position, then $\varphi(t) \equiv a$ is a solution of the equation (a process starting in state a always remains in that state). Figure 6 shows an equilibrium position a. It can be seen that this equilibrium position is unstable: if the initial condition deviates by a small amount from the equilibrium position, the phase point moves away from the equilibrium position as time goes on.

Figure 6 also depicts the direction field for this equation. Since v is independent of t, the field maps into itself under translations along the t-axis.

According to the theorem of Sect. 3, the problem of constructing the integral curves of this equation can be solved by integration alone (in a region where the field is not parallel to the t-axis, i.e., where there are no equilibria, so that $v(x) \neq 0$). Assume that the function $v(x)$ is continuous and never vanishes. We shall write out an explicit formula that determines the integral curves.

The tangent of the angle between our field and the x-axis equals $1/v(x)$. Consequently *the direction field of the equation* $dx/dt = v(x)$ *coincides with the direction field of the equation* $dt/dx = 1/v(x)$. Hence the integral curves of these fields also coincide. But an integral curve of the second equation is given by Barrow's formula; in the present case it has the form

$$t - t_0 = \int_{x_0}^{x} \frac{d\xi}{v(\xi)}. \tag{3}$$

Thus we have proved the following theorem.

Theorem. *The solution $x = \varphi(t)$ of the equation $\dot{x} = v(x)$ with a continuous nonvanishing right-hand side which satisfies the initial condition (t_0, x_0) is given by formula (3). Conversely the function $x = \varphi(t)$ defined by formula (3) is a solution and satisfies this initial condition.*

Remark. The following is a mnemonic device for remembering formula (3). Write the original equation in the form $dx/dt = v(x)$. Even though the student is taught that dx/dt is a single symbol and not a fraction when the derivative is introduced in courses of analysis, we shall operate with this symbol as if it were a fraction and rewrite the equation, gathering all the x's on the left-hand side and all the t's on the right, in the form $dx/v(x) = dt$. Integrating both sides, we obtain the relation $t = \int dx/v(x)$, i.e., (3).

Actually this device is of course more than a mnemonic rule. Leibniz would not have introduced the complicated notation $\frac{dx}{dt}$ if he had not had in mind a real fraction: *dx divided by dt*. The point is that dx and dt are not at all mysterious "infinitely small" quantities, but perfectly finite numbers – functions of a vector, to be more precise.

Consider (Fig. 7) a velocity vector $\mathbf{A}$ in the plane attached at some point in a plane in which the coordinates (t, x) are fixed. The rate of change of the t-coordinate in this motion is a function of this vector. It is linear. It is this linear function of the vector that is denoted dt. For example, the value of this function at the vector $\mathbf{A}$ with components $(10, 20)$ is $dt(\mathbf{A}) = 10$. In exactly the same way one can define $dx(\mathbf{A}) = 20$, the rate of change of the x-coordinate under the motion with velocity vector $\mathbf{A}$, so that $\mathbf{A}$ has components $dt(\mathbf{A}), dx(\mathbf{A})$. The following proposition is obvious.

Proposition 1. *For any vector $\mathbf{A}$ tangent to the graph of a smooth function $x = \varphi(t)$ the ratio $dx(\mathbf{A})/dt(\mathbf{A})$ is equal to the derivative dx/dt of the function φ at the corresponding point.*

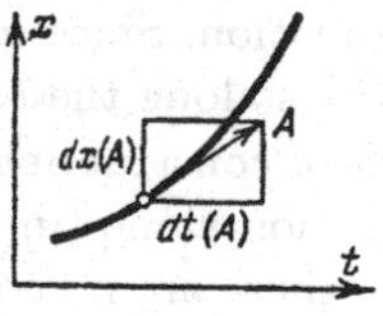

Fig. 7. The numerator and denominator of the fraction dx/dt

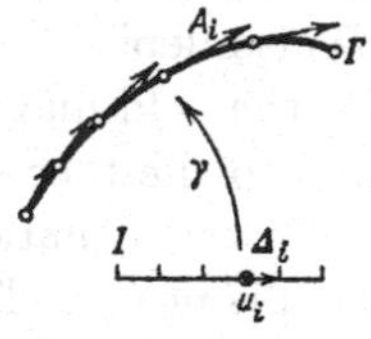

Fig. 8. The definition of the integral of a 1-form

Thus the equation $dx/v(x) = dt$ is *a relation between two linear functions of a vector tangent to an integral curve.*

Functions of the attached vector that are linear at each given point where they are attached are called *differential 1-forms.*

Every differential 1-form in the (t, x)-plane can be written in the form $\omega = a\,dt + b\,dx$, where a and b are functions in the plane.

Differential forms can be integrated along oriented closed segments of curves. On a segment Γ of a curve in the plane we choose an orienting parameter u, i.e., we represent Γ as the image of a smooth mapping $\gamma : I \to \mathbf{R}^2$ from a segment of the u-axis into the plane (Fig. 8). The integral of the form ω along Γ is defined to be the number

$$\int_{\Gamma} \omega = \int_{I} \omega(\gamma')\, du, \quad \text{where } \gamma' = d\gamma/du.$$

In other words the integral is the limit of the integral sums $\sum \omega(\mathbf{A}_i)$, where $\mathbf{A}_i = \gamma'(u_i)\Delta_i$; here u_i are points that divide the interval I into intervals of length $\Delta_i = u_{i+1} - u_i$. The vector $\mathbf{A}_i$ is tangent to Γ and differs from the chord joining the successive points of division on Γ only by infinitesimals of higher order with respect to Δ_i (Fig. 8).

The following proposition is a consequence of the theorem on the change of variable in a definite integral.[3]

Proposition 2. *The integral of a 1-form over an oriented closed segment of a curve is independent of the parameter for parameters giving the same orientation. (When the orientation is reversed, the integral changes sign.)*

The following proposition is also obvious.

Proposition 3. *The integral of a 1-form $f(x)\,dx$ over a segment of a curve on which x can be taken as parameter coincides with the usual definite integral of the function f.*

Let us now return to the proof of formula (3).

The values of the two differential forms $dx/v(x)$ and dt coincide on vectors tangent to an integral curve. Hence their integrals over a segment of the curve are equal. According to Proposition 3 the integral of the first form is equal to the right-hand side and that of the second is equal to the left-hand side of formula (3).

6. Example: The Equation of Normal Reproduction

Assume that the size of a biological population (for example the number of bacteria in a Petri dish or fish in a pond) is x and that the rate of reproduction is proportional to the number of organisms present. (This assumption holds approximately while there is sufficient food.)

Our assumption is expressed by the *differential equation of normal reproduction*

$$\dot{x} = kx, \quad k > 0.$$

From the conditions of the problem $x > 0$, so that the direction field is defined on a half-plane; it is depicted in Fig. 9. From the form of the direction field it is clear that x increases with t, but is not clear whether infinite values of x will be reached in finite time (whether an integral curve will have a vertical asymptote) or whether the curve will remain finite for all t. Along with the future the past is also unclear: will an integral curve tend to the axis $x = 0$ as t tends to a finite or infinite negative limit?

Fortunately the reproduction equation can be solved explicitly by the preceding theorem: according to formula (3),

[3] It was in solving the simplest differential equations, those now known as equations with separable variables, that Barrow came to discover this theorem.

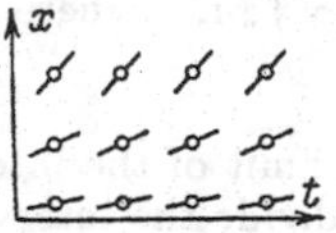

Fig. 9. The reproduction equation $\dot{x} = kx$

$$t - t_0 = \int_{x_0}^{x} \frac{d\xi}{k\xi}, \quad k(t - t_0) = \ln(x/x_0), \quad x = e^{k(t-t_0)}x_0.$$

Consequently the solutions of the normal reproduction equation increase exponentially as $t \to +\infty$ and decrease exponentially as $t \to -\infty$; *neither infinite nor zero values of x are attained for finite values of t.* Thus according to the equation the same time period is always required to double the population, regardless of its size (the doubling period of the earth's population is of the order of 40 years at present). Up to the middle of the twentieth century science also increased exponentially (Fig. 10).

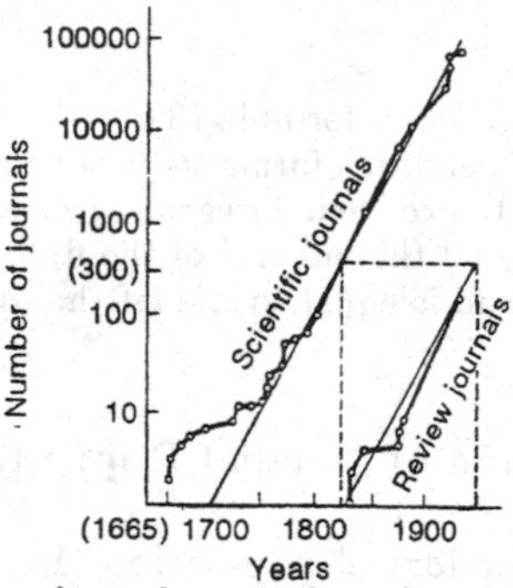

Fig. 10. Growth of the number of research and review journals (according to the book of V. V. Nalimov and A. M. Mul'chenko *Scientometry* (Nauka, Moscow 1969 [Russian])

The same differential equation with negative k describes radioactive decay. To reduce the amount of a radioactive substance by one-half requires time $T = k^{-1}\ln 2$, regardless of the initial amount of the substance. This time is called the *half-life*. The half-life of the widely known isotope radium-226 is 1620 years, and that of the commonest isotope of uranium (U-238) is 4.5×10^9 years.

The same equation is also encountered in a large number of other problems. (We shall see below that this is not coincidental, but a manifestation of a natural law according to which "every" function is approximately linear locally.)

Problem 1. At what altitude is the density of the air one-half of that at the surface of the earth? Regard the temperature as constant. One cubic meter of air at the surface of the earth weighs 1250g.

Answer. $8 \ln 2\,\text{km} \approx 5.6\,\text{km}$ – the height of Mt. Elbrus.

7. Example: The Explosion Equation

Now let us assume that the rate of reproduction is proportional not to the number of individuals in the population but to the number of pairs of individuals:

$$\dot{x} = kx^2. \tag{4}$$

In this case for large x reproduction takes place much faster than in the case of normal reproduction, and for small x it goes much more slowly (this situation occurs more often in problems of physical chemistry, where the rate of a reaction is proportional to the concentration of the two reagents; however at present whales of certain species are having such difficulty finding a mate that the reproductive rate of the whale population is subject to Eq. (4), and with small x).

The direction field appears to differ little from that of the ordinary reproduction equation (Fig. 9), but computations show that the integral curves behave quite differently. Suppose for simplicity that $k = 1$. By Barrow's formula we find the solution $t = \int \frac{dx}{x^2} + C$, i.e., $x = -\frac{1}{t - C}$ for $t < C$. An integral curve is one branch of a hyperbola (Fig. 11). The hyperbola has a vertical asymptote.

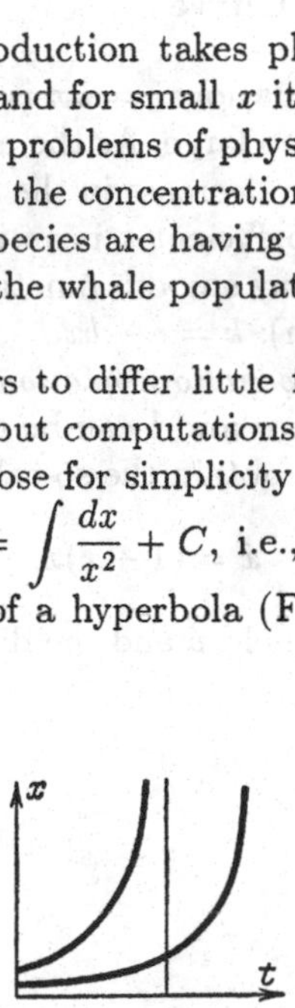

Fig. 11. The explosion equation $\dot{x} = x^2$

Thus *if the growth of the population is proportional to the number of pairs of individuals in the population, then the size of the population becomes infinitely large in a finite time.* Physically this conclusion corresponds to the explosive nature of the process. (Of course for t too close to C the idealization adopted in describing the process by a differential equation is inapplicable, so that the actual size of the population does not attain infinite values in finite time.)

It is interesting to observe that the other branch of the hyperbola $x = (C - t)^{-1}$ is also an integral curve of our equation (if we extend it from the semiaxis $x > 0$ to the entire x axis). The solutions corresponding to the two branches of the hyperbola are both given by the same formula, but have no connection with each other. The connection between these solutions is recovered if we consider time to be a complex variable or if we compactify the affine x-axis to form the projective line (cf. Chapt. 5).

Problem 1. [4] Which of the differential equations $\dot{x} = x^n$ determine on an affine line a phase velocity field that can be extended without singular points to the projective line?

Answer. $n = 0, 1, 2$.

8. Example: The Logistic Curve

The ordinary reproduction equation $\dot{x} = kx$ is applicable only as long as the number of individuals is not too large. As the number of individuals increases competition for food leads to a decrease in the rate of reproduction. The simplest hypothesis is that the coefficient k is an inhomogeneous linear function of x (when x is not too large any smooth function can be approximated by an inhomogeneous linear function): $k = a - bx$.

We thus arrive at the *reproduction equation taking account of competition* $\dot{x} = (a - bx)x$. The coefficients a and b can be taken as 1 by a change of scale on the t- and x-axes. We thus obtain the so-called *logistic equation*

$$\dot{x} = (1 - x)x.$$

The phase velocity vector field v and the direction field in the (t, x)-plane are depicted in Fig. 12.

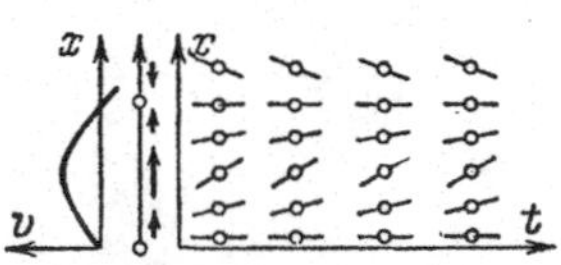

Fig. 12. The vector field and the direction field of the equation $\dot{x} = (1 - x)x$

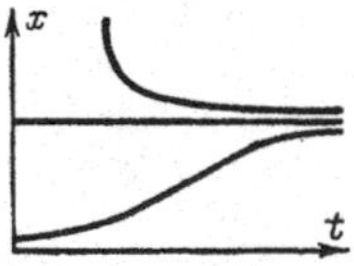

Fig. 13. The integral curves of the equation $\dot{x} = (1 - x)x$

We conclude from this that the integral curves look as depicted in Fig. 13. More precisely, we see that

1) the process has two equilibrium positions $x = 0$ and $x = 1$;

2) between the points 0 and 1 the field is directed from 0 to 1, and for $x > 1$ to the point 1.

Thus the equilibrium position 0 is unstable (as soon as a population arises it begins to grow), while the equilibrium position 1 is stable (a smaller population increases, and a larger one decreases).

[4] Here and in the sequel problems marked with an asterisk are more difficult than the others.

For any initial state $x > 0$, as time passes the process moves toward the stable equilibrium state $x = 1$.

It is not clear from these considerations, however, whether this passage takes place in a finite or infinite time, i.e., whether integral curves starting in the region $0 < x < 1$ can have points in common with the line $x = 1$.

It can be shown that there are no such common points and that these integral curves tend asymptotically to the line $x = 1$ as $t \to +\infty$ and to the line $x = 0$ as $t \to -\infty$. These curves are called *logistic curves*. Thus a logistic curve has two horizontal asymptotes ($x = 0$ and $x = 1$) and describes the passage from one state (0) to another (1) in an infinite time.

Problem 1. Find the equation of a logistic curve.

Solution. By formula (3) $t = \int dx/(x(1-x)) = \ln(x/1-x)$, or $x = e^t/(1+e^t)$. This formula proves the asymptotic property of the logistic curve mentioned above.

Problem 2. Prove that the integral curves of the equation $\dot{x} = (1-x)x$ in the region $x > 1$ tend asymptotically to the line $x = 1$ as $t \to +\infty$ and have the vertical asymptote $t = \text{const}$.

For small x the logistic curve is practically indistinguishable from the exponential curve, i.e., competition has little influence on reproduction. However, as x increases the reproduction becomes nonexponential, and near $x = 1/2$ the exponential curve diverges sharply upward from the logistic curve; subsequently logistic growth describes the saturation of the system, i.e., the establishment of an equilibrium mode in it ($x = 1$).

Up to the middle of the twentieth century science grew exponentially (cf. Fig. 10). If such growth were to continue, the entire population of the earth would consist of scientists by the end of the twenty-first century and there would not be enough forests on the earth to print all the scientific journals. Consequently saturation must set in before that point: we are nearing the point where the logistic curve begins to lag behind the exponential curve. For example, the number of mathematical journal articles increased at a rate of 7% per year from the end of the Second World War until the 1970's but the growth has been slower for the past several years.

9. Example: Harvest Quotas

Up to now we have considered a free population developing according to its own inner laws. Assume now that we harvest a part of the population (for example, we catch fish in a pond or in the ocean). Let us assume that the rate of harvesting is constant. We then arrive at the differential equation for harvesting

$$\dot{x} = (1-x)x - c.$$

The quantity c characterizes the rate of harvesting and is called the *quota*. The form of the vector field and the phase velocity field under different values of the harvest rate c is shown in Fig. 14.

We see that for a harvesting rate that is not too large ($0 < c < 1/4$) there exist two equilibrium positions (A and B in Fig. 14). The lower equilibrium position $x = A$) is unstable. If for any reason (overharvesting or disease) the

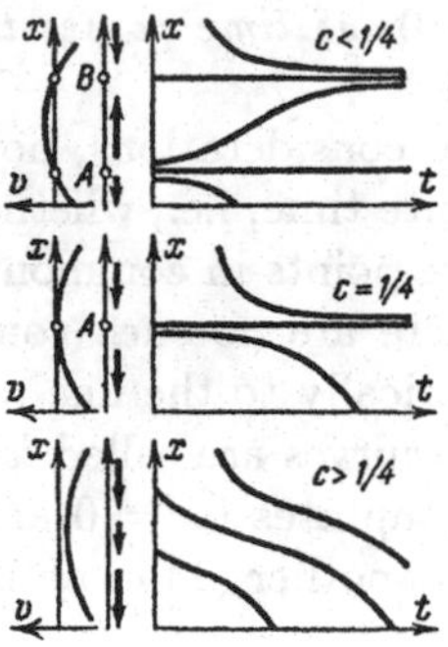

Fig. 14. The harvest equation $\dot{x} = (1 - x)x - c$

size of the population x drops below A, the population will subsequently die out in a finite time.

The upper equilibrium position B is stable – this is the steady state toward which the population converges under constant harvest c.

If $c > 1/4$ there is no equilibrium, and the entire population will be harvested in a finite time (for example, Steller's sea-cow).

For $c = 1/4$ there is one unstable equilibrium state ($A = B = 1/2$). It is mathematically possible to continue harvesting indefinitely at such a rate if the initial population is sufficiently large. However, an arbitrarily small fluctuation in the established equilibrium size of the population will lead to a complete harvest of the population in a finite time.

Thus, although any quotas up to the maximal ($c \leq 1/4$) are theoretically admissible, *the maximal quota* $c = 1/4$ *leads to instability and is therefore inadmissible.* Moreover *as a practical matter quotas close to* $1/4$ *are also inadmissible* since with these values the threshold of danger A is close to the steady state B (small random fluctuations throw the population below the threshold A, after which it perishes.)

It turns out, however, that the harvest can be organized so as to obtain in a stable manner a harvest at the rate $1/4$ for 1 unit of time (more than this cannot be achieved since $1/4$ is the maximal reproductive rate of the unharvested population).

10. Example: Harvesting with a Relative Quota

Instead of an absolute rate of harvest, let us fix a relative rate, i.e., let us fix the fraction of the population to be harvested in unit time: $\dot{x} = (1-x)x - px$. The form of the vector field and the integral curves (for $p < 1$) is depicted in Fig. 15.

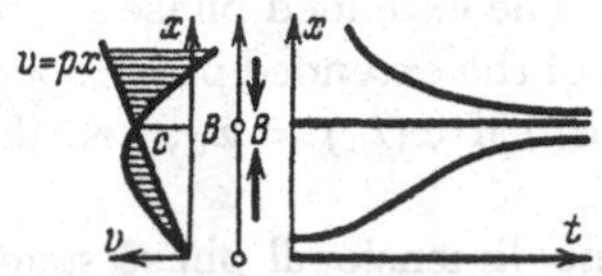

Fig. 15. The harvest equation $\dot{x} = (1 - x)x - px$

The lower, unstable equilibrium point is now at the point $x = 0$, and the second equilibrium position B is now stable for any p, $0 < p < 1$.

After a certain period of adjustment the population converges to the steady state $x = B$. When this happens, the absolute rate of harvest becomes established at the value $c = pB$. This is the ordinate of the point of intersection of the graphs of the functions $v = (1 - x)x$ and $v = px$ (Fig. 15, left). Let us study the behavior of the quantity c as p varies. Under small relative harvests (small p) the steady rate of harvest is also small; as $p \to 1$ it also tends to 0 (overharvesting). The highest absolute value of the rate c is the largest ordinate of the graph of the function $v = (1 - x)x$. It is attained when the line $v = px$ passes through the vertex of the parabola (i.e., when $p = 1/2$), and is equal to $c = 1/4$.

Let us take $p = 1/2$ (i.e., assign the relative quota so that the steady population will be half of the unharvested population). We will then have achieved the maximum possible steady harvest rate $c = 1/4$, and the system will remain stable (return to its steady state under small fluctuations of the initial population from the steady population).

11. Equations with a Multidimensional Phase Space

In the examples considered above the phase space was one-dimensional. In more complicated cases (for example, taking account of the interaction between several populations) a point of the phase space is determined by several numbers (two for two populations, etc.). The definitions of a differential equation, solutions, etc., in this case are analogous to those given above. We shall repeat these definitions.

Let $\boldsymbol{v}$ be a vector field in a region U of an n-dimensional phase space. The *autonomous differential equation* defined by the field $\boldsymbol{v}$ is the equation

$$\dot{\boldsymbol{x}} = \boldsymbol{v}(\boldsymbol{x}), \quad \boldsymbol{x} \in U \subset \mathbf{R}^n.$$

A *solution* of this equation is a smooth mapping $\varphi : I \to U$ of an interval of the time axis into the phase space for which $d\varphi/dt = v(\varphi(t))$ for all t in I.

The image of the mapping φ is called a *phase curve* and the graph[5] of the mapping φ is called an *integral curve.* The integral curve lies in the direct

[5] The *graph* of a mapping $f : X \to Y$ is the subset of the direct product $X \times Y$ consisting of all pairs of the form $(x, f(x))$, where $x \in X$; the *direct product* $X \times Y$ is the set of all ordered pairs (x, y), where $x \in X$ and $y \in Y$.

product of the time axis and the phase space. This direct product is called the *extended phase space*. The extended phase space has dimension $n+1$.

Let $(t_0, \boldsymbol{x}_0)$ be a point of the extended phase space. The solution φ satisfies the *initial condition* $(t_0, \boldsymbol{x}_0)$ if $\varphi(t_0) = \boldsymbol{x}_0$, i.e., if the integral curve passes through the point $(t_0, \boldsymbol{x}_0)$.

As in the case of a one-dimensional phase space, the integral curves can be described using a direction field in the extended phase space. The slope is replaced by the following construction.

Assume that a direction field is given in the region V of the direct product $\mathbf{R} \times \mathbf{R}^n$ and that the direction of the field is nowhere vertical (parallel to $\mathbf{R}^n$). Let t be a coordinate in $\mathbf{R}$, and let $\boldsymbol{x} = (x_1, \ldots, x_n)$ be coordinates in $\mathbf{R}^n$. Then at each point there exists (a unique) vector of the direction attached at this point having horizontal coordinate (t-component) equal to 1. Thus this vector has the form $(1, \boldsymbol{v}(t, \boldsymbol{x}))$, where $\boldsymbol{v}(t, \boldsymbol{x})$ is a vector in $\mathbf{R}^n$ depending on a point of the extended phase space. In other words a nonvertical direction field in the extended phase space determines a time-dependent vector field in the phase space.

Each integral curve of the given direction field obviously satisfies the differential equation

$$\dot{\boldsymbol{x}} = \boldsymbol{v}(t, \boldsymbol{x}),$$

i.e., is the graph of the mapping φ of an interval of the time axis into the phase space for which $d\varphi/dt = \boldsymbol{v}(t, \varphi(t))$ for all t. Conversely the graph of any solution is an integral curve of this field.

The solution satisfies the initial condition $(t_0, \boldsymbol{x}_0)$ if and only if the integral curve passes through this point.

Remark. In coordinate notation a vector field in n-dimensional space is defined by n functions of n variables. Our differential equation therefore assumes the form of a "system of n first-order equations":

$$\dot{x}_1 = v_1(t; x_1, \ldots x_n), \ldots, \quad \dot{x}_n = v_n(t; x_1, \ldots, x_n).$$

A solution is defined by a vector-valued function $(\varphi_1, \ldots, \varphi_n)$ of the variable t for which $d\varphi_k/dt = v_k(t; \varphi_1(t), \ldots, \varphi_n(t))$, $k = 1, \ldots, n$, for all t. An initial condition is given by $n+1$ numbers $(t_0; x_{1,0}, \ldots x_{n,0})$.

12. Example: The Differential Equation of a Predator-Prey System

The simplest and crudest model describing the struggle of two species – the predator-prey model – consists of the following. Consider a pond in which two species of fish live, say carp and pike. If there were no pike, the carp would multiply exponentially at a rate proportional to their numbers x, say $\dot{x} = kx$ (we assume that the total mass of the carp is much less than the mass of the pond). If y is the number of pike, one must take account of the number of carp eaten by the pike. We shall assume that the number of encounters of carp with pike is jointly proportional to the numbers of carp and pike; then for the rate of fluctuation in the number of carp, we obtain the equation $\dot{x} = kx - axy$.

As for the pike, they die out in the absence of carp: $\dot{y} = -ly$, while in the presence of carp they begin to increase at a rate proportional to the number of carp eaten: $\dot{y} = -ly + bxy$.

We thus arrive at a system of differential equations for the simplest model of a predator-prey system:

$$\begin{cases} \dot{x} = kx - axy, \\ \dot{y} = -ly + bxy. \end{cases}$$

This model is called the *Lotka-Volterra* model, after its creators. The right-hand side defines a vector field in the plane: the vector attached at the point (x, y) has components $(kx - axy, -ly + bxy)$. This is a phase velocity field.

The phase space is the sector $x \geq 0$, $y \geq 0$.

The phase velocity vector field is not difficult to sketch by following the changes of sign of the two components (Fig. 16). The critical point ($x_0 = l/b$, $y_0 = k/a$) corresponds to an equilibrium number of carp and pike, when the increase in the carp population is balanced by the activity of the pike and the increase in the pike is balanced by their natural mortality.

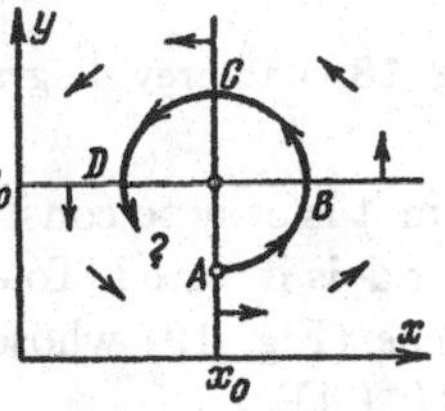

Fig. 16. The phase velocity field of the predator-prey model

If the initial number of pike is less than y_0 (the point A in the figure), then the numbers of carp and pike both increase until the increasing numbers of pike begin to eat carp faster than the latter can increase (point B). At that point the number of carp begins to decrease while the number of pike continues to increase until the shortage of food causes the pike also to begin to die out (point C). The number of pike will then decrease so much that the carp will again begin to multiply (point D). This new increase in the number of carp will lead to an increase in the number of pike also after a time. In this way the populations of carp and pike will oscillate about their equilibrium numbers.

The question arises, however, whether these oscillations are periodic or not. Our approximate picture of the phase velocity field does not provide an answer to this question, and different cases can be imagined, as depicted in Fig. 17, for example.

In order to investigate these cases, consider the line segment joining the critical point to the x-axis. Each point A of this line segment (except the point on the x-axis) determines a phase curve that again intersects the line segment

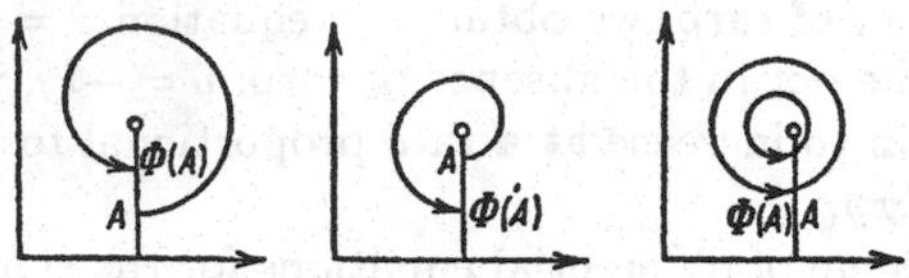

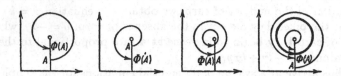

Fig. 17. The first return function

in some point $\Phi(A)$. The function Φ is called the *first return function* (or the *Poincaré mapping*, as well as the *monodromy* or *holonomy*).

Consider the graph of the first return function. It is called the *Lamerey diagram*. The Lamerey diagrams for the four cases of Fig. 17 are depicted in Fig. 18.

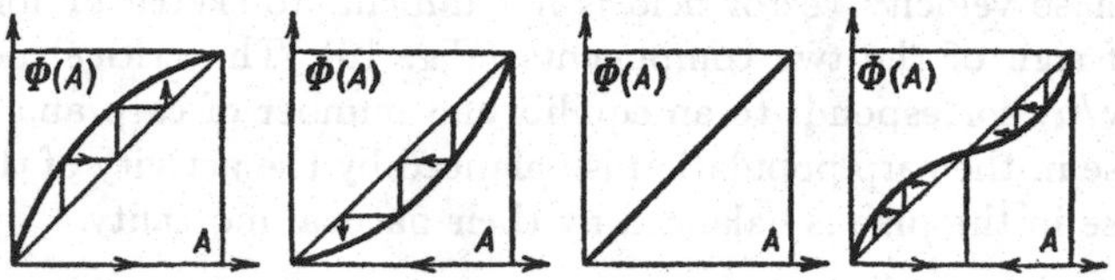

Fig. 18. Lamerey diagrams

From the Lamerey diagram it is easy to construct the sequence of images of the point A when the mapping Φ is iterated. To do this one needs to construct the so-called *Lamerey staircase* (Fig. 19) whose vertices have abscissas and ordinates $A, \Phi(A), \Phi^2(A) = \Phi(\Phi(A)), \ldots$.

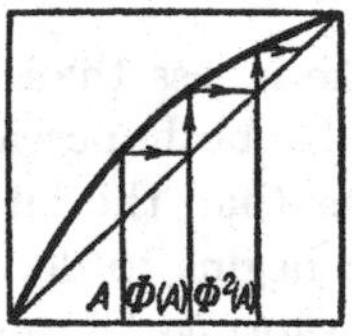

Fig. 19. The Lamerey staircase

The points of intersection of the graph of the return function with the diagonal (the graph of $\Phi \equiv A$) correspond to closed phase curves (*cycles*) in the phase plane.

A cycle is demonstrably stable (resp. unstable) if at the corresponding point A we have $\Phi'(A) < 1$ (resp. $\Phi'(A) > 1$). As for our four Lamerey diagrams (Fig. 18), in the first case the curves are spirals winding in toward the critical point; in the second case they wind outward from this point; in the third case they are closed. In the fourth case the phase curves wind toward a stable cycle from both within and without.

Accordingly in the first case an equilibrium in the population of the pond becomes established as time passes and the oscillations die out. In the second case the equilibrium is unstable, and the oscillations increase. In this case there will come a time when the number of carp (or pike) is less than 1; before that time our model will become inapplicable, and *the population of the pond will become extinct.*

In the third case periodic oscillations about the equilibrium state will be observed in the numbers of carp and pike; the amplitude of the oscillations is determined by the initial conditions.

In the fourth case periodic oscillations in the numbers of carp and pike are also observed, *but the amplitude of the steady oscillations is independent of the initial conditions*: any phase spiral winds toward a limit cycle. In this case it is said that *self-sustaining oscillations* are established in the system.

Which of these cases holds for the Lotka-Volterra system? We cannot yet answer this question (for the solution see § 2).

13. Example: A Free Particle on a Line

According to Newton's First Law the acceleration of a material point not subject to any external force is zero: $\ddot{x} = 0$. If the point x belongs to $\mathbf{R}$, we speak of a *free particle on the line* (one may imagine a bead on a wire).

The phase space has dimension 2 since the whole motion is determined by the initial position and the initial velocity. On the phase space with coordinates $x_1 = x$ and $x_2 = \dot{x}$ a phase velocity vector field arises:

$$\dot{x}_1 = x_2, \quad \dot{x}_2 = 0,$$

and consequently the components of the field are equal to $(x_2, 0)$ (Fig. 20).

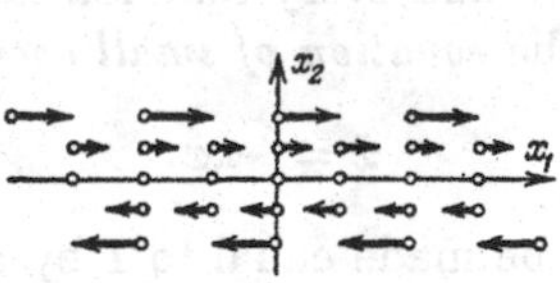

Fig. 20. The phase velocity field of a free particle

All the points of the x_1-axis are equilibrium positions. This kind of equilibrium is called *neutral equilibrium* in physics and *unstable equilibrium* in mathematics (a suitable arbitrarily small change in the initial phase point causes a change in the position that is not small after a sufficiently long time).

The phase curves are the horizontal lines $x_2 = \text{const}$ and all the points of the x_1-axis.

Problem 1. Find the solution with initial condition (a, b) for $t_0 = 0$.

Solution . $\varphi_1(t) = a + bt$, $\varphi_2(t) = b$.

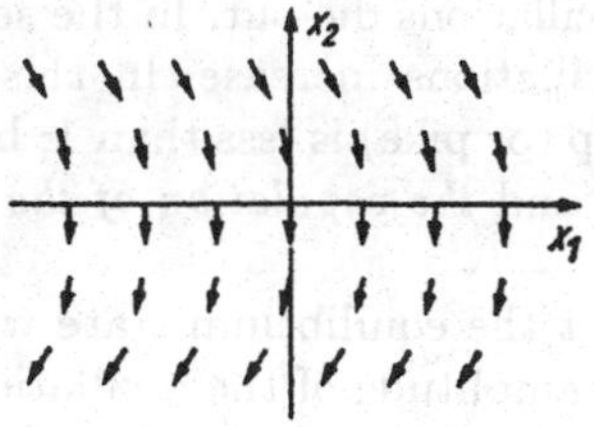

Fig. 21. The phase velocity field for a falling particle

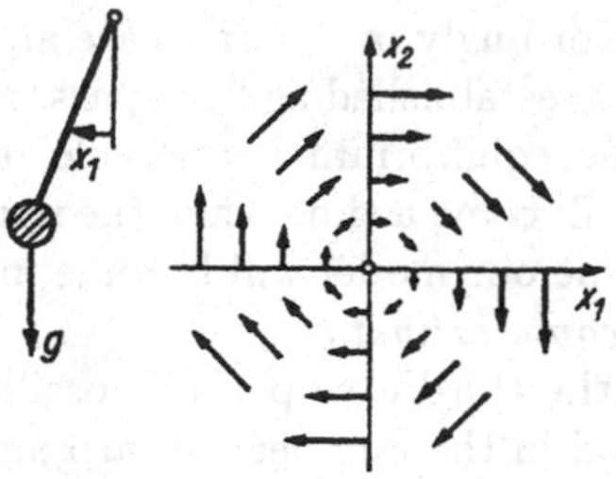

Fig. 22. The phase velocity field for small oscillations

14. Example: Free Fall

According to Galileo the acceleration g of bodies falling near the surface of the earth is constant. If x is the height, then $\ddot{x} = -g$. Introducing coordinates on the phase plane, as in the preceding example, we obtain the system

$$\dot{x}_1 = x_2, \quad \dot{x}_2 = -g.$$

The vector field defined by the right-hand side is depicted in Fig. 21.

Problem 1. Prove that the phase curves are parabolas.

15. Example: Small Oscillations

In many cases the force that restores a system to its equilibrium position is more or less proportional to the displacement from the equilibrium position (Hooke's law, and the like; the essence of the matter is that the force is zero in the equilibrium position, and every function is linear on the infinitesimal level). We thus arrive at the *equation of small oscillations*:

$$\ddot{x} = -kx.$$

The coefficient $k > 0$ can be made equal to 1 by a choice of time scale. The equation then assumes the form

$$\ddot{x} = -x.$$

Introducing the coordinates $x_1 = x$ and $x_2 = \dot{x}$ on the phase plane as before, we rewrite this equation as the system

$$\dot{x}_1 = x_2, \quad \dot{x}_2 = -x_1.$$

The right-hand sides of these equations define a vector field on the phase plane. This field is depicted in Fig. 22.

Problem 1. Prove that the phase curves are a set of concentric circles and their common center.

Solution. The phase velocity vector is perpendicular to the radius-vector.

Problem 2. Prove that the phase point moves along a circle with constant angular velocity equal to 1.

Solution. The length of the phase velocity vector equals the length of the radius vector.

Problem 3. Find the solution with the initial condition $x(0) = a$, $\dot{x}(0) = b$.

Solution. According to the two preceding problems the vector of the initial condition must be rotated by an angle t. We obtain

$$x_1(t) = a\cos t + b\sin t, \quad x_2(t) = -a\sin t + b\cos t.$$

Remark. Thus we have proved that x executes harmonic oscillations and we have proved the "law of conservation of energy": the quantity $x_1^2/2 + x_2^2/2$ is constant along a phase curve.

Problem 4. Prove the law of conservation of energy $x_2^2/2 + kx_1^2/2$ for the system $\dot{x}_1 = x_2$, $\dot{x}_2 = -kx_1$.

Remark. The quantity $x_2^2/2$ is called the ***kinetic energy***, and $kx_1^2/2$ is called the ***potential energy***.

Problem 5. Prove that the integral curves of the system (with $k = 1$) are helices.

16. Example: The Mathematical Pendulum

Consider a weightless rod of length l attached at one end and bearing a point mass m at the other end. We denote by θ the angle by which this pendulum deviates from the vertical. According to the laws of mechanics, the angular acceleration $\ddot{\theta}$ of the pendulum is proportional to the torque of the weight (Fig. 23):

$$I\ddot{\theta} = -mgl\sin\theta,$$

where $I = ml^2$ is the moment of inertia (the sign is negative because the torque tends to decrease the deviation).

Thus the equation of the pendulum has the form $\ddot{\theta} = -k\sin\theta$, $k = g/l$. The coefficient k can be made equal to 1 by a choice of the time scale. The equation then assumes the form $\ddot{\theta} = -\sin\theta$.

The phase space has dimension 2. The coordinates can be taken as the displacement angle $x_1 = \theta$ and the angular velocity $x_2 = \dot{\theta}$. The equation assumes the form of the system

$$\dot{x}_1 = x_2, \quad \dot{x}_2 = -\sin x_1.$$

Fig. 23. The mathematical pendulum

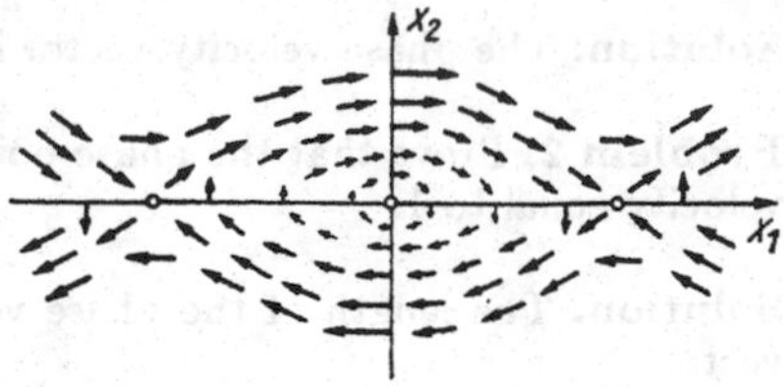

Fig. 24. The phase velocity field of the pendulum

The right-hand side defines a phase velocity vector field. It is depicted in Fig. 24.

Problem 1. Prove that the origin ($x_1 = x_2 = 0$) and the point ($x_1 = \pi$, $x_2 = 0$) are phase curves.

We shall make a detailed study of the form of the other phase curves below (§ 12).

Remark. For infinitesimal angles of deviation $\sin\theta$ is equivalent to the angle θ. Replacing $\sin\theta$ by the approximate value θ, we reduce the equation of the pendulum to the equation of small oscillations (Sect. 15). The question as to the extent to which conclusions drawn from the study of this very simple equation carry over to the full pendulum equation requires a special investigation, which we shall carry out below (§ 12).

17. Example: The Inverted Pendulum

Consider the behavior of a pendulum turned upside down. In this case the angle θ is close to π, and so it is natural to introduce the angle of deviation from the upper position $\psi = \theta - \pi$. Then $\ddot{\psi} = \sin\psi$, and for small ψ approximately

$$\ddot{\psi} = \psi.$$

This equation is called the *equation of "small oscillations" of the inverted pendulum.* The phase space is two-dimensional. For coordinates we shall take $x_1 = \psi$ and $x_2 = \dot{\psi}$. We then obtain the system

$$\dot{x}_1 = x_2, \quad \dot{x}_2 = x_1.$$

The phase velocity vector field is depicted in Fig. 25. We shall study its phase curves in detail in § 2.

18. Example: Small Oscillations of a Spherical Pendulum

The deviation from the vertical is characterized by two numbers x and y.

It is known from mechanics that the equations of small oscillations have the form

$$\ddot{x} = -x, \quad \ddot{y} = -y.$$

The phase space is 4-dimensional. As coordinates in the phase space we take $x_1 = x$, $x_2 = \dot{x}$, $x_3 = y$, $x_4 = \dot{y}$. The equations can be written in the form

$$\dot{x}_1 = x_2, \quad \dot{x}_2 = -x_1, \quad \dot{x}_3 = x_4, \quad \dot{x}_4 = -x_3.$$

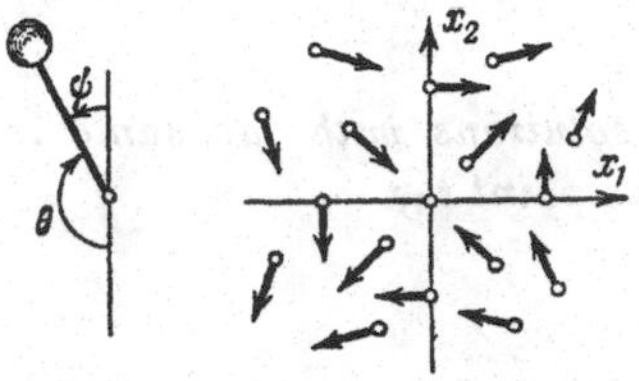

Fig. 25. The phase velocity field of the inverted pendulum

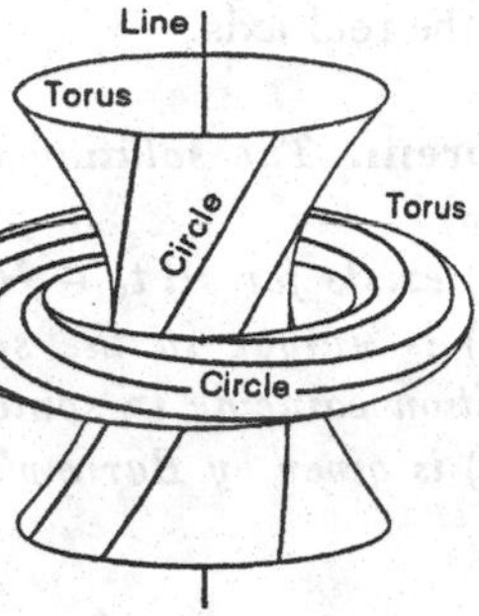

Fig. 26. The phase curves of the spherical pendulum on a hypersurface of constant energy

The right-hand sides define a vector field on $\mathbf{R}^4$.

Problem 1. Prove that the phase curves of this field lie on the three-dimensional spheres $x_1^2 + \cdots + x_4^2 = \text{const}$.

Problem 2. Prove that the phase curves are great circles of these spheres.

However, not every great circle of a sphere is a phase curve.

Problem 3. * Prove that the set of all phase curves on each three-dimensional sphere itself forms a two-dimensional sphere.

A three-dimensional sphere S^3 can be thought of as the three-dimensional space $\mathbf{R}^3$ completed by one "point at infinity." Consequently a partition of S^3 into circles determines a partition of $\mathbf{R}^3$ into circles and one nonclosed circle ("with both ends going off to infinity"). This partition is depicted in Fig. 26.

Problem 4. * Verify that any two of the circles of this partition are linked with a linking number equal to 1 (the linking number tells how many times one of the curves crosses a film stretched over the other, points of intersection being counted positive or negative depending on the direction of crossing.).

§ 2. Vector Fields on the Line

In this section we study a differential equation defined by a vector field on the line and equations with separable variables that reduce to such an equation.

1. Existence and Uniqueness of Solutions

Let v be a smooth (continuously differentiable) function defined on an interval U of the real axis.

Theorem. *The solution φ of the equation $\dot{x} = v(x)$ with initial condition (t_0, x_0)*

1) *exists for all $t_0 \in \boldsymbol{R}$ and $x_0 \in U$;*

2) *is unique in the sense that any two solutions with the same initial condition coincide in some neighborhood of the point t_0;*

3) *is given by Barrow's formula:*

$$t - t_0 = \int_{x_0}^{\varphi(t)} \frac{d\xi}{v(\xi)}, \quad \text{if } v(x_0) \neq 0,$$
$$\varphi(t) \equiv x_0, \quad \text{if } v(x_0) = 0.$$

Proof. Suppose x_0 is not an equilibrium position. We have seen in § 1 that: 1) the solution is given by Barrow's formula in a neighborhood of the point t_0 and 2) the function φ defined by this formula is a solution and satisfies the initial condition.

In the case when x_0 is an equilibrium position the function $\varphi(t) \equiv x_0$ is also a solution, and the theorem is proved. □

Problem 1. Find the gap in this proof.

2. A Counterexample

Let $v = x^{2/3}$ (see Fig. 27). The two solutions $\varphi_1 = 0$ and $\varphi_2 = (t/3)^3$ satisfy the same initial condition $(0, 0)$, contrary to the assertion of uniqueness.

To be sure, the function v is not differentiable, so that this example does not refute the assertion of the theorem. However, the proof given made no use of the smoothness of v: it goes through even in the case when the function v is merely continuous. Consequently this proof cannot be correct. And indeed, the assertion of uniqueness was proved only under the assumption $v(x_0) \neq 0$. We see that if the field v is merely continuous (and not differentiable), the solution with initial condition in the equilibrium position may fail to be unique. It turns out that *the smoothness of v guarantees the uniqueness in this case* (cf. Sect. 3 below).

This example can also be described as follows: under a motion with velocity $v(x) = x^{2/3}$ the equilibrium point ($x = 0$) can be reached from another point in a finite time.

In § 1 we have studied the motion in a linear field (with velocity $v(x) = kx$). In this case it required an infinite time to reach the equilibrium position (for example, if $v(x) = -x$, then the phase point approaches equilibrium so slowly that at every instant it would require one more unit of time to reach equilibrium if its velocity were constant from that instant on).

The reason for nonuniqueness in the case $v(x) = x^{2/3}$ is that the velocity decreases too slowly when approaching the equilibrium position. As a result the solution manages to reach the singular point in a finite time.

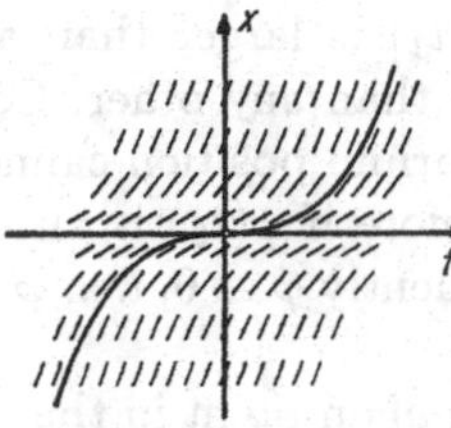

Fig. 27. An example of nonuniqueness

3. Proof of Uniqueness

Let us assume that φ is a solution of the equation $\dot{x} = v(x)$ with a smooth right-hand side v. We shall suppose that $\varphi(t_0) = x_0$ is an equilibrium position and $\varphi(t_1) = x_1$ is not (Fig. 28). On the interval between t_0 and t_1 consider the instant t_2 closest to t_1 at which $v(\varphi(t_2)) = 0$. By Barrow's formula for any point t_3 between t_2 and t_1 we have

$$t_3 - t_1 = \int_{x_1}^{x_3} \frac{d\xi}{v(\xi)}, \quad x_3 = \varphi(t_3).$$

If the function v is smooth, then the integral tends to infinity as x_3 tends to x_2. Indeed, the slope of the chord of the graph of a smooth function on an interval is bounded (Fig. 29), so that $|v(\xi)| \leq k|\xi - x_2|$, where the constant k is independent of the point ξ of the interval $[x_1, x_2]$ (the condition that the slope of the chord of the graph be bounded is called a *Lipschitz condition* and the constant k a *Lipschitz constant*). Thus

$$|t_3 - t_1| \geq \left| \int_{x_1}^{x_3} \frac{d\xi}{k(\xi - x_2)} \right|.$$

The latter integral is easily calculated; it tends to infinity as x_3 tends to x_2. It is easy to verify this without even calculating the integral: it must be equal

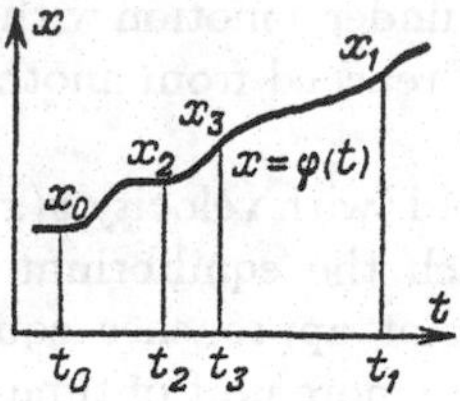

Fig. 28. Proof of uniqueness

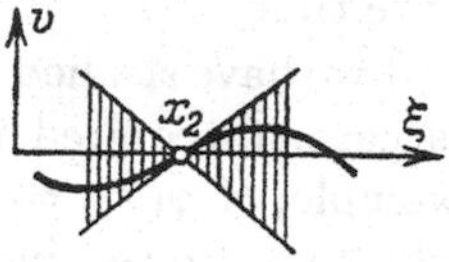

Fig. 29. The Lipschitz condition

to the time of transit between the two points in the linear field, and this time tends to infinity when one of the points tends to the equilibrium position.

Thus the number $|t_2 - t_1|$ is larger than any preassigned number. But there are no numbers larger than any other. Consequently the solution with initial condition in an equilibrium position cannot assume values that are not equilibrium positions. Therefore if $\varphi(t_0)$ is an equilibrium position, we have $v(\varphi(t)) \equiv 0$ for all t. Consequently $\dot{\varphi} \equiv 0$, i.e., φ is a constant. The uniqueness is now proved.

We now remark that the main point in the proof was the comparison of a motion in a smooth field with a more rapid motion in a suitable linear field. For the latter motion the time to enter an equilibrium position is infinite, and consequently it is *a fortiori* infinite for the slower motion in the original field.

Problem 1. Can the integral curves of a smooth equation $\dot{x} = v(x)$ approach each other faster than exponentially as $t \to \infty$?

Answer. If one of them corresponds to an equilibrium position, no; otherwise, yes.

Problem 2. Does the uniqueness theorem hold in the case when the derivative of the function v exists but is discontinuous?

Answer. Yes.

Problem 3. Show that a sufficient condition for uniqueness of the solution with initial value x_0 is that the integral $\displaystyle\int_{x_0}^{x} \frac{d\xi}{v(\xi)}$ diverge at x_0.

Problem 4. Show that a sufficient condition for uniqueness is that the function v satisfy a Lipschitz condition $|v(x) - v(y)| \le k|x - y|$ for all x and y.

Problem 5. Prove that the solution of the equation $\dot{x} = v(t, x)$ with initial condition $\varphi(t_0) = x_0$ is unique, where v is a smooth function.

Hint: Reduce the equation to an equation with right-hand side zero by the change of variable $x \mapsto x - \varphi(t)$, and then compare the direction field with a suitable linear field. This comparison proves the uniqueness in a phase space of any dimension.

Problem 6. Prove that the phase curves of the predator-prey system (§ 1, Sect. 12) do not intersect the coordinate axes (for example, what was originally a positive number of carp cannot become negative at a later time).

Problem 7. Prove that every two solutions of the equation $\dot{x} = v(x)$ with smooth v satisfying the same initial condition coincide everywhere that they are both defined.

4. Direct Products

Consider two differential equations:

$$\dot{x}_1 = v_1(x_1), \quad x_1 \in U_1; \tag{1}$$
$$\dot{x}_2 = v_2(x_2), \quad x_2 \in U_2. \tag{2}$$

The *direct product* of these equations is the system

$$\begin{cases} \dot{x}_1 = v_1(x_1), \\ \dot{x}_2 = v_2(x_2), \end{cases} \tag{3}$$

whose phase space is the direct product U of the phase spaces of Eqs. (1) and (2). The following result is an immediate consequence of the definition.

Theorem. *The solution φ of the differential equation* (3), *which is the direct product of Eqs.* (1) *and* (2) *is a mapping $\varphi : I \to U$ of the form $\varphi(t) = (\varphi_1(t), \varphi_2(t))$, where φ_1 and φ_2 are the solutions of Eqs.* (1) *and* (2), *defined on the same interval.*

In particular, suppose the phase spaces U_1 and U_2 are one-dimensional. Then we know how to solve each of Eqs. (1) and (2). Consequently we can also solve the system (3) of two equations explicitly.

To be specific, by the theorem of Sect. 5 of § 1 the solution φ with the condition $\varphi(t_0) = x_0$ can be found in a neighborhood of the point $t = t_0$ from the relations

$$\int_{x_{1,0}}^{\varphi_1(t)} \frac{d\xi}{v_1(\xi)} = t - t_0 = \int_{x_{2,0}}^{\varphi_2(t)} \frac{d\xi}{v_2(\xi)} \quad (x_0 = (x_{1,0}, x_{2,0})).$$

if $v_1(x_{1,0}) \neq 0 \neq v_2(x_{2,0})$.

If $v_1(x_{1,0}) = 0$, then the first relation is replaced by $\varphi_1 \equiv x_{1,0}$ and if $v_2(x_{2,0}) = 0$, then the second is replaced by $\varphi_2 = x_{2,0}$. Finally, if $v_1(x_{1,0}) = v_2(x_{2,0}) = 0$, then x_0 is a singular point of the vector field v and an equilibrium position of the system (3): $\varphi(t) \equiv x_0$.

5. Examples of Direct Products

Consider the system of two equations

$$\dot{x}_1 = x_1, \quad \dot{x}_2 = kx_2.$$

Problem 1. Sketch the corresponding vector fields in the plane for $k = 0, \pm 1, 1/2, 2$.

We have already solved each of these equations separately. Thus the solution φ with initial condition $\varphi(t_0) = x_0$ has the form

$$\varphi_1 = x_{1,0}e^{(t-t_0)}, \quad \varphi_2 = x_{2,0}e^{k(t-t_0)}. \tag{4}$$

Consequently along each phase curve $x = \varphi(t)$ we have either

$$|x_2| = C|x_1|^k, \tag{5}$$

where C is a constant independent of t, or $x_1 \equiv 0$.

Problem 2. Is the curve in the phase plane (x_1, x_2) given by Eq. (5) a phase curve?

Answer. No.

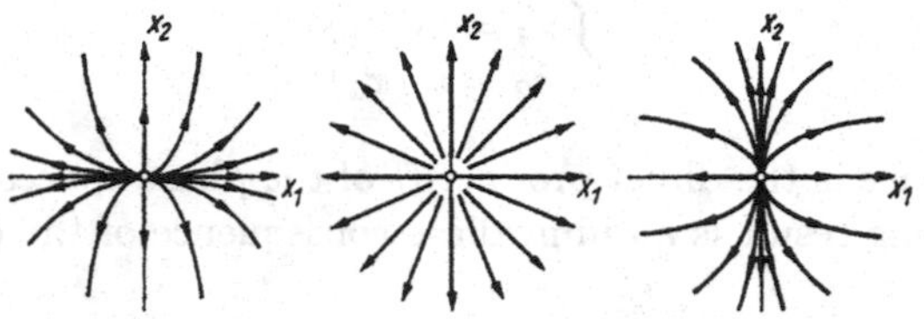

Fig. 30. Nodes: The phase curves of the systems $\dot{x}_1 = x_1$, $\dot{x}_2 = kx_2$, for $k > 1$, $k = 1$, and $0 < k < 1$

The family of curves (5), where $C \in \boldsymbol{R}$, has various forms depending on the value of the parameter k. If $k > 0$ this is a family of "parabolas[6] with exponent k." Such parabolas are tangent to the x_1-axis if $k > 1$ or to the x_2-axis if $k < 1$. (Fig. 30; for $k = 1$ a family of lines passing through the origin is obtained). The distribution of phase curves depicted in Fig. 30 is called a *node*. For $k < 0$ the curves (5) have the form of hyperbolas (Fig. 31)[7] and in a neighborhood of the origin they form a *saddle*. For $k = 0$ the curves (5) become straight lines (Fig. 32).

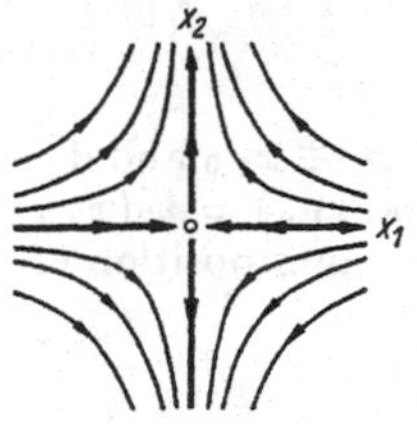

Fig. 31. A saddle; the phase curves of the system $\dot{x}_1 = x_1$, $\dot{x}_2 = kx_2$, $k < 0$

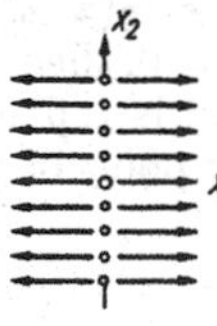

Fig. 32. The phase curves of the system $\dot{x}_1 = x_1$, $\dot{x}_2 = 0$

It can be seen from Eq. (4) that each phase curve lies entirely in one quadrant (or on a coordinate semiaxis or coincides with the origin, which is

[6] True parabolas are obtained only for $k = 2$ and $k = 1/2$.
[7] True hyperbolas are obtained only for $k = -1$.

a phase curve for all k). The arrows in the figures indicate the direction of motion of the point $\varphi(t)$ as t increases.

Problem 3. Prove that each of the parabolas $x_2 = x_1^2$ ($k = 2$) consists of three phase curves. Describe all the phase curves for other values of k ($k > 1$, $k = 1$, $0 < k < 1$, $k = 0$, $k < 0$).

It is interesting to trace the transition from one figure to another as k varies continuously.

Problem 4. Describe the node corresponding to $k = 0.01$ and the saddle corresponding to $k = -0.01$.

Problem 5. Solve the equation for the inverted pendulum $\dot{x}_1 = x_2$, $\dot{x}_2 = x_1$ and sketch the phase curves.

Solution. In the phase plane we introduce new coordinates: $X = x_1 + x_2$, $Y = x_1 - x_2$. The system breaks up into a direct product: $\dot{X} = X, \dot{Y} = -Y$. In the (X, Y)-plane the phase curves form a saddle, as in Fig. 31. Consequently we also obtain a saddle in the (x_1, x_2)-plane (Fig. 33). Hence, in particular, it follows that *for a given displacement of the pendulum from the vertical there exists one and only one initial velocity under which the pendulum approaches asymptotically the upper equilibrium position as* $t \to +\infty$ (the corresponding phase curve is a straight ray that reaches 0). With a smaller or larger initial velocity the pendulum falls, either without reaching the upper equilibrium position or having passed through it (the corresponding phase curves are halves of hyperbolas).

The solutions have the form $X = X_0 e^t$, $Y = Y_0 e^{-t}$, whence $x_1 = Ae^t + Be^{-t} = a\cosh t + b\sinh t$, $x_2 = Ae^t - Be^{-t} = a\sinh t + b\cosh t$.

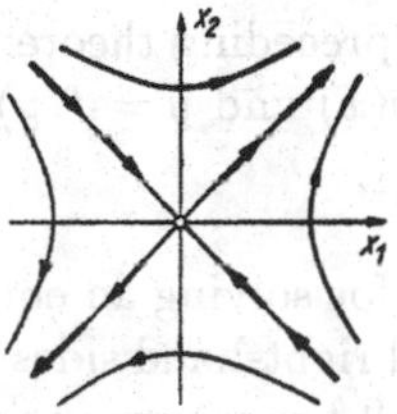

Fig. 33. The phase curves of the inverted pendulum

6. Equations with Separable Variables

Definition. An *equation with separable variables* is an equation

$$\frac{dy}{dx} = \frac{f(y)}{g(x)}. \tag{6}$$

We shall assume that f and g are smooth functions that do not vanish in the region under consideration.

Along with this equation we consider the system

$$\dot{x} = g(x), \quad \dot{y} = f(y). \tag{7}$$

Theorem. *The phase curves of the system* (7) *are integral curves of Eq.* (6)*; and, conversely, the integral curves of Eq.* (6) *are phase curves of the system* (7).

Proof. The slope of the phase velocity vector is $f(y)/g(x)$. Hence a phase curve of the system is tangent to the direction field of the equation at each of its points.

Conversely suppose an integral curve of Eq. (6) is given. Then the parameter t on the curve can be chosen so that the parametric equation of the curve is $x = \varphi(t)$, $y = \psi(t)$, and the function φ is a solution of the equation $\dot{x} = g(x)$ (this is where the condition $g \neq 0$ is used). The second coordinate ψ of the point with parameter t then satisfies the relation $(d\psi/dt)/(d\varphi/dt) = f(\psi(t))/g(\varphi(t))$, i.e., is a solution of the equation $\dot{y} = f(y)$. Consequently our curve is a phase curve of the system. □

Theorem. *The solution of Eq.* (6) *with initial condition* (x_0, y_0) *exists, is unique,*[8] *and is given by the formula*

$$\int_{x_0}^{x} \frac{d\xi}{g(\xi)} = \int_{y_0}^{y} \frac{d\eta}{f(\eta)}.$$

Proof. This follows from the preceding theorem and the formulas for the solution of the equations $\dot{x} = g(x)$ and $\dot{y} = f(y)$ with intial conditions (t_0, x_0) and (t_0, y_0) respectively. □

Remark. A "mnemonic" rule for solving an equation with separable variables is to regard both the left- and right-hand sides as fractions and transpose "all the x terms to one side and all the y terms to the other side":

$$\frac{dx}{g(x)} = \frac{dy}{f(y)}. \tag{8}$$

When this has been done, "equating the integrals" gives the desired relation between x and y in the form of the equality $\int \frac{dx}{g(x)} = \int \frac{dy}{f(y)} + C$ for the primitives or in the form shown in the theorem for the definite integrals.

Of course this "mnemonic" rule, properly understood, is a completely rigorous deduction of the formulas for the solution. Indeed relation (8) expresses the *equality of values of two differential forms at any vector tangent to an integral curve of Eq.* (6) (and conversely a curve all of whose tangent vectors satisfy relation (8) is an integral curve for Eq. (6)).

---

[8] In the sense that any two such solutions coincide wherever both are defined.

The integrals of the forms on the left- and right-hand sides of Eq. (8) over the same segment of an integral curve of Eq. (6) are equal (since only the values of the form at the vectors tangent to a curve occur in the definition of the integral over a curve, and on these vectors the values of the two forms coincide). Finally the integral of the form $dx/g(x)$ along a segment of the curve equals the ordinary integral of the function $1/g$ along the projection of this curve on the x-axis, and similarly for the form $dy/f(y)$.

Formula (8) is sometimes called the *symmetric form* for writing Eq. (6).

Problem 1. Sketch the integral curves of the equations $dy/dx = y/x$, $dy/dx = x/y$, $dy/dx = -y/x$, and $dy/dx = -x/y$.

Problem 2. Sketch the integral curves of the equations $dy/dx = kx^\alpha y^\beta$, $dy/dx = \sin y/\sin x$, and $dy/dx = \sin x/\sin y$.

Problem 3. Sketch the phase curves of the pendulum equation $\dot{x} = y$, $\dot{y} = -\sin x$.

Hint: Consider the variables-separable equation $dy/dx = -(\sin x)/y$.

7. An Example: The Lotka-Volterra Model

In Sect. 12 of § 1 we studied the simplest model for the interaction of y predators (pike) and x prey (carp):

$$\dot{x} = kx - axy, \quad \dot{y} = -ly + bxy. \tag{9}$$

But we were not able to sketch the phase curves.

Theorem. *The phase curves of the system* (9) *are closed* (Fig. 34).

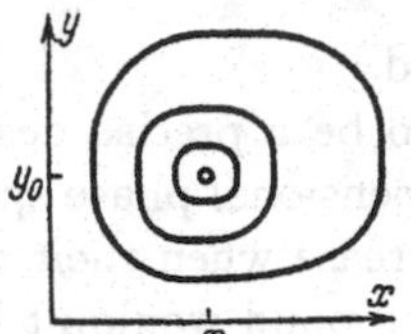

Fig. 34. The phase curves of the Lotka-Volterra model

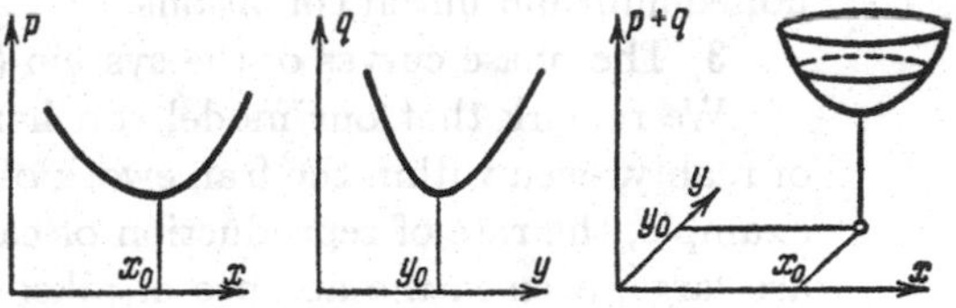

Fig. 35. The structure of the phase curves of the Lotka-Volterra model

Proof. The phase curves of the system (9) coincide with the integral curves of the variables-separable equation $\dfrac{dy}{dx} = \dfrac{y(bx-l)}{x(k-ay)}$ or with the phase curves of the product-equation

$$\frac{dx}{d\tau} = \frac{x}{bx-l}, \quad \frac{dy}{d\tau} = \frac{y}{k-ay}$$

(in the region where x, y, $bx-l$, and $k-ay$ are nonzero).

Consequently $\int \frac{k-ay}{y}\,dy = \int \frac{bx-l}{x}\,dx + C$ or $p(x)+q(y)=C$, where $p = bx - l\ln x$, $q = ay - k\ln y$. The graphs of the functions p and q are convex upward. Therefore the graph of the function $p+q$ is also convex upward (Fig. 35). Consequently the level lines of the function $p+q$ are closed curves. It is easy to verify that the phase curves of Eq. (9) are not only among the level lines of $p+q$ but actually coincide with them; the theorem is now proved. □

It follows from the fact that the phase curves are closed that the number of carp and pike in the Lotka-Volterra model vary periodically with time. The period of oscillation depends on the initial condition.

Problem 1. Prove that the period of oscillation in the Lotka-Volterra model (9) tends to infinity as the initial condition tends to the point $(0,0)$.

Remark. Mathematical approach to infinity must be distinguished from physical approach to infinity. For example as $\varepsilon \to 0$ the function $1/\varepsilon$ does indeed tend to infinity (for example for $\varepsilon = 10^{-6}$ the quantity $1/\varepsilon$ is indeed large). At the same time $|\ln\varepsilon|$ virtually remains bounded as $\varepsilon \to 0$ (for example when $\varepsilon = 10^{-6}$ this quantity is of the order of 10). In practice logarithms in asymptotic relations can often be treated as constants.

Problem 2. How does the period of oscillation in the Lotka-Volterra model (9) tend to infinity when the initial condition has the form (x_0, ε) and $\varepsilon \to 0$?

Answer. Logarithmically.

Consider certain deductions from our computations.

For the Lotka-Volterra system (9).

1) There exists an equilibrium position (x_0, y_0) (which is unique for $x > 0$, $y > 0$).

2) The numbers of carp and pike vary periodically with time under nonequilibrium initial conditions.

3) The phase curves of the system (9) are closed.

We remark that our model can hardly claim to be a precise description of reality, even within the framework of the two-dimensional phase space. For example, the rate of reproduction of carp must decrease when their numbers are large, even in the absence of pike; otherwise the pond wouldn't be large enough for the carp, etc. We may think, therefore that a more precise model has the form

$$\begin{cases} \dot{x} = x(k - ay + \varepsilon f(x,y)), \\ \dot{y} = y(-l + bx + \varepsilon g(x,y)), \end{cases} \tag{9_ε}$$

where $x\varepsilon f$ and $y\varepsilon g$ are small corrections to our model that were neglected in the idealization (the correction to $\dot{x}$ is divisible by x since the rate of reproduction of carp is 0 when their number is 0; for the same reason the correction to $\dot{y}$ is divisible by y). We shall assume that f and g are smooth functions (strictly speaking, here and below we are considering a bounded portion of the phase plane since there are no grounds for believing the corrections are small for very large values of the coordinates).

We shall call a property of the model (9) a *robust* property if it (or a closely similar property) also holds for every system (9_ε) for sufficiently small ε.

Let us consider the deductions 1)–3) from this point of view.

Theorem. *The system (9_ε) has an equlibrium position $x(\varepsilon)$, $y(\varepsilon)$ depending smoothly on small ε and such that $x(0) = x_0$, $y(0) = y_0$ is an equilibrium position of the system* (9).

Proof. By the implicit function theorem the system of equations

$$F(x,y,\varepsilon) = 0, \quad G(x,y,\varepsilon) = 0$$

in x and y has a solution $(x(\varepsilon), y(\varepsilon))$ that becomes (x_0, y_0) for $\varepsilon = 0$ if the Jacobian $J = D(F,G)/D(x,y)|_{(x_0,y_0,0)}$ is nonzero.

In our case $F = k - ay + \varepsilon f$, $G = -l + bx + \varepsilon g$, so that $J = \begin{vmatrix} 0 & -a \\ b & 0 \end{vmatrix} \neq 0$, which was to be proved. □

Hence the conclusion 1) is robust: the equilibrium position exists not only for the system (9) but for any nearby system (9_ε).

In contrast, conclusions 2) and 3) are nonrobust. Indeed, the return function for the system (9) has the form $\Phi(A) \equiv A$. For the nearby system (9_ε) the graph of the return function is near the diagonal, but does not necessarily coincide with it. Depending on the form of the perturbations of f and g the Lamerey diagram may be located above or below the diagonal or intersect it in one or more points corresponding to stable or unstable cycles.

Consequently *the conclusions that the phase curves are closed and that the numbers of pike and carp vary periodically with an amplitude depending on the initial conditions are not robust.* Although for the nearby system (9_ε) every coil of a phase curve is also near a closed cycle, it does not close precisely, and after a long time (of the order of $1/\varepsilon$) a self-oscillatory mode establishes itself (the phase curve winds onto a limit cycle).

The property of having a limit cycle is stable with respect to small perturbations of the system of equations. More precisely, assume that a cycle corresponds to a fixed point $A = \Phi(A)$ of the return function Φ and that $\Phi'(A) \neq 1$. In such a situation the cycle is called *nondegenerate.*

If a system defined by a vector field v_0 has a nondegenerate limit cycle passing through A_0, then every nearby system (defined by the field v_ε for small ε) has a nearby cycle (passing through a point $A_{(\varepsilon)}$ near A_0).

To prove this it is necessary to apply the implicit function theorem to the equation $\Phi(A,\varepsilon) = A$, $A(0) = A_0$.

Consequently *the conclusion that the system has self-oscillations described by a nondegenerate limit cycle is robust:* in every nearby system there are nearby self-oscillations.

We remark that the degenerate limit cycles can disappear under a small deformation of the system. Nevertheless they occur in a way that cannot be removed by a small deformation in the case when a family of systems depending on a parameter, rather than an individual system, is considered. In this case distinct cycles may coalesce for particular values of the parameter, and a similar coalescence may take place for some nearby value of the parameter in any nearby family. At the instant when two nondegenerate cycles coalesce a degenerate cycle appears. In this situation, in general, one of the two coalescing cycles is stable and the other unstable. The degenerate cycles that arise when two nondegenerate cycles coalesce are of interest because they always occur on the boundary of the region in which the oscillatory mode exists in the parameter space.

For example, Fig. 36 depicts the Lamerey diagrams for three very near values of the parameter (curves 1, 2, and 3). Diagram 1 intersects the bisector in two points: in this case the system has two limit cycles, the stable cycle being inside the unstable cycle (Fig. 37). The equilibrium position is unstable; the whole region inside the unstable cycle is a region of attraction ("sink") of the stable cycle. When the initial conditions are in this region (except for the equilibrium position) self-oscillations are set up in the system depicted by the stable cycle.

Curve 2 corresponds to a critical value of the parameter: a stable cycle coalesces with an unstable cycle and becomes degenerate. The phase curves that originate in the region bounded by the cycle tend to the cycle as time passes. However the oscillatory mode that is thereby set up is unstable: an arbitrarily small random deviation can throw the phase point outside the cycle.

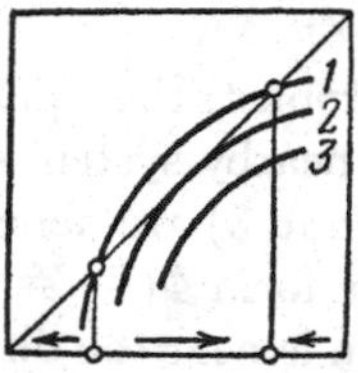

Fig. 36. Metamorphosis of Lamerey diagrams

When the parameter is changed even more (curve 3) the cycle disappears entirely. Thus the coalescence of cycles leads to a spasm in the behavior of the system: the stable self-oscillatory mode with a finite region of attraction suddenly disappears. Motions whose initial conditions lie in the sink of a disappearing cycle move to other regions of the phase space (Fig. 37) after the disappearance. In our example, after the parameter has passed through a critical value in the populations of predators and prey, an arbitrarily small deviation of the initial conditions from equilibrium leads to unbounded increase in the oscillations, and consequently to extinction.

The metamorphosis of the qualitative picture of the motion as the parameter varies is studied by *bifurcation theory* (bifurcation = branching), and the application of bifurcation theory to the study of spasmodic reactions of mechanical, physical, chemical, biological, economic, and other systems to smooth variation of the external conditions has lately come to be known as *catastrophe theory*.

It can be seen from Fig. 36 that when the value of the parameter differs from the critical value by a small amount Δ, the distance between the stable and unstable cycles is of order $\sqrt{\Delta}$. Consequently the rate of approach of the cycles as the parameter varies grows rapidly as the parameter approaches a critical value. At the instant of a catastrophe both cycles are moving toward each other with infinite velocity. This explains why it is so difficult to avert the impending catastrophe of loss of stability of a system by the time signs of it have become noticeable.

Problem 3. Study the bifurcation of cycles when the parameter c varies in the system given in polar coordinates by the equations

$$\dot{r} = cr - r^3 + r^5, \quad \dot{\varphi} = 1.$$

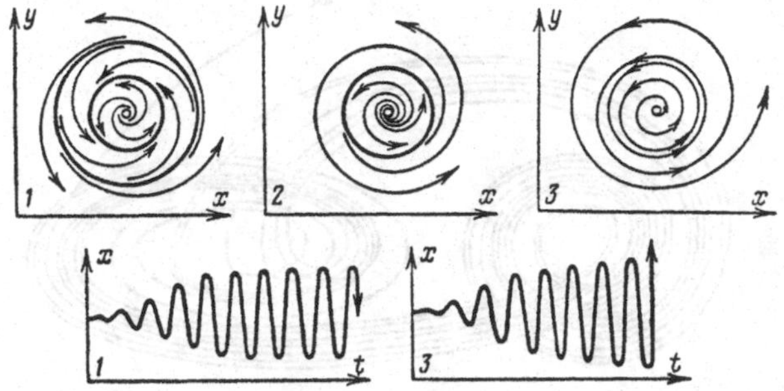

Fig. 37. Metamorphosis of the phase portrait and the behavior of the solutions

Solution. For $c = 0$ a stable cycle whose radius is of the order $\sqrt{c}$ is generated from the equilibrium position $r = 0$. It disappears for $c = 1/4$, coalescing with an unstable cycle.

Remark. It can be shown that the birth or death of a cycle at an equilibrium position, like the birth or death of a pair of cycles, is a generic phenomenon encountered in the variation of the parameter in general one-parameter families of differential equations.

The stable limit cycles describe stationary periodic oscillations of a system under stationary external conditions. The oscillations describable by stable cycles are called *self-oscillations*, in contrast to the *forced* oscillations caused by periodic external action and oscillations of the same type as the free oscillations of a pendulum. The occurrence of self-oscillations is quite marvelous in itself, but they are encountered, for example, in such systems as a clock, a steam engine, an electric bell, the heart, a radio transmitter, and Cepheid variable stars. The functioning of each of these mechanisms is described by a limit cycle in a suitable phase space.

It should not be thought, however, that all oscillatory processes are described by limit cycles: in multidimensional phase space much more complicated behavior of the phase curves is possible. Examples are furnished by the precession of a gyroscope, the motion of planets and their satellites and their revolution about their axes (the aperiodicity of these motions is responsible for the complexity of the calendar and the difficulty of predicting the tides), and the motion of charged particles in magnetic fields (which causes the aurora). We shall study the simplest motions of this type in § 24 and Sect. 6 of § 25. In systems with a multidimensional phase space the phase curves can even approach a set on which all nearby trajectories diverge rapidly from one another (Fig. 38). Such attracting sets have lately come to be known as *strange attractors*: they are connected with phenomena of same type as turbulence and are responsible, for example, for the impossibility of long-range weather forecasting.

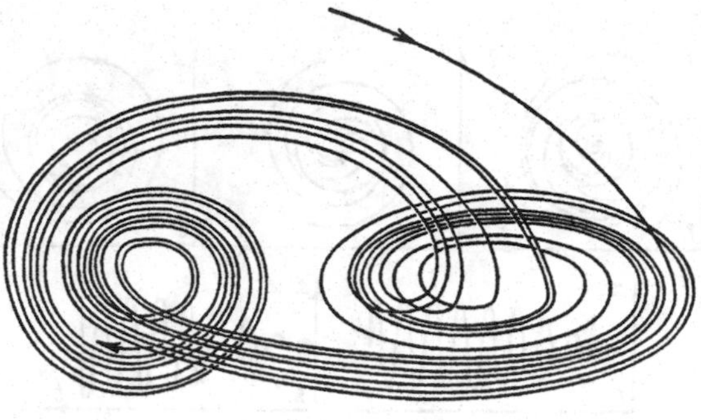

Fig. 38. An attractor with scattering of its phase curves

§ 3. Linear Equations

Linear equations describe the influence that small variations in the initial conditions or in the right-hand sides of arbitrary equations exert on the solutions. In this section linear homogeneous and inhomogeneous equations with one dependent variable are solved explicitly and studied: the monodromy operator, the δ-function, the Green's function, and forced oscillations appear.

1. Homogeneous Linear Equations

Definition. A *first-order homogeneous linear equation* is an equation

$$\frac{dy}{dx} = f(x)y, \tag{1}$$

whose right-hand side is a (homogeneous) linear function of the one-dimensional dependent variable y.

This is a special case of an equation with separable variables. Solving it according to the general rule, we find $dy/y = f(x)\,dx$, $\ln(y/y_0) = \int_{x_0}^{x} f(\xi)\,d\xi$. A consequence of this is the following result.

Theorem. *Every solution of Eq.* (1) *can be extended to the entire interval on which the function f is defined; the solution with initial condition* (x_0, y_0) *is given by the formula* $y = y_0 e^{\int_{x_0}^{x} f(\xi)\,d\xi}$.

Remark 1. Let $y = \varphi(x)$ be a solution of Eq. (1). Then for any constant c the function $y = c\varphi(x)$ is also a solution. The sum of two solutions (defined on the entire interval of definition of f) of Eq. (1) is also a solution. Therefore all such solutions of the homogeneous linear equation (1) form a *vector space.* The dimension of this vector space is 1 (why?).

Remark 2. Dilating the extended phase space (x, y) along the y-axis maps the direction field of the homogeneous linear equation (1) into itself. Therefore *the integral curves map into one another under the action of dilations along the y-axis.* They can all be obtained from any one of them by dilations (Fig. 39).

Linear equations occupy a special place in the theory of differential equations because, by one of the basic principles of analysis, every smooth function is well approximated in a neighborhood of each point by a linear function. The operation of *linearization* that thereby arises leads to linear equations as a first approximation in the study of an arbitrary equation near any solution.

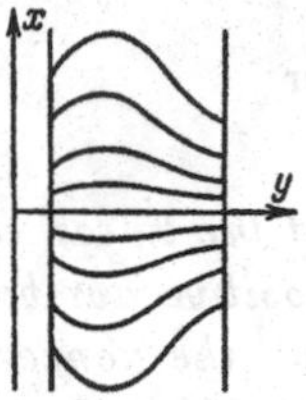

Fig. 39. The integral curves of a linear equation

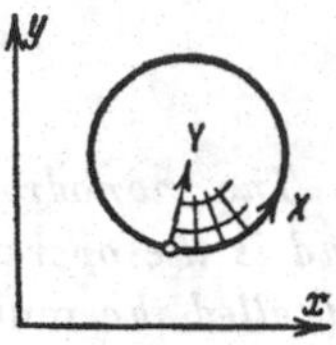

Fig. 40. A coordinate system near a cycle

Consider, for example, an autonomous system with a two-dimensional phase plane (x, y) having a limit cycle (Fig. 40). Introduce coordinates $(X \bmod T,\ Y)$ in a neighborhood of this cycle in such a way that the equation of the cycle assumes the form $Y = 0$, and a traversal of the cycle in the direction of the phase velocity corresponds to increasing X by T. Then under the mapping $(x, y) \mapsto (X, Y)$ the phase curves of the initial system transform into integral curves of an equation of the form

$$\frac{dY}{dX} = a(X, Y), \quad \text{where } a(X, 0) \equiv 0, \quad a(X + T, Y) \equiv a(X, Y). \tag{2}$$

Linearization of this equation on Y at the point $Y = 0$ leads to a linear equation

$$\frac{dY}{dX} = f(X)Y, \quad \text{where } f(X) = \partial a/\partial Y|_{Y=0}.$$

We remark that the function f has period T.

We thus arrive at the problem of investigating a linear equation with a periodic coefficient f.

2. First-order Homogeneous Linear Equations with Periodic Coefficients

Definition. A *first-order homogeneous linear equation with T-periodic coefficient* is an equation

$$\frac{dY}{dX} = f(X)Y, \text{ where } f(X+T) \equiv f(X). \tag{3}$$

The solutions of Eq. (3) determine a linear mapping of the Y-axis into itself, assigning to the value $\varphi(0)$ at $X = 0$ the value $\varphi(T)$ of the same solution for $X = T > 0$. This mapping $A : \boldsymbol{R} \to \boldsymbol{R}$ is called a *monodromy* (Fig. 41). (We plan to use a similar operator in the multidimensional case.)

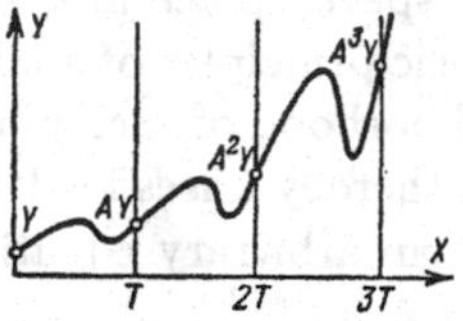

Fig. 41. A monodromy operator

Theorem. *The monodromy operator $A : \boldsymbol{R} \to \boldsymbol{R}$ of the linear equation* (3) *is linear and is the operator of multiplication by a positive number λ. If this number λ (called the* multiplier*) is greater than* 1, *all the nonzero solutions tend to infinity as $x \to +\infty$; if $\lambda < 1$, they tend to 0; if $\lambda = 1$, all solutions are bounded.*

Proof. The linearity of A follows from the fact that dilations along the Y-axis map integral curves into integral curves and that $\lambda > 0$, i.e., the X-axis is an integral curve. Translation by T along the X-axis also takes integral curves into integral curves (because of the periodicity of f). It follows from this that the values of the solution with initial condition $\varphi(0) = Y$ are equal to λY, $\lambda^2 Y$, $\lambda^3 Y$,... for $X = T$, $2T$, $3T$,...; therefore $\varphi(NT) \to \infty$ as $N \to +\infty$ if $\lambda > 1$ and $\varphi(NT) \to 0$ as $N \to +\infty$ if $\lambda < 1$. Moreover translating the extended phase space by NT along the X-axis, we find

$$\varphi(NT + S) = \lambda^n \varphi(S),$$

from which all the assertions being proved follow (why?). □

Remark. A formula for the multiplier follows from the theorem of Sect. 1:

$$\ln \lambda = \int_0^T f(\xi)\, d\xi.$$

Thus *the multiplier is greater or less than one according as the average value of the function f is positive or negative.*

In the first case the zero solution of the linear equation (3) is unstable, and in the second case it is stable (moreover solutions with initial conditions near

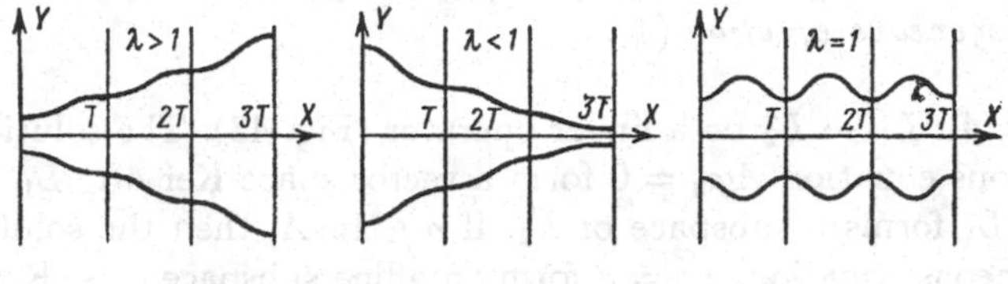

Fig. 42. Stability of the zero solution

zero tend to 0); in the case $\lambda = 1$ the solutions with nonzero initial conditions are periodic (Fig. 42).

A natural question arises: what relation does our theorem on the solutions of the linearized equation (3) have to the original problem on the behavior of the solutions of the nonlinear equation (2), i.e., to the problem of phase curves near a cycle?

Problem 1. Prove that if $\lambda > 1$, then the cycle is unstable and the phase curves originating near a cycle are unwinding spirals diverging from the cycle; if $\lambda < 1$, then the cycle is stable and the phase curves originating in a neighborhood of it are spirals winding onto the cycle.

In other words in the cases when the multiplier is different from 1 linearization leads to a correct judgment about the stability of the cycle. On the other hand if $\lambda = 1$, then, even though the solutions of Eq. (3) are periodic, it would be incorrect to extend this inference from the linearized equation (3) to the original equation (2), for which the solutions near $Y = 0$ are, in general, not periodic, and no judgment about the stability of the cycle can be made from the linearized equation.

Hint: Consider the return function Φ defined by the solutions φ of Eq. (2) and assigning to the initial condition $Y = \varphi(0)$ for $X = 0$ the value $\Phi(Y) = \varphi(T)$. Prove that the linearization of Φ at the point $Y = 0$ is a monodromy operator.

Problem 2. Study the stability of the limit cycle $r = 1$ for the system given in polar coordinates by the equations

$$\dot{r} = (r^2 - 1)(2x - 1), \quad \dot{\varphi} = 1 \quad (\text{where } x = r\cos\varphi).$$

3. Inhomogeneous Linear Equations

Definition. A *first-order inhomogeneous linear equation* is an equation

$$\frac{dy}{dx} = f(x)y + g(x). \tag{4}$$

The term *solution* is taken to mean a solution defined on the entire interval of definition of the functions f and g.

Theorem. *If one particular solution of an inhomogeneous equation $y = \varphi_1(x)$ is known, all other solutions have the form $y = \varphi_1(x) + \varphi_0(x)$, where φ_0 is a*

solution of the homogeneous equation (1); ***every function of this type satisfies the inhomogeneous equation*** (4).

Proof. Let $A : L_1 \to L_2$ be a linear operator (Fig. 43). The solutions φ_0 of a homogeneous equation $A\varphi_0 = 0$ form a vector space $\operatorname{Ker} A \subset L_1$. The image $\operatorname{Im} A = AL_1$ forms a subspace of L_2. If $g \in \operatorname{Im} A$, then the solutions of the inhomogeneous equation $A\varphi = g$ form an affine subspace $\varphi_1 + \operatorname{Ker} A$ parallel to $\operatorname{Ker} A$. In our case $A\varphi = d\varphi/dx - f\varphi$. This is a linear operator[9], so that the assertion of our theorem follows from the algebraic theorem on the solution of an inhomogeneous linear equation. □

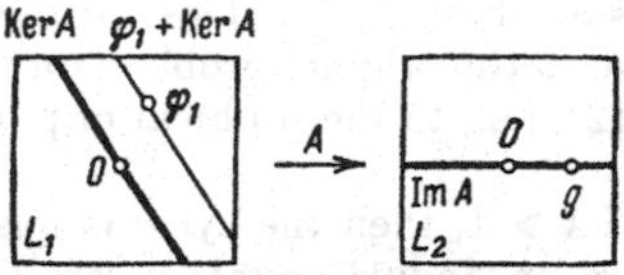

Fig. 43. The solution space of an inhomogeneous linear equation

To find a particular solution one may use the method of "variation of parameters."

The method of variation of parameters is often used in studying the influence of all possible perturbations. Consider, for example, the motion of the planets about the Sun. In first approximation, not taking account of the attraction of the planets on one another, we arrive at the independent motion of planets in Keplerian ellipses. This is the solution of the unperturbed equations of motion.

To take account of the perturbing influence of the planets on one another we may proceed as follows: assume that the planets make a Keplerian motion, but the parameters of the Keplerian ellipse vary slightly with time.[10] Thus quantities that had been constant in the unperturbed motion are now regarded as functions of time.

The differential equations describing the variation of these parameters are sometimes simpler to solve or study than the original equations. In particular, in the application to inhomogeneous linear equations, where the role of the unperturbed problem is played by the homogeneous equation and the role of the perturbation is played by the inhomogeneity, the method of variation of parameters leads to an explicit formula for the solution. In this case the perturbation is not required to be small.

We already know that every solution of the homogeneous equation (1) has the form $y = c\varphi(x)$, where c is an arbitrary constant and φ is any nonzero

[9] The spaces L_1 and L_2 can be chosen in various ways. For example, we may assume that L_1 consists of the once continuously differentiable functions and L_2 of continuous functions.

[10] For example, the oscillation of the eccentricity of the Earth's orbit is one of the causes of Ice Ages.

solution. We shall try to choose the function $c = c(x)$ so that $y = c(x)\varphi(x)$ is a solution of the inhomogeneous equation (4).

Theorem. *The solution of the inhomogeneous linear equation* (4) *with initial condition $y(x_0) = 0$ exists, is unique, and is given by the formula*

$$y = \int_{x_0}^{x} e^{\int_{\xi}^{x} f(\zeta)\, d\zeta} g(\xi)\, d\xi. \tag{5}$$

Proof. Substituting $y = c(x)\varphi(x)$ into (4) gives

$$c'\varphi + c\varphi' = fc\varphi + g.$$

But φ is a solution of the homogeneous equation (1). Hence $\varphi' = f\varphi$ and

$$c' = g/\varphi, \quad c(x) = \int_{x_0}^{x} g(\xi)/\varphi(\xi)\, d\xi.$$

Substituting the known solution of the homogeneous equation in place of φ, we obtain formula (5) (after introducing φ into the integral), which was to be proved. □

4. The Influence Function and δ-shaped Inhomogeneities

Formula (5) has a simple "physical meaning," which can be made clear as follows. The following principle is obvious.

The Principle of Superposition. *If φ_1 and φ_2 are solutions of the inhomogenous linear equations $A\varphi_1 = g_1$ and $A\varphi_2 = g_2$, then $\varphi_1 + \varphi_2$ is a solution of the equation $A\varphi = g_1 + g_2$.*

This principle makes it possible to separate the various perturbations when taking account of all possible perturbations, calculate their influence individually, and add the effects of the perturbations (for example, if two stones are thrown into water, the waves from each of them can be calculated independently and the perturbations added; in the flight of a missile one can introduce independent corrections for wind and the deviation of the density of the atmosphere from the tabular density, etc.).

In applications to our inhomogeneous equation (4) the function g plays the role of the perturbation. We shall attempt to represent the function g in the form of a linear combination of "elementary perturbations"; then the solution is the same linear combination of solutions of equations with these elementary perturbations as the inhomogeneity g.

Definition. A *δ-shaped sequence* is a sequence h_N of nonnegative smooth functions equal to 0 outside neighborhoods that tend to 0 as $N \to \infty$ and each possessing an integral equal to 1.

An example of such a sequence is easy to construct (Fig. 44). Physicists say that "the limit of the sequence h_N is the Dirac delta-function equal to zero everywhere except at the point 0 and having integral 1."

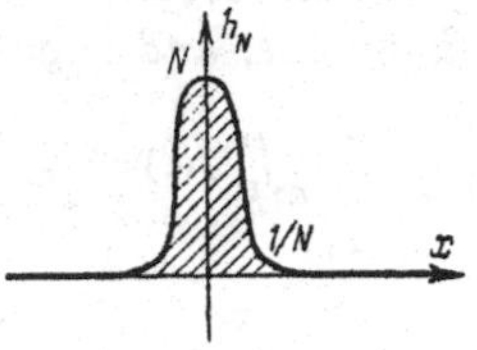

Fig. 44. A δ-shaped sequence

Of course no function δ with these properties exists.

Nevertheless many quantities in whose definitions the functions h_N occur tend to definite limits as $N \to \infty$, which are said to be the corresponding quantities calculated for the δ-function. For example, for any continuous function g

$$\lim_{N\to\infty} \int_{-\infty}^{+\infty} h_N(x)g(x)\,dx = g(0).$$

(Prove this!) Therefore by definition

$$\int_{-\infty}^{+\infty} \delta(x)g(x)\,dx = g(0).$$

In exactly the same way, translating all h_N by ξ along the x-axis, we find

$$\int_{-\infty}^{+\infty} \delta(x-\xi)g(x)\,dx = g(\xi),$$

i.e., $\delta(\cdot - \xi)$ is the "δ-function concentrated at the point ξ."

The last formula can also be interpreted as the representation of any smooth function g as a "continuous linear combination" of δ-functions concentrated at different points x with coefficients equal to the values of g at these points.

Thus an arbitrary inhomogeneity g in Eq. (4) can be decomposed into a continuous linear combination of inhomogeneities "each concentrated at one point" and having the form of shifted δ-functions. According to the principle of superposition, to find a particular solution of Eq. (4) with an arbitrary inhomogeneity it suffices to know this solution for a δ-shaped inhomogeneity.

Definition. The solution of the equation

$$\frac{dy}{dx} = f(x)y + \delta(x-\xi), \quad \xi > 0,$$

with initial condition $y(0) = 0$, is called the *influence function of the perturbation at the instant ξ on the solution at the instant x* (or the *Green's function*[11]) and is denoted $y = G_\xi(x)$.

Theorem. *The Green's function is given by the formula*

$$G_\xi(x) = \begin{cases} 0 & \text{if } x < \xi, \\ e^{\int_\xi^x f(\zeta)\,d\zeta} & \text{if } x > \xi. \end{cases} \tag{6}$$

Remark. As explained above, we are talking about the limit of a sequence of solutions of the equations

$$\frac{dy}{dx} = f(x)y + h_N(x - \xi), \tag{7}$$

where $\{h_N\}$ is a δ-shaped sequence, as $N \to \infty$.

Heuristic proof: For $x < \xi$ the solution is zero, since the inhomogeneity disappears. For $x > \xi$ the solution coincides with some solution of the homogeneous equation, since the inhomogeneity disappears. For x near ξ the second term on the right-hand side of Eq. (7) is large in comparison with the first term, so that the integral of dy/dx over a small neighborhood of the point ξ is almost equal to

$$\int_{-\infty}^{+\infty} h_N(x - \xi)\,dx = 1.$$

Passing to the limit as $N \to \infty$, we see that the jump in the solution $y(x)$ as x passes through the point ξ equals 1, i.e., for $x > \xi$ the function G_ξ of the variable x is the solution of the homogeneous equation with initial condition $y(\xi) = 1$, which was to be proved. □

This reasoning can be made completely rigorous, but it is simpler to carry out the following argument.

Mathematical proof: Substituting the function h_N translated by ξ for g in formula (5) for the solution of Eq. (4) and passing to the limit as $N \to \infty$, we obtain what was required:

$$G_\xi(x) = \lim_{N\to\infty} \int_{x_0}^{x} e^{\int_\nu^x f(\zeta)\,d\zeta} h_N(\nu - \xi)\,d\nu = e^{\int_\xi^x f(\zeta)\,d\zeta},$$

if $x_0 < \xi < x$. □

Corollary. *The solution of the inhomogeneous equation* (4) *with inhomogeneity g and with zero initial condition is expressed in terms of the influence function by the formula $y(x) = \int_0^x G_\xi(x)g(\xi)\,d\xi$ for $x > 0$.*

---

[11] This function is also called the *retarded Green's function* in order to avoid confusion with the Green's functions of boundary-value problems for higher-order equations, which we do not consider here.

Of course this formula is equivalent to formula (5) (by (6)).

Problem 1. Solve the equation $dy/dx = y+h_N$, where $h_N(x) = N$ for $|x-1| < 1/2N$ and 0 for $|x-1| \geq 1/2N$ with initial condition $y(0) = 0$, and find the limit of the solution as $N \to \infty$.

5. Inhomogeneous Linear Equations with Periodic Coefficients

Theorem. *If the equation*

$$\frac{dy}{dx} = f(x)y + g(x)$$

with right-hand side of period T in x is such that the mean value of f over a period is nonzero, then the equation has a solution of period T, and moreover exactly one such solution (stable if the average value is negative and unstable if it is positive, cf. Fig. 45*).*

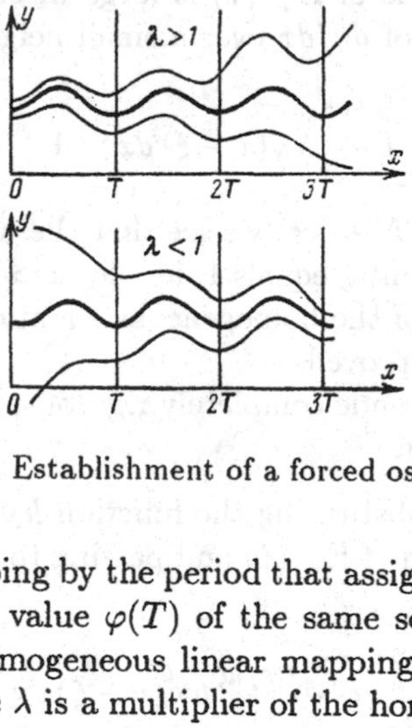

Fig. 45. Establishment of a forced oscillation

Proof. Consider the mapping by the period that assigns to the initial condition $\varphi(0)$ of a solution φ the value $\varphi(T)$ of the same solution at the instant T. This mapping is an inhomogeneous linear mapping (why?); it has the form $\varphi(T) = \lambda\varphi(0) + C$, where λ is a multiplier of the homogeneous equation. The logarithm of λ equals the integral of f over a period. Consequently $\lambda \neq 1$ if the mean value of f is not 0, and the assertion being proved follows from this fact. □

Thus *for $\lambda < 1$ after a certain "transition process" has been carried out a completely definite oscillatory mode is established in the system independently of the initial condition.* The oscillations that arise are called *forced oscillations.* They are caused by a periodic external action on the system, i.e., by the function g.

Problem 1. Find a periodic solution of the equation

$$\frac{dy}{dx} = -y + \sin x$$

and study its stability.

Remark. Inhomogeneous linear equations arise naturally in the cases when we study the influence exerted on the solution by small perturbations of the initial condition simultaneously with small perturbations of the *right-hand side* of the differential equation (neglecting infinitesimals of higher order than first with respect to the perturbations). The inhomogeneity g in Eq. (4) corresponds precisely to a perturbation of the *equation*.

For example under a small perturbation of the vector field in a neighborhood of a limit cycle with multiplier different from 1 the cycle does not disappear, but merely deforms slightly; the periodic solution of the corresponding inhomogeneous linear equation gives a first approximation to this deformation of the cycle.

Problem 2. Suppose the smooth function $\varphi(t,\varepsilon)$ is a solution of the equation $\dot{x} = v(t,x;\varepsilon)$ depending on the parameter ε and becomes a solution $\varphi_0(t)$ of the equation $\dot{x} = v(t,x;0)$ for $\varepsilon = 0$. Prove that the derivative of the solution with respect to the parameter, $\psi(t) = \partial\varphi/\partial\varepsilon|_{\varepsilon=0}$ satisfies the inhomogeneous linear equation $\dot{\psi} = f(t)\psi + g(t)$, where f and g are the values of $\partial v/\partial x$ and $\partial v/\partial\varepsilon$ for $\varepsilon = 0$, $x = \varphi_0(t)$. This equation is called an (inhomogeneous) ***equation in variations***, since ψ describes a small variation in the solution under the action of a small change in the equation corresponding to $\varepsilon = 0$.

§ 4. Phase Flows

The mathematical formalization of the concept of a deterministic process leads to the concept of a one-parameter transformation group.

In this section we define and study one-parameter diffeomorphism groups and their connections with vector fields. We shall need some algebraic terminology. All the theorems in this section are essentially obvious.

1. The Action of a Group on a Set

A *transformation* of a set is a one-to-one mapping of the set onto itself.

Problem 1. Which of the three following mappings is a transformation?

$$1)\ \boldsymbol{R} \to \boldsymbol{R},\ x \mapsto e^x;\quad 2)\ \boldsymbol{R} \to \boldsymbol{R},\ x \mapsto x^3;\quad 3)\ \boldsymbol{C} \to \boldsymbol{C},\ z \mapsto z^2.$$

Answer. Only the second.

The *product* fg of the transformations f and g of a set is the transformation obtained by applying first g, then f, i.e., $(fg)(x) = f(g(x))$.

Problem 2. Give an example in which fg is not the same as gf.

The transformation f^{-1} *inverse* to f is defined by the condition that if f takes x to y, then f^{-1} takes y to x.

A collection of transformations of a set is called a *transformation group* if it contains the inverse of each of its transformations and the product of any two of its transformations.

Problem 3. Is the set of the three reflections about the vertices of an equilateral triangle a transformation group?

Problem 4. How many elements are there in the group of isometries[12] of an equilateral triangle? In the group of rotations of a tetrahedron?

Answer. 6, 12.

The concept of a transformation group is one of the most fundamental in all of mathematics and at the same time one of the simplest: the human mind naturally thinks in terms of invariants of transformation groups (this is connected with both the visual apparatus and our power of abstraction).

Let A be a transformation group on the set X. Multiplication and inversion define mappings $A \times A \to A$ and $A \to A$ (the pair (f, g) goes to fg, and the element g to g^{-1}). A set A endowed with these two mappings is called an *abstract group* (or briefly, simply a *group*). Thus a *group* is obtained from a *transformation group* by simply *ignoring* the set that is transformed.

Problem 5. Prove that the set $\boldsymbol{R}$ of all real numbers becomes a group when equipped with the operations of ordinary addition and changing the sign.

Algebraists usually define a group as a set with two operations satisfying a collection of axioms such as $f(gh) = (fg)h$. These axioms automatically hold for transformation groups. Actually these axioms mean simply that the group is formed from some transformation group by ignoring the set that is transformed. Such axioms, together with other unmotivated definitions, serve mathematicians mainly by making it difficult for the uninitiated to master their subject, thereby elevating its authority.

Let G be a group and M a set. We say that an *action of the group G on the set M* is defined if to each element g of the group G there corresponds a transformation $T_g : M \to M$ of the set M, to the product of any two elements of the group corresponds the product of the transformations corresponding to these elements, and to any two mutually inverse elements correspond mutually inverse transformations: $T_{fg} = T_f T_g$, $T_{g^{-1}} = (T_g)^{-1}$.

[12] An isometry is a transformation that preserves distances (so that the distance between the images of two points equals the distance between the points themselves).

Each transformation group of a set naturally acts on that set ($T_g \equiv g$), but may also act on other sets. For example, consider an equilateral triangle. The group of its six isometries acts on the set of its two orientations: the reflections reverse the orientation, the rotations do not.

Problem 6. Which permutations of the three coordinate axes are realized by the action of the group of isometries of the cube $\max(|x|, |y|, |z|) \le 1$ on the set of axes?

Answer. All six.

Problem 7. How does the group of linear changes of coordinates act on the set of matrices of linear operators from a space into itself?

Answer. $T_g m = gmg^{-1}$.

The transformation T_g is also called the *action of the element* g of the group G on M. The action of the group G on M defines another mapping $T : G \times M \to M$ assigning to the pair $g \in G$, $m \in M$ the point $T_g m$.

If the action T is fixed, then the result $T_g m$ of the action of the element g of the group G on a point m of the set M is denoted by gm for short. Thus $(fg)m = f(gm)$, and so the parentheses are usually omitted.

Let us fix a point m of the set M and act on it by all the elements of the group G. We thereby obtain a subset $\{gm, g \in G\}$ of the set M. This subset is called the *orbit of the point* m (for the given group action), and is denoted Gm.

Problem 8. Find the orbits of the group of rotations of the plane about zero.

Problem 9. Prove that any two orbits of an action are either disjoint or coincident.

Problem 10. How many orbits are there in the action of the group of isometries of the tetrahedron on the set of unordered pairs of its edges?

Problem 11. How many colorings of the six faces of a cube by six colors $1, \dots, 6$ are essentially different (cannot be transformed into one another by rotations of the cube)?

Answer. $6!/24 = 30$.

A mapping $\varphi : G \to H$ of the group G into the group H is called a *homomorphism* if it takes products into products and inverses into inverses:

$$\varphi(fg) = \varphi(f)\varphi(g); \quad \varphi(g^{-1}) = (\varphi(g))^{-1}.$$

The action of a group G on a set M is a homomorphism of the group G into the group of all transformations of the set M.

2. One-parameter Transformation Groups

A group is called *commutative* (or *Abelian*) if the product is independent of the order of the factors: $fg = gf$ for any two elements of the group.

Example 1. The group of isometries of an equilateral triangle is not Abelian.

Example 2. The group of translations of the real axis is Abelian.

The operation in an Abelian group is usually denoted $+$.

For example, the successive application of translations by a and b in either order is a translation by $a + b$. Therefore the set of all real numbers with the operation of addition is an Abelian group; the natural action of this group on the line assigns to the number a the translation by a.

Definition. A *one-parameter group of transformations of a set* is an action on the set by the group of all real numbers.

Remark. Actions by the group of integers $\boldsymbol{Z}$ are sometimes called "one-parameter groups with discrete time." For such an action $T_n = (T_1)^n$, so that the whole group consists of powers of one transformation.

A one-parameter group of transformations of the set M is usually denoted $\{g^t\}$. Here $g^t : M \to M$ is the transformation corresponding to the point t of $\boldsymbol{R}$.

Thus a one-parameter group of transformations of the set M is a collection of transformations g^t parametrized by the real parameter t such that for any real numbers s and t

$$1) \quad g^{s+t} = g^s g^t, \quad 2) \quad g^{-t} = (g^t)^{-1}.$$

The parameter t is usually called *time* and the transformation g^t is called the *transformation in time t*.

Example 1. $M = \boldsymbol{R}$, g^t is translation by $2t$ (i.e., $g^t x = x + 2t$). Properties 1) and 2) are obvious.

Example 2. $M = \boldsymbol{R}$, g^t is dilation by a factor e^t (i.e., $g^t x = e^t x$). Properties 1) and 2) are obvious. The notation g^t derives from this example.

Example 3. $M = \boldsymbol{R}$, $g^t x = x + \sin t$. Property 2) holds, but Property 1) does not; $\{g^t\}$ is not a one-parameter group.

Remark. It follows obviously from Property 1) that g^0 is the identity transformation, which leaves each point fixed. Therefore Property 2) follows from Property 1). Property 1) is called the *group property*.

A one-parameter transformation group is the mathematical equivalent of the physical concept of a "two-sided deterministic process." Let M be the phase space of the process. A point of this space is a definite state of the process. Assume that at the instant $t = 0$ the process was in state x. Then at another moment t the process will be in another state. Denote this new state of the process by $g^t x$. We have defined for each t a mapping $g^t : M \to M$ from the phase space of the process into itself. The mapping g^t takes the state at the instant 0 to the state at the instant t. It is called the *transformation in time t*.

The mapping g^t really is a transformation (a mapping that is one-to-one and onto). This follows from the fact that, by the definition of determinacy, each state uniquely determines both the future and the past of the process. The group property also holds. Indeed, suppose at the initial instant the process was in state x. One may either pass to the state at the instant $t + s$ directly ($x \mapsto g^{t+s}x$) or first study the intermediate state $g^t x$ at which the process arrives in time t, then see where this intermediate state moves in time s. The agreement of the results ($g^{t+s}x = g^s g^t x$) means that the transition from the initial state to the final state in a fixed time always takes place in the same way, independently of the instant of time at which we leave the initial state.

A one-parameter group of transformations of the set M is also called a *phase flow* with the phase space M (the phase space can be thought of as filled with a fluid, particle x of which passes to the point $g^t x$ during time t).

The orbits of a phase flow are called its *phase curves* (or *trajectories*).

Example. Let g^t be a rotation of the plane about the point 0 through angle t. The group property obviously holds. The orbits of the phase flow $\{g^t\}$ are the point 0 and circles with center at 0.

The points that are phase curves are called *fixed points* of the flow.

3. One-parameter Diffeomorphism Groups

Assume now that the set M under consideration is endowed with the structure of a smooth manifold. Examples of smooth manifolds are: 1) any open domain of Euclidean space; 2) a sphere; 3) a torus. The general definition is given in Chapt. 5. For the time being we may assume that we are dealing with an open domain of Euclidean space.

A *diffeomorphism* is a mapping that is smooth, along with its inverse. (A mapping is called *smooth* if the coordinates of the image-point are smooth functions of the coordinates of the pre-image and vice versa.)

Problem 1. Which of the functions x, $-x$, x^2, x^3, $\arctan x$ define a diffeomorphism of the line onto itself?

Answer. Only the first two.

Definition. A *one-parameter diffeomorphism group* is a one-parameter transformation group whose elements are diffeomorphisms satisfying the additional condition that $g^t x$ depends smoothly on both of the arguments t and x.

Example 1. $M = \boldsymbol{R}$, g^t is multiplication by e^{kt}.

Example 2. $M = \boldsymbol{R}^2$, g^t is rotation about 0 by the angle t.

Remark. The condition of smooth dependence on the time t is needed in order to eliminate pathological examples such as the following: let $\{\alpha\}$ be a basis of the group $\boldsymbol{R}$, i.e., a set of real numbers such that each real number has a unique representation in the form of a finite linear combination of numbers of the set with integer coefficients. To each number α of the basis we assign the translation of the line by some distance, paying no attention to other elements of the basis. Setting $g^{\alpha_1+\cdots+\alpha_k} = g^{\alpha_1}\cdots g^{\alpha_k}$, we obtain a one-parameter transformation group each of whose elements is a translation of the line and consequently a diffeomorphism; but in general g^t is not a smooth function of t and is even discontinuous.

Instead of smoothness with respect to t one may require only continuity (from which smoothness is a consequence) but we have no need to do this.

Definition. A *one-parameter group of linear transformations* is a one-parameter diffeomorphism group whose elements are linear transformations.

Example. On the plane with coordinates (x, y) consider the transformation $g^t(x, y) = (e^{\alpha t}x, e^{\beta t}y)$.

It is clear that g^t is a linear transformation (in time t the x-axis is dilated by the factor $e^{\alpha t}$ and the y-axis by the factor $e^{\beta t}$).

The group property $g^{t+s} = g^t g^s$ follows from the exponential property ($e^{u+v} = e^u e^v$), and the smoothness of this dependence on t is also obvious. Thus $\{g^t\}$ is a one-parameter group of linear transformations of the plane.

In particular let $\alpha = 1$, $\beta = 2$ (Fig. 46). In this case the phase curves are the fixed point $(0, 0)$, the coordinate semi-axes, and semiparabolas; the action of one of the transformations of the phase flow on a domain E is depicted in Fig. 46. The areas of domains are increased by a factor of e^{3t} under the action of g^t.

Consider also the case $\alpha = 1$, $\beta = -1$ (Fig. 47). In this case the transformation g^t consists of a compression by a factor of e^t in the direction of the y-axis and a dilation by a factor of e^t in the direction of the x-axis. Such a transformation is called a *hyperbolic rotation*, since the phase curves of the flow $\{g^t\}$ are halves of the hyperbolas $xy = \text{const}$ (of course the equilibrium position 0 and the coordinate semi-axes are also phase curves). Hyperbolic rotations preserve area, although they strongly distort the shape of figures (Fig. 47).

We remark that our one-parameter group of linear transformations of the plane decomposes into the "direct product" of two one-parameter groups of linear transformations of lines (namely dilations of the axes).

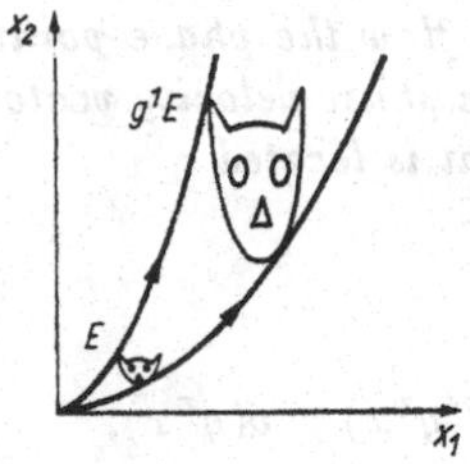

Fig. 46. The action of the phase flow on a domain

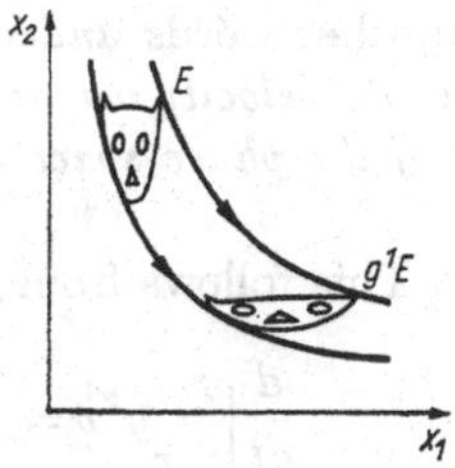

Fig. 47. A hyperbolic rotation

Problem 2. Does every one-parameter group of linear transformations of the plane decompose similarly?

Hint: Consider rotations or shifts $(x, y) \mapsto (x + ty, y)$.

4. The Phase Velocity Vector Field

Consider a one-parameter group $\{g^t\}$ of diffeomorphisms of a domain M.

Definition. The *phase velocity vector of the flow* $\{g^t\}$ at the point x in M is the velocity with which the point $g^t x$ leaves x, i.e.,

$$v(x) = \left.\frac{d}{dt}\right|_{t=0} (g^t x).$$

The phase velocity vectors of a flow at all points of the domain M form a smooth vector field (since $g^t x$ depends smoothly on t and x). It is called the *phase velocity field.*

Problem 1. Find the phase velocity fields of the following flows on the line: $g^t x = x + t$, $g^t x = e^t x$, $g^t x = e^{-t} x$

Answer. $v(x) = 1, x, -x$.

Problem 2. The fixed points of the flow are singular points of the phase velocity field, i.e., the phase velocity vector vanishes at these points. Is the converse true?

Answer. Yes. See Sect. 3 of § 2.

Let us fix a point x_0 and study its motion under the action of the phase flow g^t. In other words consider the mapping $\varphi : \boldsymbol{R} \to M$ defined as follows: $\varphi(t) = g^t x_0$.

Theorem. *The mapping* φ *is a solution of the equation* $\dot{x} = v(x)$ *with initial condition* $\varphi(0) = x_0$.

In other words *under the action of the phase flow the phase point moves so that its velocity vector at any instant equals the phase velocity vector at the point of the phase space at which the moving point is located.*

Proof. This follows from the group property:

$$\left.\frac{d}{dt}\right|_{t=\tau} g^t x = \left.\frac{d}{d\varepsilon}\right|_{\varepsilon=0} g^{\tau+\varepsilon} x = \left.\frac{d}{d\varepsilon}\right|_{\varepsilon=0} g^{\varepsilon}(g^{\tau} x) = v(g^{\tau} x).$$

□

Thus *with each one-parameter diffeomorphism group there is associated a differential equation* (defined by the phase velocity vector field); the solutions of this equation are the motions of the phase points under the action of the phase flow.

Problem 3. Is the converse true, i.e., is every solution given by the formula $\varphi(t) = g^t x_0$?

Answer. Yes, by the uniqueness theorem (§ 2, Sect. 3).

If the phase flow describes the course of a process with arbitrary initial conditions, then the differential equation defined by its phase velocity vector field determines the *local law of evolution of the process*; the theory of differential equations is supposed to reconstruct the past and predict the future knowing this law of evolution.

The statement of a law of nature in the form of a differential equation reduces any problem about the evolution of a process (physical, chemical, ecological, etc.) to a *geometric* problem of the behavior of the phase curves of the given vector field in the corresponding phase space.

Definition. The *phase flow of the differential equation* $\dot{x} = \boldsymbol{v}(x)$ is the one-parameter diffeomorphism group for which $\boldsymbol{v}$ is the phase velocity vector field.

To find the phase flow of an equation it suffices to solve the equation: $g^t x_0$ is the value of the solution φ at the instant t with the initial condition $\varphi(0) = x_0$.

Examples. The phase flow of the equation $\dot{x} = kx$ is the group $\{e^{kt}\}$. The phase flow of the equation of small oscillations of a pendulum ($\dot{x}_1 = x_2$, $\dot{x}_2 = -x_1$) consists of rotations of the plane through the angle t. The phase flow of the equation of small oscillations of the inverted pendulum ($\dot{x}_1 = x_2$, $\dot{x}_2 = x_1$) consists of hyperbolic rotations.

Problem 4. Find the phase flows of the differential equations $\dot{x} = 0$, $\dot{x} = 1$, $\dot{x} = x-1$, $\dot{x} = \sin x$ $(0 < x < \pi)$.

Answer. $g^t x = x$, $g^t x = x + t$, $g^t x = (x-1)e^t + 1$, $g^t x = 2\operatorname{arccot}(e^{-t}\cot x/2)$.

Problem 5. Find the phase flows of the systems

$$\begin{cases} \dot{x} = y, \\ \dot{y} = 0; \end{cases} \quad \begin{cases} \dot{x} = y, \\ \dot{y} = 1; \end{cases} \quad \begin{cases} \dot{x} = \sin y, \\ \dot{y} = 0. \end{cases}$$

Answer. $(x + ty, y)$, $(x + ty + t^2/2, y + t)$, $(x + t \sin y, y)$.

The question arises, *is every smooth vector field the phase velocity vector field of a flow?*

The answer to this question is negative.

Example 1. Consider the differential equation $\dot{x} = 1$ with phase space $0 < x < 1$. It is clear that the transformation g^t can be only a translation by t, but for $t \neq 0$ such a translation does not take the phase space into itself.

Example 2. Consider the case $v(x) = x^2$, $x \in \boldsymbol{R}$. The solution of the equation $\dot{x} = v(x)$ with initial condition x_0 for $t = 0$ can easily be found explicitly:

$$\frac{dx}{x^2} = dt, \quad -\frac{1}{x} = t + C, \quad C = -\frac{1}{x_0}, \quad x = \frac{x_0}{1 - x_0 t}.$$

Thus $g^t x = \dfrac{x}{(1 - tx)}$. It is not difficult to verify that $g^{t+s} = g^t g^s$, so that at first glance we appear to have found the phase flow.

Unfortunately the mapping g^t is not a diffeomorphism of the line for any value of t except $t = 0$ (it is not even defined everywhere). Therefore *the field $v(x) = x^2$ is not the phase velocity vector field of any one-parameter group of diffeomorphisms of the line.*

Remark. The reason why the two fields just given have no phase flows lies in the noncompactness of the phase space. We shall see below that a smooth vector field on a compact manifold always defines a phase flow. In particular the field $v(x) = x^2$ on the affine line can be extended to a smooth vector field on the entire projective line (including the point at infinity). The projective line is compact (a topological circle), and a smooth vector field on it defines a phase flow. The formulas we have found for the mappings g^t describe just this flow: g^t is a diffeomorphism of the projective line, not the affine line!

Problem 6. Prove that every smooth vector field on the line that has at most linear growth at infinity ($|v(x)| \leq a + b|x|$) is the phase velocity field of a one-parameter group of diffeomorphisms of the line.

Hint: By comparing the motion with a faster motion in a suitable linear field prove that the solution cannot become infinite in a finite time and consequently can be continued to the entire axis t.

Problem 7. Does the equation $\dot{x} = e^x \sin x$ define a phase flow on the line?

Answer. Yes.

Problem 8. Consider the vector space of all polynomials p of degree less than n in the variable x. Define a transformation in time t as the translation of the argument of the polynomial by t (i.e., $(g^t p)(x) \equiv p(x+t)$). Prove that $\{g^t\}$ is a one-parameter group of linear transformations and find its phase velocity vector field.

Solution. The vector field at the point p is the polynomial dp/dx.

§ 5. The Action of Diffeomorphisms on Vector Fields and Direction Fields

The main method of solving and studying differential equations is to choose a suitable change of variables, i.e., in geometric terms, a suitable diffeomorphism that simplifies the given vector field or direction field. In this section we introduce formal definitions of the necessary concepts. We begin by recalling certain simple facts from differential calculus.

1. The Action of Smooth Mappings on Vectors

In studying all possible mathematical objects it is useful to study mappings[13] along with objects. We recall the definition of the action of smooth mappings on vectors.

Let $f : M \to N$ be a smooth mapping of the domain M of a vector space into the domain N of a vector space, and let $\boldsymbol{v}$ be a vector attached at the point x of the pre-image domain M, i.e., an arrow with tail at the origin x (Fig. 48). Then at the image point $f(x)$ of the domain N there also arises a vector denoted by $f_{*x}\boldsymbol{v}$ and called the *image of the vector* $\boldsymbol{v}$ *under the mapping* f, defined as below.

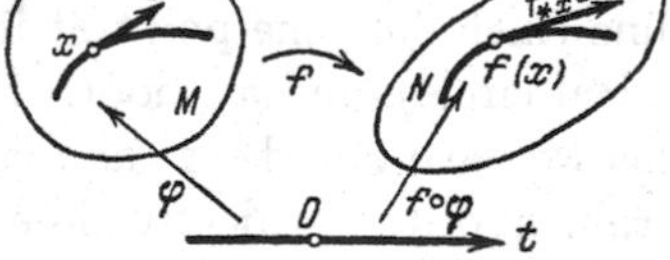

Fig. 48. The action of a smooth mapping on a vector

Definition. The *image of the vector* $\boldsymbol{v}$ *under the mapping* f is the velocity vector with which the moving point $f(\varphi(t))$ leaves the point $f(x)$ when the moving point $\varphi(t)$ leaves the point x with velocity $\boldsymbol{v}$:

[13] This is the essence of the so-called "category" point of view. Roughly speaking a category is a collection of objects and mappings (example: the category of all vector spaces and linear mappings of them).

$$f_{*x}\boldsymbol{v} = \left.\frac{d}{dt}\right|_{t=0} f(\varphi(t)), \text{ where } \varphi(0) = x, \quad \left.\frac{d}{dt}\right|_{t=0} \varphi(t) = \boldsymbol{v}. \tag{1}$$

In other words the arrow $\boldsymbol{v}$ is shortened by a factor of 1000, then under the action of f becomes a bent arrow, then the latter is lengthened by a factor of 1000, and finally 1000 tends to infinity.

Problem 1. Prove that the image of the vector $\boldsymbol{v}$ does not depend on the choice of the motion φ provided the point $\varphi(t)$ leaves x with velocity $\boldsymbol{v}$.

Solution. Let ψ be another motion leaving x with the same velocity. Then the distance between the points $\varphi(t)$ and $\psi(t)$ for small $|t|$ is $o(|t|)$. Since the mapping f is smooth, the distance between the image-points $f(\varphi(t))$ and $f(\psi(t))$ at N is also $o(|t|)$, which was required.

Problem 2. Suppose $\boldsymbol{v}$ is a positive unit vector of the line attached at the point a and let $f(x) = x^2$. Find $f_{*a}\boldsymbol{v}$.

Answer. $2a$ · unit vector.

Problem 3. Can two points on a plane moving along different coordinate axes leave the origin with the same velocity vector?

Answer. Yes, if the velocity is zero. For example: $\varphi(t) = (t^2, 0)$, $\psi(t) = (0, t^2)$.

The set of velocity vectors of motions leaving point x of a domain M is a vector space: this is simply the space of vectors attached at the point x. Its dimension is the dimension of the domain M. This space is called the *tangent space* to the domain M at the point x and is denoted T_xM.

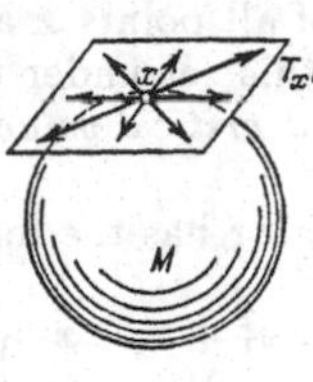

Fig. 49. The tangent space

Everyone encountering this for the first time finds it difficult to distinguish the tangent space to a vector space from the vector space itself. The following generalization is presented as an aid in dealing with this difficulty. Consider a smooth surface M in $\boldsymbol{R}^3$, for example a sphere. The velocity vectors with which a point moving over the sphere can leave a given point of the sphere obviously form a plane (the two-dimensional tangent space of the sphere at the given point x); this tangent plane T_xM (Fig. 49) is clearly distinct from the sphere M itself.

The mapping f_{*x} defined above maps the tangent space to the pre-image domain M at the point x into the tangent space to the image-domain at the point $f(x)$.

Problem 4. Prove that the mapping $f_{*x}: T_x M \to T_{f(x)} N$ is linear.

Solution. By Taylor's formula

$$f(x+vt) = f(x) + (\partial f/\partial x)vt + o(|t|),$$

and consequently $f_{*x} = \partial f/\partial x$ is a linear operator.

If Cartesian coordinates $(x_1, \ldots, x_m)$ and $(y_1, \ldots, y_n)$ have been chosen respectively in the pre-image and image spaces of the mapping f, so that f is given by a set of n functions f_i of m variables x_j, then the components of the vector $f_{*x}v$ are expressed in terms of the components of the vector v by the formula

$$(f_{*x}v)_i = \sum_i \frac{\partial f_i}{\partial x_j} v_j.$$

In other words the *matrix of the operator f_{*x} is composed of the partial derivatives $(\partial f_i/\partial x_j)$*.

Definition. The linear operator f_{*x} is called the *derivative of the mapping f* at the point x.

Problem 5. Consider the mapping f of the line into the plane $f(x) = (\sin x, \cos x)$. Find the value of its derivative on the vector v of length 10 attached at the point α and giving the positive orientation of the x-axis.

Answer. $f_{*\alpha}v = (10\cos\alpha, -10\sin\alpha)$.

Problem 6. Consider the mapping f of the plane into itself $f(x_1, x_2) = (x_1^3 + x_1 x_2, x_2)$ (Fig. 50). Find the set of all points x at which the linear operator f_{*x} is degenerate and find the image of this set under the mapping f (these two sets are called the sets of *critical points* and *critical values* respectively).

Solution. The matrix of the operator has the form

$$\begin{pmatrix} 3x_1^2 + x_2 & x_1 \\ 0 & 1 \end{pmatrix},$$

and therefore the derivative is degenerate on the parabola $x_2 = -3x_1^2$. Its image is the semicubical parabola $(y_1/2)^2 + (y_2/3)^3 = 0$.

The mapping of this problem is called the *Whitney mapping* (cusp). H. Whitney proved that the cusp singularity is typical for smooth mappings of the plane into itself (for example, every smooth mapping near f has a similar singularity near the origin.)

Remark. The linear structure (vector addition) in the tangent space to M at the point x was defined above using the linear structure of the ambient space in which M is imbedded, or in other words, using a Cartesian coordinate system.

Actually both the set $T_x M$ and the vector space structure on it can be defined independently of the choice of the coordinate system, even with curvilinear coordinate

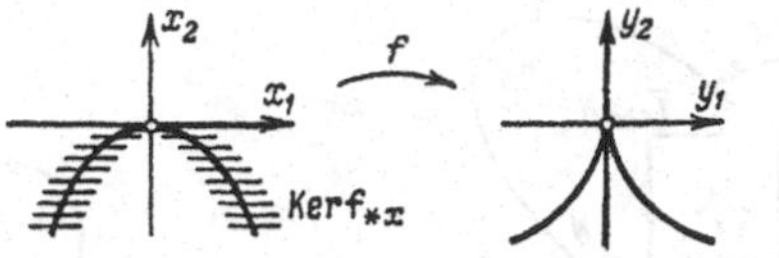

Fig. 50. The critical points and critical values of the Whitney mapping

systems, provided the system is admissible, i.e., connected to the Cartesian coordinate system by a smooth change of variables (diffeomorphism). The independence of the tangent space from the coordinate system is not entirely obvious, since the arrow sketched in the domain M (the attached vector) bends under a diffeomorphism.

A definition independent of the coordinate system for the exit velocity vector from the point x has a rather abstract appearance:

Definition. A *tangent vector* to a domain M at a point x is an equivalence class of smooth motions $\varphi : \boldsymbol{R} \to M$, for which $\varphi(0) = x$; the equivalence $\varphi \sim \psi$ is defined by the following condition: the distance between the points $\varphi(t)$ and $\psi(t)$ in some (and then every) system of coordinates is $o(|t|)$ as $t \to 0$ (Fig. 51).

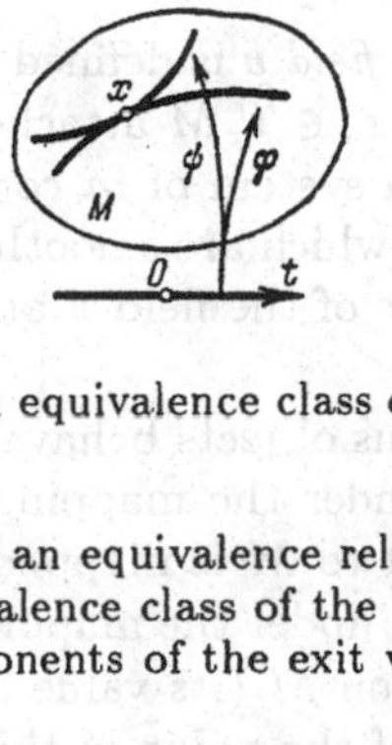

Fig. 51. An equivalence class of motions

It is clear that this is indeed an equivalence relation ($\varphi \sim \varphi$, $\varphi \sim \psi \Rightarrow \psi \sim \varphi$, $\varphi \sim \psi \sim \chi \Rightarrow \varphi \sim \chi$). The equivalence class of the motion φ is defined (for a fixed coordinate system) by the components of the exit velocity vector of $\varphi(t)$ from the point $\varphi(0)$.

Thus our vector defined without use of coordinates becomes an ordinary arrow as soon as a coordinate system is fixed. The only thing that needs to be proved is the independence of the vector operations (addition and scalar multiplication) from the coordinate systems occurring in their definition. But this independence follows immediately from the linearity of the operation of taking the derivative of a mapping at a point (the relevant mapping is a "change of variables," i.e., the diffeomorphism assigning to the set of old coordinates of a point the set of its new coordinates).

Although our definition is independent of the coordinate system, there remains a dependence on the whole class of coordinate systems connected by smooth changes of variables. This class is called a *differentiable structure* and the concepts just introduced depend essentially on it.

The derivative of the mapping f at the point x is a linear operator $f_{*x} : T_x M \to T_{f(x)} N$ that is *independent of both the coordinate system in the pre-image and the coordinate system in the image* by its very definition (1) (Fig. 52).

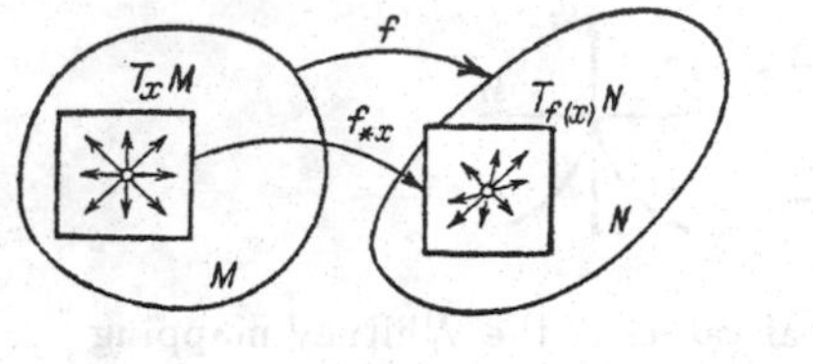

Fig. 52. The derivative of a mapping at a point

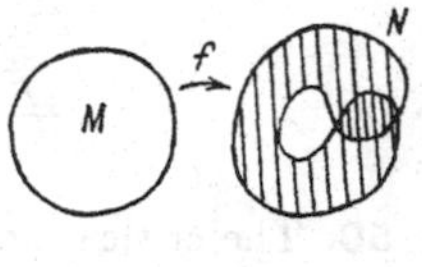

Fig. 53. A local diffeomorphism may fail to be a global diffeomorphism

Problem 7. Let f be a diffeomorphism of M onto N. Prove that the mapping f_{*x} is a vector-space isomorphism by its very definition (1) (Fig. 52).

Problem 8. Is the converse true?

Answer. No, even if f_{*x} is an isomorphism for every x (cf. Fig. 53).

2. The Action of Diffeomorphisms on Vector Fields

Definition. A smooth *vector field* $\boldsymbol{v}$ is defined in a domain M if to each point x there is assigned a vector $\boldsymbol{v}(x) \in T_xM$ attached at that point and depending smoothly on the point x (if a system of m coordinates is chosen, the field is defined by its m components, which are smooth functions of m variables). The vector $\boldsymbol{v}(x)$ is called the *value* of the field $\boldsymbol{v}$ at the point x.

Let us examine how various objects behave under smooth mappings. *Tangent vectors move forward* under the mappings $g : M \to N$ (i.e., under the action of g a vector $\boldsymbol{v}$ tangent to M is mapped into a vector $g_{*x}\boldsymbol{v}$ tangent to N). *Functions move backward* under the mappings $g : M \to N$, i.e., a function f on N generates a function on M (its value at the point $x \in M$ is equal to the value of f at the image of the point x; this function is denoted g^*f; the upper asterisk denotes *backward* motion).

Vector fields in general map neither backward nor forward. Indeed, under a mapping two points of the pre-image may map to the same point and carry different vectors with them, so that a field in the pre-image does not transform to the image. Moreover, many tangent vectors at a given point of the pre-image may have a common image, so that a field on the image does not transform to the pre-image.

Definition. The *image of a vector field under a diffeomorphism onto* is the vector field whose value at each point is the image of the vector of the original field at the pre-image of the given point. The image of the field $\boldsymbol{v}$ under the diffeomorphism g is denoted $g_*\boldsymbol{v}$.

In other words, the image $g_*\boldsymbol{v}$ of the field $\boldsymbol{v}$ in M under a *diffeomorphism* g of a domain M *onto* N is the field $\boldsymbol{w}$ in N defined by the formula $\boldsymbol{w}(y) = (g_{*x})\boldsymbol{v}(x)$, where $x = g^{-1}y$ (Fig. 54).

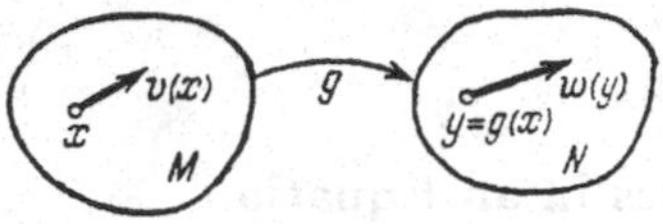

Fig. 54. The action of a diffeomorphism on a vector field

Problem 1. Find the image of the field $v(x) = 1$ on the line under the action of the diffeomorphism $g(x) = 2x$.

Answer. $(g_*v)(y) = 2$.

The vector field on the x-axis whose unique component is v is frequently denoted[14] by the symbol $v\partial/\partial x$. The convenience of this notation is that under dilations of the axis $\partial/\partial x$ behaves like $1/x$. For example, the solution of the preceding problem can be written as follows:

$$\frac{\partial}{\partial x} = \frac{\partial}{\partial(y/2)} = 2\frac{\partial}{\partial y}.$$

In this notation the formula for the action of a diffeomorphism of the line on a vector field assumes the form of the following formula for change of variable: if $y = g(x)$, then $\dfrac{\partial}{\partial y} = \dfrac{\partial}{\partial(g(x))} = \dfrac{1}{g'}\dfrac{\partial}{\partial x}$. Thus the notation $\partial/\partial x$ makes the computation of the action of diffeomorphisms on fields automatic.

Problem 2. Find the image of the field $x\partial/\partial x$ under the action of the diffeomorphism $y = e^x$.

Answer. $y\ln y\,\partial/\partial y$.

If $(x_1, \ldots, x_n)$ is a fixed coordinate system in $\boldsymbol{R}^n$, then the basis vector fields (with components $(1,0,\ldots,0),\ldots,(0,\ldots,0,1)$) are denoted $\partial/\partial x_1, \ldots, \partial/\partial x_n$. The field with components $(v_1, \ldots, v_n)$ is therefore denoted $v_1\partial/\partial x_1 + \cdots + v_n\partial/\partial x_n$.

Problem 3. Find the images of the "Euler field" $\boldsymbol{v} = x_1\partial/\partial x_1 + x_2\partial/\partial x_2$ on the plane under the action of the following diffeomorphisms: 1) a rotation about 0; 2) a hyperbolic rotation; 3) any linear transformation.

Answer. $\boldsymbol{v}$.

[14] In reality $v\partial/\partial x$ is the operator of differentiation in the direction of the field $\boldsymbol{v}$ (cf. § 10), but since the operator $v\partial/\partial x$ and the field $\boldsymbol{v}$ determine each other uniquely, they are frequently identified.

Problem 4. Prove that a diffeomorphism taking the vector field $\boldsymbol{v}$ to the field $\boldsymbol{w}$ takes the phase curves of the field $\boldsymbol{v}$ to the phase curves of the field $\boldsymbol{w}$. Is the converse true?

Answer. No, for example: $\boldsymbol{v} = x\partial/\partial x$, $\boldsymbol{w} = 2x\partial/\partial x$.

3. Change of Variables in an Equation

Let $\boldsymbol{w}$ be the image of the vector field $\boldsymbol{v}$ in M under a diffeomorphism g of a domain M onto a domain N, i.e., $\boldsymbol{w} = g_*\boldsymbol{v}$.

Theorem. *The differential equations*

$$\dot{x} = \boldsymbol{v}(x), \quad x \in M \tag{1}$$

and

$$\dot{y} = \boldsymbol{w}(y), \quad y \in N \tag{2}$$

are equivalent in the sense that if $\varphi : I \to M$ is a solution of the first, then $g \circ \varphi : I \to N$ is a solution of the second equation and conversely.

In other words: *the change of variables $y = g(x)$ takes Eq. (1) into Eq. (2)*. Or again: *substituting $g(x)$ for y turns Eq. (2) into Eq. (1)*.

Proof. This is obvious. In other words, applying successively the rule for differentiating a composite function, the definition of a solution φ, and the definition of the field $g_*\boldsymbol{v}$, we find $\frac{d}{dt} g \circ \varphi = g_{*x}\dot{\varphi}(t) = g_{*x}\boldsymbol{v}(x) = \boldsymbol{w}(y)$, where $x = \varphi(t)$, $y = g(\varphi(t))$, which was to be proved. □

Problem 1. Solve the equation of small oscillations of a pendulum

$$\dot{x}_1 = x_2, \quad \dot{x}_2 = -x_1,$$

by passing to polar coordinates[15] through the substitution $x_1 = r\cos\theta$, $x_2 = r\sin\theta$.

Solution. By carrying out the substitution we find $\dot{r} = 0$, $\dot{\theta} = -1$, whence $x_1 = r_0\cos(\theta_0 - t)$, $x_2 = r_0\sin(\theta_0 - t)$.

Problem 2. Study the phase curves of the system

$$\begin{cases} \dot{x}_1 = x_2 + x_1(1 - x_1^2 - x_2^2), \\ \dot{x}_2 = -x_1 + x_2(1 - x_1^2 - x_2^2). \end{cases}$$

[15] Of course the usual conventions are needed concerning the multivaluedness of polar coordinates: the mapping $(r,\theta) \mapsto (x_1, x_2)$ is not a diffeomorphism of the plane onto the plane. For example, one can consider separately the diffeomorphism of the domain $r > 0$, $0 < \theta < 2\pi$ onto the plane less the positive x_1-semiaxis defined by this mapping and the diffeomorphism of the domain $r > 0$, $-\pi < \theta < \pi$ onto the plane less the negative x_1-semiaxis.

Solution. Passing to polar coordinates, we obtain

$$\dot{r} = r(1 - r^2), \quad \dot{\theta} = -1.$$

The phase curves of this system in the (r, θ)-plane coincide with the integral curves of the equation $dr/d\theta = r(r^2 - 1)$. Sketching these curves (Fig. 55) and returning to Cartesian coordinates, we obtain Fig. 56. The only singular point is the origin. The phase curves that originate near this point wind outwards toward the circle $x_1^2 + x_2^2 = 1$ as time passes. This circle is a closed phase curve (limit cycle). The phase curves also wind inward to this circle from the outside.

Passing to polar coordinates makes it possible to integrate explicitly the original system also.

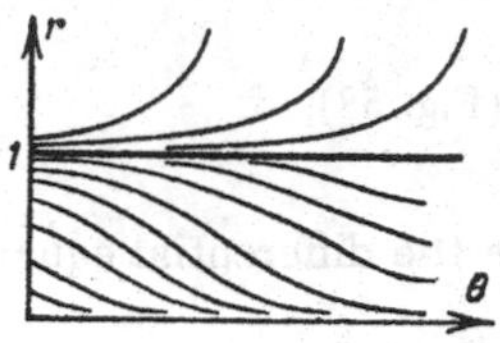

Fig. 55. Integral curves in the (r, θ)-plane

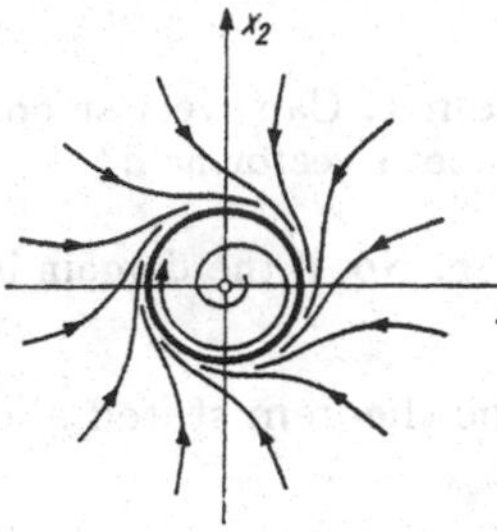

Fig. 56. Phase curves in the (x_1, x_2)-plane

4. The Action of a Diffeomorphism on a Direction Field

Let g be a diffeomorphism of a domain M onto a domain N, and suppose that a direction field is defined in the domain M. Then a direction field also arises in the domain N. It is called the *image of the original field under the action of the diffeomorphism* g and is defined as follows.

Consider some point y of the domain N (Fig. 57). In M it has a unique pre-image $x = g^{-1}M$. Consider the direction of this field at the point x. This is a line in the tangent space T_xM. Let us take any nonzero vector of this line. Its image under the action of g is a nonzero vector in the tangent space T_yN (since g is a diffeomorphism). The line defined by this vector is independent of the choice of the vector on the original line (since g_{*x} is a linear operator). This new line is a line of the new direction field at the point y. The following theorem is obvious.

Theorem. *Under the action of a diffeomorphism $g : M \to N$ the integral curves of the original direction field on M map into integral curves of the direction field on N obtained by the action of g on the original field.*

To prove this it suffices to extend the given direction field (in a neighborhood of each point of the domain M) to a vector field whose vectors lie on the lines of the given direction field and are different from zero, then apply the theorem of Sect. 3.

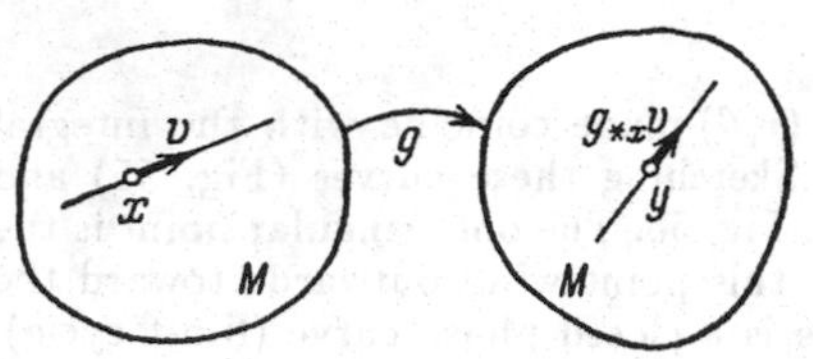

Fig. 57. The action of a diffeomorphism on a direction field

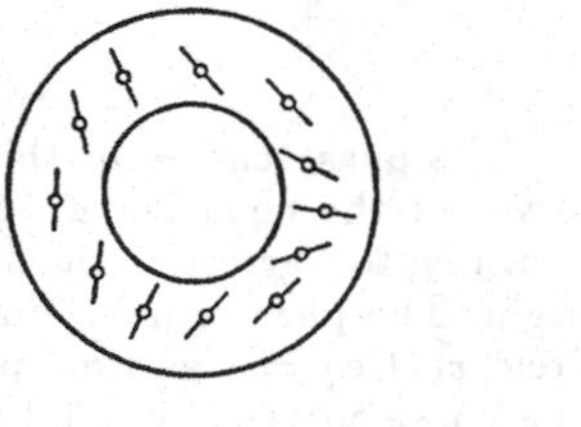

Fig. 58. A direction field that cannot be extended to a vector field

Problem 1. Can every smooth direction field in a domain of the plane be extended to a smooth vector field?

Answer. No, if the domain is not simply connected (Fig. 58).

The theorem stated above shows that to solve the differential equation

$$\frac{dx}{dt} = v(t, x)$$

it suffices to construct a diffeomorphism that maps the direction field to the direction field of an equation that we already know how to solve – for example, the field of an equation with separable variables. In other words it suffices to choose a change of variable that reduces the equation to one that has already been solved.

Problem 2. Choose a change of variables such that the variables become separable in the equation $\dfrac{dx}{dt} = \dfrac{x^2 - t^2}{x^2 + t^2}$.

Solution. Polar coordinates will do.

Problem 3. Find a diffeomorphism that takes all the integral curves of the equation $dx/dt = x - 1$ into parallel lines.

Solution. We solve the homogeneous equation: $x = Ce^t$. We find a particular solution of the inhomogeneous equation : $\dot{C}e^t = -1$, $C = e^{-t}$, $x = 1$.

Consequently each solution of the inhomogeneous equation has the form $x = 1 + ae^t$. The mapping taking (t, x) to (t, a) is the desired diffeomorphism $(a = e^{-t}(x-1))$, since a is constant along integral curves.

Another solution: Assign the point (t, y) to the point (t, x), where y is the ordinate of the point of intersection of the integral curve passing through the point (t, x) and the ordinate axis (Fig. 59).

Problem 4. Does every smooth direction field defined on the entire plane become a field of parallel lines under a suitable diffeomorphism?

Answer. No, see Fig. 60.

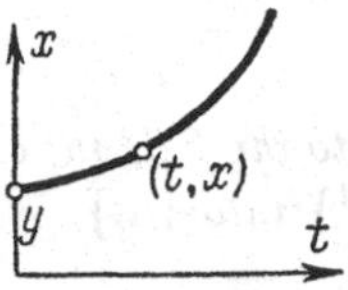

Fig. 59. Rectifying integral curves

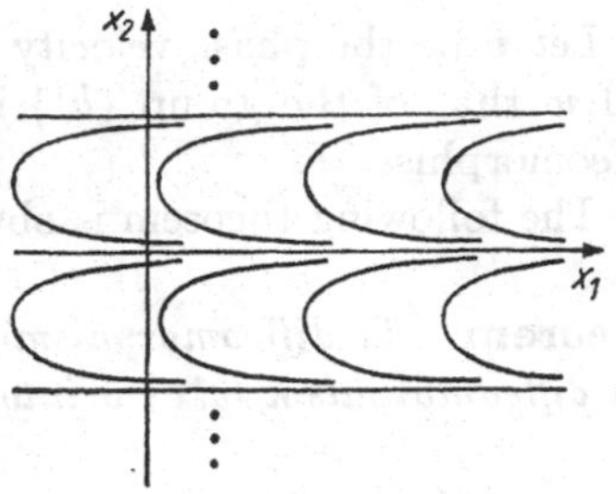

Fig. 60. A nonrectifiable direction field on the plane

Problem 5. Can a diffeomorphism of the plane map the direction field of the differential equation $\dot{x} = x^2$ into a field of parallel lines?

Answer. It is possible, though an explicit formula is difficult to write out.

5. The Action of a Diffeomorphism on a Phase Flow

Let $\{g^t : M \to M\}$ be a one-parameter diffeomorphism group, and let $f : M \to N$ be another *onto* diffeomorphism.

Definition. The *image of the flow* $\{g^t\}$ *under the action of the diffeomorphism* f is the flow $\{h^t : N \to N\}$, where $h^t = fg^tf^{-1}$.

In other words, the diagram

$$\begin{array}{ccc} M & \xrightarrow{g^t} & M \\ f\downarrow & & \downarrow f \\ N & \xrightarrow{h^t} & N \end{array}$$

is commutative for any t. It is clear that f takes the orbits of the group $\{g^t\}$ into orbits of the group $\{h^t\}$.

If we regard the diffeomorphism f as a "change of variables," then the transformation h^t is simply the transformation g^t "written in new coordinates."

Remark. The flows $\{g^t\}$ and $\{h^t\}$ are sometimes called *equivalent* (or *similar* or *conjugate*), and the diffeomorphism f is called an *equivalence* (or a *conjugating diffeomorphism*).

Problem 1. Prove that $\{h^t\}$ is a one-parameter diffeomorphism group.

Problem 2. Are the one-parameter groups of rotations of the plane and hyperbolic rotations of it conjugate?

Let $\boldsymbol{v}$ be the phase velocity vector field of the one-parameter group $\{g^t\}$ and $\boldsymbol{w}$ that of the group $\{h^t\}$ into which the first group is mapped by the diffeomorphism f.

The following theorem is obvious.

Theorem. *The diffeomorphism f takes the field $\boldsymbol{v}$ into the field $\boldsymbol{w}$; conversely, if a diffeomorphism takes $\boldsymbol{v}$ into $\boldsymbol{w}$, then it takes $\{g^t\}$ into $\{h^t\}$.*

Problem 3. Can the vector fields on the line defining the following five differential equations be transformed into one another by diffeomorphisms: $\dot{x} = \sin x$, $\dot{x} = 2\sin x$, $\dot{x} = \sin^2 x$, $\dot{x} = \sin 2x$, $\dot{x} = 2\sin x + \sin^2 x$?

Answer. The second can be transformed into the fourth and the fifth.[16]

§ 6. Symmetries

In this section we solve homogeneous and quasi-homogeneous differential equations. The solution of these equations is based on the use of one-parameter groups of symmetries of vector fields and direction fields, which we study first of all.

1. Symmetry Groups

Definition. A diffeomorphism $g: M \to M$ is called a *symmetry* of the vector field $\boldsymbol{v}$ on M if it maps the field into itself: $g_*\boldsymbol{v} = \boldsymbol{v}$. We also say that the field $\boldsymbol{v}$ is *invariant with respect to the symmetry* g.

Example. A rotation of the plane about zero is a symmetry of the Euler field $x_1\partial/\partial x_1 + x_2\partial/\partial x_2$ (whose vector at the point $\boldsymbol{x}$ is $\boldsymbol{v}(\boldsymbol{x}) = \boldsymbol{x}$ (Fig. 61)).

Problem 1. Suppose a diffeomorphism takes the phase curves of a vector field into one another. Is it a symmetry of the field?

Answer. Not necessarily.

Definition. A diffeomorphism $g: M \to M$ is called a *symmetry of a direction field on* M if it maps this direction field into itself; the field is then said to be

[16] The fifth case is extremely difficult!

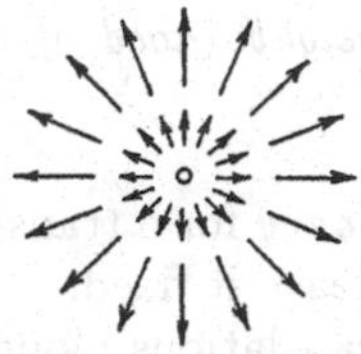

Fig. 61. The Euler field

invariant with respect to the symmetry. The integral curves of the field map into one another under the action of a symmetry.

Example. The direction field of the equation $\dot{x} = v(t)$ in the extended phase space is invariant with respect to translations along the x-axis (Fig. 4 on p. 17 above), and the field of the equation $\dot{x} = v(x)$ is invariant with respect to translations along the t-axis (Fig. 6 on p. 19).

Problem 2. Supppose a diffeomorphism maps the integral curves of a direction field into one another. Is it a symmetry of the direction field?

Answer. Yes.

A field is said to be *invariant with respect to a group* of diffeomorphisms if it is invariant with respect to each transformation of the group. In this case we say that the field *admits* this symmetry group.

Example. The Euler field on the plane admits, among others, the following four symmetry groups: the one-parameter group of dilations ($x \mapsto e^t x$), the one-parameter group of rotations through the angle t, the one-parameter group of hyperbolic rotations, and the group of all linear transformations of the plane $\mathrm{GL}(2, \boldsymbol{R})$.

All the symmetries of a given field form a group (prove this!).

Problem 3. Find the group of all symmetries of the Euler field on the plane.

Answer. $\mathrm{GL}(2, \boldsymbol{R})$.

2. Application of a One-parameter Symmetry Group to Integrate an Equation

Theorem. *Suppose a one-parameter group of symmetries of a direction field on the plane is known. Then the equation defined by this direction field can* *be integrated explicitly in a neighborhood of each nonstationary point of the symmetry group.*

A point is called *nonstationary* for a transformation group if not all of the transformations of the group leave it fixed.

If the group consists of translations along a line, the equation with this symmetry group was solved in § 1, p. 18 (Barrow's formula). We shall show that the general case reduces to this one by a suitable diffeomorphism (i.e., by an intelligent choice of local coordinates on the plane).

Lemma. *In a neighborhood of every nonstationary point of action of a one-parameter group of diffeomorphisms on the plane one can choose coordinates* (u, v) *such that the given one-parameter diffeomorphism group can be written in the form of a group of translations:*

$$g^s(u, v) = (u + s, v) \text{ for sufficiently small } |u|, |v|, |s|.$$

This formula says that the coordinate v indexes the orbits of the given group, and the coordinate u on each orbit is simply the time of the motion (measured from some line in the plane).

Proof. Through the given point O we pass a line Γ intersecting transversally (at a nonzero angle) the phase curve $\{g^s O\}$ passing through the point (Fig. 62). Let v be the coordinate of a point $\gamma(v)$ on this line, measured from the point O. Consider the mapping Φ of the (u, v)-plane into our plane that takes the point with coordinates (u, v) to the point $g^u\gamma(v)$. This mapping is a diffeomorphism of a neighborhood of the point $(0, 0)$ onto a neighborhood of the point O. Therefore (u, v) are local coordinates. In (u, v)-coordinates the action of g^s assumes the required form, since $g^s g^u = g^{s+u}$. □

The theorem follows from the lemma, since in the (u, v)-coordinate system the slope of the given direction field is independent of u.

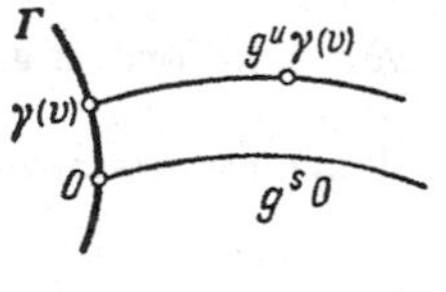

Fig. 62. Rectification of a one-parameter diffeomorphism group

Remark. This proof also gives an *explicit method of integrating* an equation; in the coordinates of the lemma the equation assumes the form $dv/du = w(v)$ (the line Γ must be taken nontangential to the direction of the given field at O). In practice it is not always convenient to use just these coordinates. It suffices that the lines $v =$ const be the orbits of the given one-parameter diffeomorphism group; as the other coordinate one can take any function of

u instead of u itself, say z. All that matters is that the transformations g^s map the lines $z = \text{const}$ into one another. In the (z, v)-coordinate system the original direction field defines an equation with separable variables $dv/dz = w(v)f(z)$, where $f(z) = du/dz$.

Problem 1. Suppose a one-parameter group of symmetries of a direction field in an n-dimensional domain is known. Reduce the problem of integrating the corresponding differential equation to finding the integral curves of a direction field on a domain of dimension $n - 1$.

Hint: The space of orbits of the symmetry group has dimension $n - 1$.

3. Homogeneous Equations

Definition. An equation is called *homogeneous* if the direction field defining it on the plane is homogeneous, i.e., invariant with respect to the one-parameter group of dilations, $g^s(x, y) = e^s(x, y)$ (Fig. 63).

The domain of definition of such a field is not necessarily the entire plane: it suffices that this field be defined on some domain that is invariant with respect to dilations (for example, in a sector).

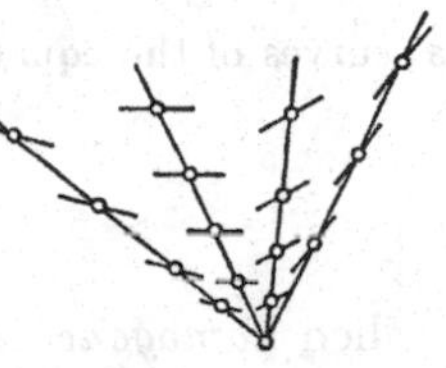

Fig. 63. The direction field of a homogeneous equation

Problem 1. Which of the following equations are homogeneous: $dy/dx = y/x$, $dy/dx = x/y$, $dy/dx = \ln x - \ln y$ $(x > 0,\ y > 0)$?

Answer. All three.

Theorem. *A homogeneous equation $dy/dx = F(x, y)$ can be reduced to an equation with separable variables by the substitution $y = vx$ (i.e., by passing to (x, v)-coordinates) in the domain $x > 0$.*

Proof. The orbits of the group of dilations are the rays passing through the origin (Fig. 64). As the line Γ we take the straight line $x = 1$ with the usual parameter y on it. The coordinates u and v exhibited in the lemma are $u = \ln x$, $v = y/x$.

The variables are separable in (x, v)-coordinates by the remark in Sect. 2. □

Problem 2. Solve the equation $dy/dx = y/x + y^2/x^2$, $x > 0$.

Solution. $dy = v\,dx + x\,dv$, $dy/dx = v + x\,dv/dx$, $x\,dv/dx = v^2$, $-1/v = \ln x + C$, $y = -x/(\ln x + C)$.

If K is an integral curve of a homogeneous equation, then the curve $e^s K$ homothetic to it is also an integral curve (Fig. 65). Thus to study all the integral curves of a homogeneous equation it suffices to sketch one curve in each sector of the plane.

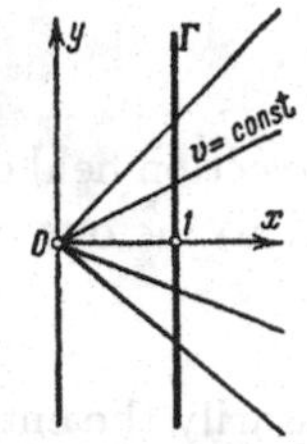

Fig. 64. Coordinates for solving a homogeneous equation

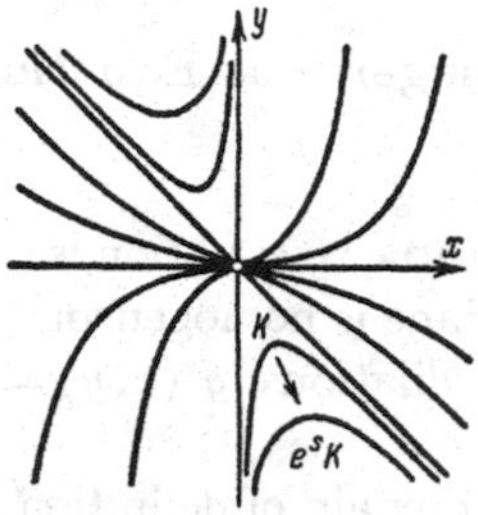

Fig. 65. The integral curves of a homogeneous equation

Problem 3. Sketch the integral curves of the equation $dy/dx = 2y/x + y^2/x^2$.

Answer. See Fig. 65.

Definition. A function f is called *homogeneous of degree r* if it satisfies the relation

$$f(e^s x) \equiv e^{rs} f(x) \tag{1}$$

identically.

In other words a *homogeneous function of degree r* is a common eigenvector of all the linear operators $(e^s)^*$ with eigenvalues e^{rs}.

The operator g^* (action of the diffeomorphism g on a function) is defined in § 5, p. 58.

Example. Let us draw the line $p + q = r$ in the (p, q)-plane. A polynomial $\sum a_{p,q} x^p y^q$ is homogeneous of degree r if and only if the exponents of each monomial occurring in it with nonzero coefficient lie on this line (called a *Newton diagram*).

Theorem (Euler). *A necessary and sufficient condition for a function f to be homogeneous of degree r is that it satisfy the Euler relation* $\sum x_i \partial f/\partial x_i = rf$.

The Euler relation says that f is an eigenvector of the operator of differentiation along the Euler field depicted in Fig. 61 (the phase velocity field of the group of dilations e^s) with eigenvalue r.

Proof. The Euler relation is obtained by differentiating the definition (1) of a homogeneous function with respect to s at $s = 0$. Relation (1) is obtained from Euler's relation by integrating the differential equation with separable variables that is defined by the Euler relation on each orbit of the group of dilations: $df/dx = rf/x$. □

A necessary and sufficient condition for the direction field of the differential equation $dy/dx = F(x,y)$ to be homogeneous is that the right-hand side be a homogeneous function of degree zero. For example the ratio of any two homogeneous polynomials of the same degree will do.

Remark. Passing from (x, y)-coordinates to the coordinates $(x, v = y/x)$ in a domain where $x \neq 0$ and to coordinates $(u = x/y, y)$ in a domain where $y \neq 0$ is called the *σ-process* or *inflating the point* 0.

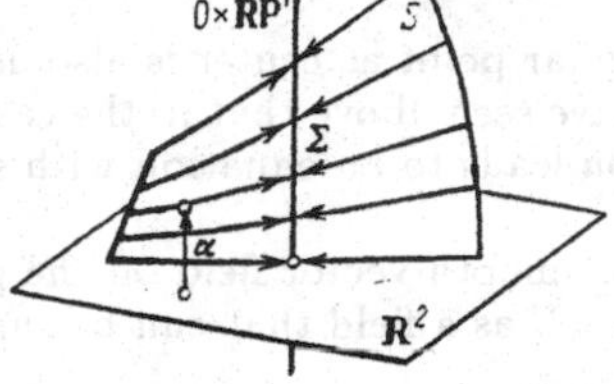

Fig. 66. The σ-process

This construction has a simple geometric meaning: it denotes the passage from the plane to the surface obtained from it by removing the origin and gluing in its place the whole projective line. Here is how this is done. Consider the mapping (fibration) $\alpha : (\boldsymbol{R}^2 \setminus 0) \to \boldsymbol{RP}^1$ that defines the projective line[17].

The mapping α assigns to a point of the plane the line joining that point to the origin. The graph of the mapping α (Fig. 66) is a surface S in the space $(\boldsymbol{R}^2 \setminus 0) \times \boldsymbol{RP}^1$. The imbedding of $\boldsymbol{R}^2 \setminus 0$ in $\boldsymbol{R}^2$ imbeds this graph in the product $M = \boldsymbol{R}^2 \times \boldsymbol{RP}^1$ (differomorphic to the interior of an anchor ring).

Problem 4. Prove that the closure of a graph in M is a smooth surface.

Hint: The equations $y = vx$ and $x = uy$ define smooth surfaces.

This surface Σ (the closure of the graph) consists of the graph itself and the line $0 \times \boldsymbol{RP}^1$ (diffeomorphic to a circle). The projection of M onto the first factor $\boldsymbol{R}^2$ defines a smooth mapping of the surface Σ onto the plane. This mapping is called a *deflation*. It takes the entire circle $0 \times \boldsymbol{RP}^1$ to the point 0 and maps the remaining part of Σ (i.e., the graph) diffeomorphically onto the plane with a point removed.

Problem 5. Prove that the surface Σ is diffeomorphic to a Möbius band.

[17] The *projective line* is the set of all lines in the plane passing through the origin. In general the projective space $\boldsymbol{RP}^{m-1}$ is the set of lines passing through the origin in $\boldsymbol{R}^m$.

Any geometric object having a singularity at the point 0 can be lifted from the plane with a point removed to Σ using the diffeomorphism exhibited above. When this is done, it turns out that the singularities become simplified in the lifting (to Σ).

Repeating the inflation process, it is possible to *resolve* the singularity. For example, one can turn any algebraic curve with a singularity at the point 0 into a curve having no singularities except ordinary self-intersections.

Problem 6. Resolve the singularity of the semicubical parabola $x^2 = y^3$.

Answer. See Fig. 67.

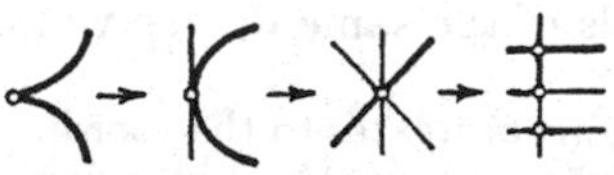

Fig. 67. Resolving a singularity

Inflation about a singular point as center is also useful in studying vector fields and direction fields. We have seen above that in the case of a homogeneous direction field the very first inflation leads to an equation with separable variables.

Problem 7. Prove that a smooth vector field on the plane equal to 0 at the origin can be lifted to the surface Σ as a field that can be smoothly extended to the circle glued in during the σ-process.

4. Quasi-homogeneous Equations

Let us fix a system of linear coordinates (x, y) in the plane and two real numbers α and β.

Definition. A *group of quasi-homogeneous dilations of the plane* is a one-parameter group of linear transformations

$$g^s(x, y) = (e^{\alpha s}x, e^{\beta s}y).$$

The numbers α and β are called the *weights* of the variables x and y. (the terms *weighted-homogeneous* and *generalized-homogeneous* are also used along with *quasi-homogeneous*.) Notation: $\alpha = \deg x$, $\beta = \deg y$.

If $\alpha = \beta = 1$, then $\{g^s\}$ is the usual dilation group.

Definition. An equation is called *quasi-homogeneous* (*with weights* α *and* β) if the direction field that defines it in the plane is invariant with respect to the group of quasi-homogeneous dilations.

Problem 1. Choose the weights so that the direction field of the equation $dy/dx = -x/y^3$ is quasi-homogeneous.

Answer. $\alpha = 2$, $\beta = 1$.

Theorem. *A quasi-homogeneous equation* $dy/dx = F(x,y)$ *with weights* $\deg \alpha$ *and* $\deg \beta$ *can be reduced to an equation with separable variables by passing to the coordinates* $(x, y^\alpha/x^\beta)$ *in the domain* $x > 0$.

Proof. The orbits of the group of quasi-homogeneous dilations are halves of "parabolas" $y^\alpha = Cx^\beta$ (Fig. 30, p. 40). We choose as the line Γ (Sect. 2) the line $x = 1$ with the parameter y on it. Quasi-homogeneous dilations map the lines parallel to Γ into parallel lines. Therefore the theorem follows from the lemma of Sect. 2 and the remark following it. □

We now explain how to tell whether an equation is quasi-homogeneous or not from its right-hand side.

Definition. A function f is called *quasi-homogeneous of degree* r if it satisfies the identity $f(e^{\alpha s}x, e^{\beta s}y) \equiv e^{rs}f(x,y)$.

In other words f is a common eigenvector of the operators $(g^s)^*$ (where g^s is a quasi-homogeneous dilation) with eigenvalues e^{rs}.

Example. A polynomial is quasi-homogeneous of degree r (with weights α and β) if and only if the exponents of the monomials $x^p y^q$ occurring in it lie on the Newton diagram $\alpha p + \beta q = r$ (Fig. 68).

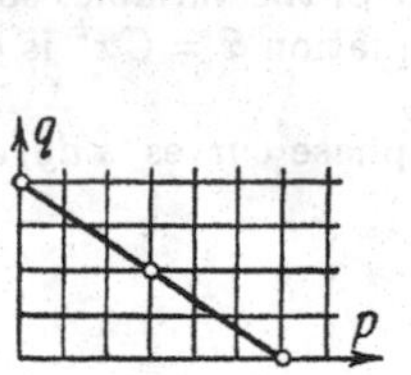

Fig. 68. The Newton diagram of a quasi-homogeneous function

The quasi-homogeneous degree of a quasi-homogeneous polynomial is also called its *weight*. For example, the weight of x being α and that of y being β, the weight of x^2y^3 will be $2\alpha + 3\beta$, etc. The assignment of weights is also called *grading*.

Problem 2. Choose the weights of the variables so that the polynomial $x^2y + y^4$ will be quasi-homogeneous of degree 1.

Answer. $\deg y = 1/4$, $\deg x = 3/8$.

Problem 3. Prove that a function f of the variables x_i with weights α_i is quasi-homogeneous of degree r if and only if it satisfies the Euler relation $\sum \alpha_i x_i \partial/\partial x_i = rf$.

Remark. A vector field $\sum \alpha_i x_i \partial/\partial x_i$ is called a *quasi-homogeneous Euler field* (it is the phase velocity field of a group of quasi-homogeneous dilations). The Euler relation says that f is an eigenvector of the operator of differentiation along the Euler field with eigenvalue r.

Theorem. *A necessary and sufficient condition for the direction field of the equation $dy/dx = F(x, y)$ to be quasi-homogeneous is that the right-hand side be quasi-homogeneous and its quasi-homogeneous degree be equal to the difference of the degrees of y and x:*

$$\deg F = \deg y - \deg x = \beta - \alpha.$$

Proof. Under the action of quasi-homogeneous dilations g^s the quantity y, and consequently dy also, is increased by a factor of $e^{\beta s}$ and x (and consequently also dx) by a factor of $e^{\alpha s}$. For a direction field to map into itself under such a dilation it is necessary that the value of F at the new point be obtained from the value at the old point by multiplying by the same factor by which dy/dx (or y/x) is multiplied, i.e., $e^{(\beta-\alpha)s}$. □

Remark. Thus in calculating the weights one may deal with dy/dx as if it were a fraction, regarding d as a "coefficient" of weight zero. Then the weight of dx is α, that of dy is β, and the weight of dy/dx is $\beta - \alpha$.

The condition for quasi-homogeneity of an equation is that the weights of the left- and right-hand sides be the same.

Problem 4. Choose the weights of the variables so that the differential equation of the phase curves of Newton's equation $\ddot{x} = Cx^k$ is quasi-homogeneous.

Solution. The equation of the phase curves is $dy/dx = Cx^k/y$. Consequently $2\beta = (k+1)\alpha$.

5. Similarity and Dimensional Considerations

The *quasi-homogeneous equations* with phase spaces of any dimension are defined in analogy with what was done above for the two-dimensional case. *Quasi-homogeneous vector fields* are defined by the condition $\deg \partial/\partial x_i = -\deg x_i$. For example the Euler field has degree 0.

Problem 1. Prove that if f is a quasi-homogeneous function of degree r and $\boldsymbol{v}$ a quasi-homogeneous field of degree s, then the derivative of f along $\boldsymbol{v}$ is a quasi-homogeneous function of degree $r + s$.

Problem 2. Let $\dot{x} = P$, $\dot{y} = Q$, where P and Q are homogeneous polynomials of degree m. Prove that if any of the phase curves is closed and is traversed in time

T, then when dilated by a factor of e^s it yields a closed phase curve with period of revolution $e^{s(1-m)}T$.

Problem 3. Let $\dot{x} = \boldsymbol{v}(x)$, where $\boldsymbol{v}$ is a quasi-homogeneous field of degree r. Prove that if T is the period of revolution over a closed curve γ and g^s is a quasi-homogeneous dilation, then $g^s\gamma$ is also a closed phase curve and its period of revolution is $e^{-sr}T$.

Problem 4. How does the period of oscillation of the "soft pendulum" $\dot{x} = y$, $\dot{y} = -x^3$ depend on the amplitude $x_{\max}$?

Answer. It is inversely proportional to the amplitude.

In applying similarity considerations one frequently encounters second derivatives as well as first derivatives. Let us see how they behave under quasi-homogeneous dilations. The following result is obvious.

Theorem. *Under a quasi-homogeneous dilation* $(x, y) \mapsto (e^{\alpha s}x, e^{\beta s}y)$ *the graph of the function* $y = \varphi(x)$ *transforms into the graph of the function* $y = \Phi(x)$, *for which*

$$\frac{d^k\Phi}{dx^k}(\textit{new point}) = e^{(\beta - k\alpha)s}\frac{d^k\varphi}{dx^k}(\textit{old point}).$$

In other words $d^ky/(dx)^k$ transforms like y/x^k (which explains the convenience of the Leibniz notation for the derivative). Consequently *in order to tell whether an equation containing derivatives of higher order is quasi-homogeneous, it suffices to ascribe a weight of* 0 *to the letter* d *and require that the weights of the right- and left-hand sides be the same.*

Problem 5. Prove that if a particle in a force field with homogeneous degree m traverses a trajectory γ in time T, then the same particle will traverse a homothetic trajectory $\lambda\gamma$ in time $T' = \lambda^{(1-m)/2}T$.

Solution. Newton's equation $d^2x/dt^2 = F(x)$, in which F is homogeneous of degree m transforms into itself under suitable quasi-homogeneous dilations: the weights α (for x) and β (for t) must be taken so that $\alpha - 2\beta = m\alpha$. We take $\alpha = 2$, $\beta = 1 - m$. To the dilation $x' = \lambda x$ corresponds the time $T' = \lambda^{(1-m)/2}T$.

Problem 6. Prove that the squares of the periodic times of similar trajectories in a gravitational field have the same ratio as the cubes of their linear dimensions[18].

Solution. From the preceding problem with $m = -2$ (the law of universal gravitation) we obtain $T' = \lambda^{3/2}T$.

Problem 7. Explain how the period of oscillation depends on the amplitude in the case of a restoring force proportional to the displacement (a "linear oscillator") and

[18] This is a special case of Kepler's Third Law, in which the trajectories are not assumed to be similar. The law of universal gravitation was found from the two preceding problems; Kepler's law was known earlier.

in the case of a restoring force proportional to the cube of the displacement (a "soft" force).

Answer. For the linear pendulum the period is independent of the amplitude; for the soft pendulum it is inversely proportional to the amplitude.

Problem 8. The equation of heat conduction has the form $\frac{\partial u}{\partial t} = a\frac{\partial^2 u}{\partial x^2}$ (t is time, x is distance, and u is the temperature). It is known that as a result of annual temperature variations the earth freezes to a depth of one meter in a certain place. To what depth would it freeze as a result of daily temperature variations of the same amplitude?

Solution. The equation transforms into itself under the quasi-homogeneous dilations $(t, x) \mapsto (e^{2s}t, e^{s}x)$. Consequently reducing the period by a factor of 365 leads to a reduction in the depth of freezing by a factor of $\sqrt{365}$. Hence the answer is 5 cm.

The use of similarity considerations originated with Galileo, who explained the limitations in size of land animals with it. The weight grows in proportion to the cube of the linear dimension and bone strength in proportion to the square. For marine animals there is no such restriction, and whales attain a much greater size than, say, elephants. Numerous applications of these considerations in various areas of natural science bear such names as: *similarity theory, dimension theory, scaling, self-modelling*, and others.

6. Methods of Integrating Differential Equations

There are several other techniques that sometimes make it possible to integrate a differential equation explicitly. For example, consider the equation

$$\frac{dy}{dx} = \frac{P(x,y)}{Q(x,y)}.$$

Let us rewrite it in the form

$$Q\,dy - P\,dx = 0$$

(a one-form equal to zero on vectors tangent to the integral curves). If the form is the total differential of a function,

$$Q\,dy - P\,dx = dF,$$

then the function F is constant along each integral curve.

Knowing the level lines of the function F, one can find the integral curves. It even suffices that the form $Q\,dy - P\,dx$ become an exact differential when multiplied by a suitable function (after all, multiplying P and Q simultaneously by the same function does not alter the original equation). Such a function is called an *integrating factor*. An integrating factor always exists (in a neighborhood of a point where Q is nonzero), but finding it is not any easier than solving the original equation.

The basic method of solving and studying differential equations is to select diffeomorphisms (changes of variables) that reduce the corresponding direction field, vector field, or phase flow to the simplest form. For example, for homogeneous and quasi-homogeneous equations such changes of variable were exhibited above.

There is a variety of techniques for finding changes of variables to integrate differential equations of special forms. Lists of such equations and techniques can be found in problem books (cf., for example, *Problems in Differential Equations*, by A. F. Filippov, (Freeman, San Francisco, 1963), §§ 4, 5, 6, 8, 9, 10) and in handbooks (cf., for example, the book of E. Kamke *Differentialgleichungen reeller Funktionen*, dritte Auflage, (Chelsea, New York, 1959), which contains about 1600 equations). Anyone can enlarge these lists as follows: take any equation that has already been solved and carry out any change of variables in it. The masters of integrating differential equations (Jacobi, for example) attained great success in the solution of specific applied problems using this technique. For the last decade we have witnessed an unexpected revival of interest in certain particular equations that can be integrated exactly, which have turned out to be connected with delicate questions of algebraic geometry on the one hand and with questions of the physics of particle solutions of partial differential equations (solitons, instantons, and the like) on the other.

However all these methods of integration have two essential defects. First, even an equation as simple as $dx/dt = x^2 - t$ cannot be solved by quadratures, i.e., the solution cannot be expressed as a finite combination of elementary and algebraic functions and their integrals[19]. Second, a cumbersome formula that gives a solution in explicit form is frequently less useful than a simple approximate formula. For example, the equation $x^3 - 3x = 2a$ can be solved explicitly by Cardan's formula $x = \sqrt[3]{a + \sqrt{a^2 - 1}} + \sqrt[3]{a - \sqrt{a^2 - 1}}$. However, if we wish to solve this equation for $a = 0.01$, it is more useful to observe that for small a it has approximately the root $x \approx -(2/3)a$ - a circumstance that is not at all obvious from Cardan's formula. Similarly the pendulum equation $\ddot{x} + \sin x = 0$ can be solved in explicit form using (elliptic) integrals. However it is simpler to solve the majority of questions about the behavior of the pendulum starting from the approximate equation for small oscillations ($\ddot{x} + x = 0$) and from qualitative considerations that make no use of the explicit formula (cf. § 12).

The exactly solvable equations are useful as examples, since one can often observe phenomena on them that also hold in more complicated cases. For example, studying the exact solution of the equation $\dot{x} = kx$ makes it possible

[19] The proof of this theorem of Liouville is similar to the proof of the unsolvability of equations of degree 5 in radicals (Ruffini-Abel-Galois); it is deduced from the fact that a certain group is not solvable. In contrast to the usual Galois theory, the discussion here involves an unsolvable Lie group rather than a finite group. The study of these questions is called *differential algebra*.

to prove a uniqueness theorem for the general equation with smooth right-hand side (cf. § 2, Sect. 3). Other examples are provided by the so-called self-modelling solutions of the equations of mathematical physics.

Problem 1. Find solutions of Laplace's equation[20] in $\boldsymbol{R}^2$ and $\boldsymbol{R}^3$ depending only on the distance from the point to the origin.

Answer. $C \ln 1/r + \text{const}$ and $C/r + \text{const}$ (Newtonian potentials; strictly speaking $\Delta(\ln 1/r) = -2\pi\delta$ in $\boldsymbol{R}^2$ and $\Delta(1/r) = -4\pi\delta$ in $\boldsymbol{R}^3$ (why?)).

Whenever an exactly solvable problem is found, it becomes possible to carry out an approximate study of nearby problems using the methods of perturbation theory.

However it is dangerous to extend results obtained in the study of an exactly solvable problem to nearby problems of a general form: frequently an exactly integrable equation can be integrated only because its solutions behave more simply than those of nearby nonintegrable problems. For example, we were able to integrate the equation of the phase curves of the Lotka-Volterra model (Sect. 7 of § 2) only because they are all closed curves (whereas for the majority of nearby nonintegrable models the majority of phase curves are nonclosed spirals).

[20] The Laplacian operator in Euclidean space $\boldsymbol{R}^n$ is the operator $\Delta = \operatorname{div}\operatorname{grad} = \sum \partial^2/\partial x_i^2$ (x_i are Cartesian coordinates). Laplace's equation has the form $\Delta u = 0$. The solutions of this equation are called *harmonic functions.* For example a steady-state temperature distribution is given by a harmonic function. The Laplacian operator measures the difference between the average value of a function in a small sphere and its value at the center of the sphere. The average of a harmonic function over any sphere is exactly equal to its value at the center of the sphere (prove this!).

Chapter 2. Basic Theorems

In this chapter we state theorems about the existence and uniqueness of solutions and first integrals and about the dependence of the solutions on the initial data and parameters. The proofs are discussed in Chapt. 4; in the present chapter we discuss only the connections among these results.

§ 7. Rectification Theorems

In this section we state the fundamental theorem on rectification of a direction field and deduce from it theorems on existence, uniqueness, and differentiable dependence of the solution on parameters and initial conditions, theorems on extension, and theorems on local phase flows.

1. Rectification of a Direction Field

Consider a smooth direction field in a domain U of n-dimensional space.

Definition. A *rectification* of a direction field is a diffeomorphism mapping it into a field of parallel directions (Fig. 69). A field is said to be *rectifiable* if there exists a rectification of it.

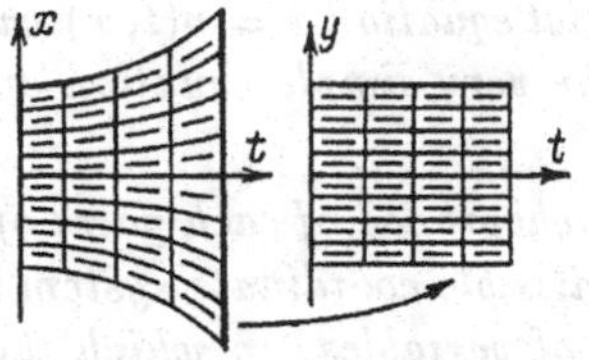

Fig. 69. Rectification of a direction field

Theorem 1 (Fundamental). *Every smooth direction field is rectifiable in a neighborhood of each point. If the field is r times continuously differentiable (of class C^r, $1 \leq r \leq \infty$), then the rectifying diffeomorphism can also be taken from the class C^r.*

Example. The direction field of the equation $\dot{x} = x$ (Fig. 69) can be rectified by the diffeomorphism $(t, x) \mapsto (t, y = xe^{-t})$. Indeed this diffeomorphism maps the integral curves $x = Ce^t$ in the (t, x)-plane to the parallel lines $y = C$ in the (t, y)-plane.

Problem 1. Rectify the direction fields of the equations $\dot{x} = t$ and $\dot{x} = x^2$ in a neighborhood of the origin.

Problem 2. Is every smooth direction field in the plane globally rectifiable?

Answer. No.

Problem 3. Suppose a (smooth) field of two-dimensional planes is given in $\boldsymbol{R}^3$ (a plane is attached at each point). Is it always possible to rectify this field (transform it into a field of parallel planes by a suitable diffeomorphism)?

Hint: A rectifiable field is a field of planes tangent to a family of surfaces.

Answer. No. Consider, for example, the field of planes given by the equation $y\,dx + dz = 0$ (a vector belongs to a plane of this field if this 1-form vanishes on it). There is *no* surface tangent to the planes of this field.

The proof of the fundamental theorem will be given in § 32. Here are two restatements of it.

Theorem 2. *All smooth direction fields in domains of the same dimension are locally diffeomorphic (can be mapped into each other by a diffeomorphism).*

1 ⇒ 2: By the fundamental theorem all fields are locally diffeomorphic to one standard field. 2 ⇒ 1: The property of being locally diffeomorphic to any field implies, in particular the property of being locally diffeomorphic to a standard field, i.e., local rectifiability.

Theorem 3. *The differential equation $\dot{x} = \boldsymbol{v}(t, x)$ with smooth right-hand side $\boldsymbol{v}$ is locally equivalent to the very simple equation $dy/d\tau = 0$.*

In other words, *in a neighborhood of each point of the extended phase space of (t, x) there exists an admissible coordinate system (τ, y) (transition to which is a diffeomorphic change of variables) in which the equation can be written in the very simple form $dy/d\tau = 0$.*

1 ⇒ 3: We first rectify the direction field $\boldsymbol{v}$ and then consider Cartesian coordinates in which the time axis τ is parallel to the lines of the rectified direction field. 3 ⇒ 1: Every direction field can be written locally as the direction field of a suitable differential equation. Passing to a local coordinate system in which the equation has the form $dy/d\tau = 0$ rectifies the given field.

Problem *4. Is it possible to rectify the direction field of the equation $\dot{x} = v(t,x)$ on the whole extended phase space $\boldsymbol{R} \times \boldsymbol{R}^n$ when the right-hand side is smooth and defined on this entire space?

Problem 5. Prove that the coordinate system in Theorem 3 can be chosen so that time is left fixed ($\tau \equiv t$).

Problem 6. Rectify the direction field of the equation $\dot{x} = x + t$ on the entire plane by a time-preserving diffeomorphism $(t,x) \mapsto (t, y(t,x))$.

Problem 7. Is it possible to rectify the direction field of the equation $\dot{x} = x^2$ on the entire plane by a time-preserving diffeomorphism?

Answer. No.

The fundamental theorem on rectification was essentially discovered by Newton. In his famous "second letter" to Oldenburg, the Secretary of the Royal Society (24 October 1676), Newton encoded the method of proving it in the form of a second (long) anagram (Newton preferred to conduct his correspondence with Leibniz, who lived in Germany, through Oldenburg). In modern terms Newton's method consists of the following.

Suppose given an equation $\dot{x} = v(t,x)$. We shall seek a rectifying diffeomorphism $y = h(t,x)$ for which $y = x$ when $t = 0$ (time is not changed). From the condition $\dot{y} = 0$ we obtain for h the equation $\partial y/\partial t + (\partial y/\partial x)v \equiv 0$. We expand v and h in series of powers of t:

$$h = h_0 + th_1 + \cdots, \qquad v = v_0 + tv_1 + \cdots.$$

Then $h_0(x) \equiv x$, and so $\partial h/\partial x = E + th_1 + \cdots$. We then substitute the series for h and v into the equation for h. We then expand the left-hand side in a series in t. We then set the coefficients of $t^0, t^1, \ldots$ equal to zero in this series (based on the uniqueness of the coefficients of a Taylor series). We obtain successively

$$h_1 + v_0 = 0, \quad 2h_2 + h_1 v_0 + v_1 = 0, \ldots.$$

The equation for h_k contains, besides h_k, only the derivatives of the h_m with smaller indices. Therefore we can sequentially ("recursively") find first h_1, then h_2 and hence all the terms of the series being sought.

This is *Newton's method of integrating differential equations by the use of series.* To apply this method one had to know how to expand given functions in series. To do that Newton had to discover his binomial formula $(1+t)^a = 1 + at + \cdots$.

Problem 8. Solve the equation $\dot{x} = x$ by Newton's method with the initial condition $\varphi(0) = 1$.

Solution. $\varphi = 1 + t\varphi_1 + t^2\varphi_2 + \cdots \Rightarrow \varphi_1 + 2\varphi_2 t + 3\varphi_3 t^2 + \cdots = 1 + \varphi_1 t + \varphi_2 t^2 + \cdots$, and consequently $\varphi_1 = 1$, $\varphi_2 = \varphi_1/2$, $\varphi_3 = \varphi_2/3, \ldots$, whence $\varphi_k = 1/k!$. This is the way in which the series for the exponential was originally discovered.

All the subsequent development of analysis, even today, follows the path marked out by Newton.

The proof of the convergence of the series constructed by Newton was much studied in the 19th century. The convergence of the series for h in the analytic case was proved by Cauchy[1]. Cauchy's theorem was extended to the case of finite smoothness by Picard. The proof is discussed in § 32.

The fundamental Theorem 1 is an assertion of the same character as the theorems of linear algebra on the reduction of quadratic forms or the matrices of linear operators to normal form. It gives an exhaustive description of the possible local behaviors of direction fields, reducing all questions to the trivial case of a parallel field.

A related theorem of analysis is the implicit function theorem. A smooth mapping $f : \boldsymbol{R}^m \to \boldsymbol{R}^n$ is called *nondegenerate at the point* 0 if the rank of the derivative at this point is as large as possible (i.e., the smaller of m and n). Suppose $f(0) = 0$.

Two such mappings f and g are called *locally equivalent at the point* 0 if one of them maps into the other under the action of diffeomorphisms of the domain and target spaces that leave 0 fixed: $h : \boldsymbol{R}^m \to \boldsymbol{R}^m$, $k : \boldsymbol{R}^n \to \boldsymbol{R}^n$, $f \circ h = k \circ g$.

In other words *two mappings are locally equivalent if under suitable choices of admissible local coordinate systems in the domain and target space (with origin at* 0*) they can be written by the same formulas.*

Implicit Function Theorem. *In some neighborhood of a nondegenerate point any two smooth mappings (of spaces of fixed dimensions m and n) are equivalent.*

In particular *every mapping is equivalent to its linear part at a nondegenerate point.* Therefore the theorem just stated is one of numerous theorems on linearization.

As a local normal form to which the mapping f reduces by diffeomorphisms h and k it is natural to take the following simplest one:

$$y_i = x_i \text{ for } i \le r, \quad y_i = 0 \text{ for } i > r,$$

where $r = \min(m, n)$ is the rank of the derivative of f at zero, x_i are the coordinates of a point in the domain space, and y_i are coordinates of a point in the target space. In other words, f is an imbedding if the dimension of the domain is less than that of the target space and a fibration otherwise.

The reader accustomed to more complicated statements of the implicit function theorem will easily verify that these more complicated statements are equivalent to the simple geometric statement given here.

All the theorems just listed on normal forms describe the orbits of the actions of various groups ("changes of variable") on sets (matrices, forms, fields, and mappings, respectively).

2. Existence and Uniqueness Theorems

The following corollary is a consequence of the fundamental Theorem 1 on rectification.

---

[1] Euler had already pointed out the necessity of a proof of convergence, noting that the series obtained by a similar route in other problems were sometimes divergent. Euler sought the solution of the equation $dx/dt = (x - t)/t^2$ equal to zero for $t = 0$ in the form of a series in t. The result was the everywhere-divergent series $x = \sum (k-1)!t^k$.

Corollary 1. *Through each point of a domain in which a smooth direction field is defined there passes an integral curve.*

Proof. Consider a diffeomorphism that rectifies the given field. The rectified field consists of parallel directions. In that field an integral curve passes through each point (a straight line, to be specific). The diffeomorphism inverse to the rectifying diffeomorphism maps this line into the desired integral curve. □

Corollary 2. *Two integral curves of a smooth direction field having a point in common coincide in a neighborhood of that point.*

Proof. For a rectified field this is obvious, but a rectifying diffeomorphism maps integral curves of the original field into integral curves of the rectified field. □

Corollary 3. *A solution of the differential equation $\dot{x} = v(t, x)$ with the initial condition (t_0, x_0) in the domain of smoothness of the right-hand side exists and is unique (in the sense that any two solutions with a common initial condition coincide in some neighborhood of the point t_0).*

Proof. We apply Corollaries 1 and 2 to the direction field of the equation in the extended phase space. The result is Corollary 3. □

Remark. In Corollary 3 and in what follows x is a point of a phase space of any (finite) dimension m. This corollary is called an *existence and uniqueness theorem for solutions of a system of m first-order equations.*

3. Theorems on Continuous and Differentiable Dependence of the Solutions on the Initial Condition

Consider the value of the solution φ of the differential equation $\dot{x} = v(t, x)$ with initial condition $\varphi(t_0) = x_0$ at the instant of time t as a function Φ of $(t_0, x_0; t)$ with values in the phase space.

The following result is a consequence of the fundamental Theorem 1 on rectification.

Corollary 4. *The solution of an equation with smooth right-hand side depends smoothly on the initial conditions.*

This means that *the function Φ shown above is defined, continuous, and smooth in a neighborhood of each point $(t_0, x_0; t_0)$ (of class C^r if v is of class C^r).*

Proof. For the simplest equation ($\boldsymbol{v} \equiv 0$) this is obvious ($\Phi \equiv x_0$). The general equation reduces to this one by a diffeomorphism (the details are left to the reader). □

Remark. The theorem on differentiability with respect to the initial condition provides a quite efficient method of studying the influence exerted on the solution by a small perturbation of the initial condition. If the solution is known for some initial condition, then to determine the deviation of the solution with a nearby initial condition from the given "unperturbed" solution a linear-homogeneous equation is obtained in first approximation (the equation of variations). The "perturbation theory" that arises in this way is but one of the variants of Newton's series method.

Problem 1. Find the derivative of the solution φ of the equation $\dot{x} = x^2 + x \sin t$ with respect to the initial condition $\varphi(0) = a$ for $a = 0$.

Solution. By Corollary 4 the solution can be expanded in a Taylor series in a: $\varphi = \varphi_0 + a\varphi_1 + \cdots$ (the dots stand for an infinitesimal of higher order than first in a). Here φ_0 is the unperturbed solution (with zero as initial condition), and φ_1 is the unknown derivative. For our equation $\varphi_0 \equiv 0$. Substituting the series into the equation and equating the terms of like degree in a on the left- and right-hand sides (by the uniqueness of Taylor series), we obtain for φ_1 the equation of variations $\dot{\varphi}_1 = \varphi_1 \sin t$ with initial condition $\varphi_1(0) = 1$ (why?). Hence the answer is $e^{1-\cos t}$.

Problem 2. Find the segment of a phase curve of the generalized Lotka-Volterra system $\dot{x} = x(1 - ya(x,y))$, $\dot{y} = y(x-1)$ passing through the point $x = 1$, $y = \varepsilon$ (with error of order ε^2).

Solution. The equation of the phase curves is $dy/dx = y(x-1)/(x(1-ya))$. The unperturbed solution is $y \equiv 0$. The equation of variations is $dy/dx = y(x-1)/x$. The answer is $y = \varepsilon e^{x-1}/x$ *independently of the form of the function* a.

Problem 3. Find the derivative of the solution of the pendulum equation $\ddot{\theta} = -\sin\theta$ with initial condition $\theta(0) = a$, $\dot{\theta}(0) = 0$ with respect to a at $a = 0$.

Solution. To apply Corollary 4 one must write the equation as a system of equations. The resulting system of equations of variations can be written in the form of a single second-order equation. It is convenient to write out only the second-order equations equivalent to these systems and their solutions rather than the systems themselves. The unperturbed solution is $\theta = 0$. The equation of variations is the equation for small oscillations of the pendulum: $\ddot{\theta} = -\theta$. The answer is $\cos t$.

Caution. In using the approximate formulas for the perturbed solutions obtained through the equation of variations, one must not forget that they give a good approximation for fixed t and small deviation ε of the initial condition from the unperturbed condition: the error for fixed t is $O(\varepsilon^2)$, but nonuniformly as $t \to \infty$ (the constant in O increases with t).

For example, the formula obtained in Problem 2 would give an incorrect picture of the form of the phase curves of the usual Lotka-Volterra model if we had tried to apply it to describe the form of these curves in the large (as we know from § 2, these curves are closed; the part of the curve far from the x-axis is by no means described by the solution to Problem 2).

In exactly the same way the solution of the full pendulum equation with initial condition $(a, 0)$ is near the solution of the equation of small oscillations (with the same initial condition) for fixed t: the difference between them is of order $O(a^3)$ (why?). However for any fixed $a \neq 0$ the error increases as t increases and for sufficiently large t the approximate solution loses contact with the perturbed solution (because of the difference of the periods of small and true oscillations). The limiting passages $t \to \infty$ and $a \to 0$ cannot be interchanged!

Problem 4. Find the first term (linear in a) of the Taylor series expansion of the solution of the equation of the soft pendulum $\ddot{x} = -x^3$ with initial condition $x(0) = 0$, $\dot{x}(0) = a$.

Solution. The unperturbed solution is $x \equiv 0$. The equation of variations is $\ddot{\varphi}_1 = 0$. The initial condition is $\varphi_1(0) = 0$, $\dot{\varphi}_1(0) = 1$ (why?). The answer is $x \approx at$.

It follows from the differentiability theorem that the error in this approximate formula is at most $O(a^2)$ for each fixed t. However for any fixed $a \neq 0$ the approximation becomes completely unsatisfactory for sufficiently large t. This can be seen, for example, from the fact that the approximate solution grows without bound, and the actual solution describes periodic oscillations of an amplitude that is small together with a (the size of the amplitude is of order $\sqrt{a}$ by similarity considerations).

To estimate the range of applicability of the approximate formula we can compute the following approximations: $x = at + a^2\varphi_2 + a^3\varphi_3 + \cdots$. Substituting into the equation, we obtain $a^2\ddot{\varphi}_2 + a^3\ddot{\varphi}_3 + \cdots = -a^3t^3 + \cdots$. Hence $\varphi_2 \equiv 0$, $\ddot{\varphi}_3 = -t^3$, $\dot{\varphi}_3 = -t^4/4$, $\varphi_3 = -t^5/20$, and $x \approx at - a^3t^5/20 + \cdots$. The second term is small in comparison with the first if $a^2t^4/20 \ll 1$, i.e., $t \ll a^{-1/2}$. In other words the value of the approximate solution must be small in comparison with the amplitude of the true oscillation, $at \ll \sqrt{a}$.

Problem 5. Prove that under this condition the relative error of the approximate solution is indeed small.

Solution. This follows from similarity conditions. The quasi-homogeneous dilations $X = e^s x$, $T = e^{-s}t$ transform the equation $\ddot{x} = -x^3$ into itself. The solution with initial condition $(0, a)$ becomes the solution with initial condition $(0, A = e^{2s}a)$. The approximate solution $x \approx at$ becomes $X \approx AT$. We choose s so that $A = 1$. For $A = 1$ the solution $X \approx T$ has small relative error when $T \ll 1$. But dilations do not change the relative errors. Hence the relative error of the approximate solution $x \approx at$ is also small for $T \ll 1$. But $T = e^{-s}t$, $a = e^{-2s}$. Hence $T \ll 1$ for $t \ll a^{-1/2}$. Thus for small a the approximation gives a small relative error, even for very large t, provided t is small in comparison with the large number $1/\sqrt{a}$.

In applications of the theory of differential equations it is always necessary to deal with a large number of quantities, some of which are "very small" and others "very large." To discern what is large in comparison with what (i.e., in what order to perform the passages to the limit) is not always easy; the study of this question is sometimes half of the job.

4. Transformation over the Time Interval from t_0 to t

Consider the differential equation $\dot{x} = v(t, x)$ with right-hand side defining a smooth direction field in a domain of the extended phase space (of any finite dimension $1 + m$).

Definition. *Transformation over the time interval from t_0 to t* is the mapping of a domain of the phase space into the phase space that assigns to the initial condition at the instant t_0 the value of the solution with this initial condition at the instant t (Fig. 70).

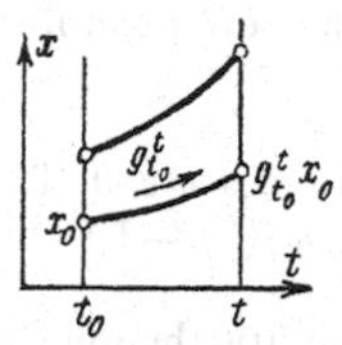

Fig. 70. The transformation over the time interval from t_0 to t

This transformation is denoted $g^t_{t_0}$.

In the notation of Corollary 4

$$g^t_{t_0}x_0 = \Phi(t_0, x_0; t).$$

The following corollary is a consequence of the fundamental theorem on rectification.

Corollary 5. *The transformations over the time interval from t_0 to t for an equation with smooth right-hand side*

1) *are defined in a neighborhood of each phase point x_0 for t sufficiently close to t_0;*

2) *are local diffeomorphisms (of class C^r if the right-hand side is of class C^r) and depend smoothly on t and t_0;*

3) *satisfy the identity $g^t_{t_0}x = g^t_s g^s_{t_0}x$ for s and t sufficiently close to t_0 (for all x in a sufficiently small neighborhood of the point x_0);*

4) *are such that for fixed ξ the function $\varphi(t) = g^t_{t_0}\xi$ is a solution of the equation $\dot{x} = v(t, x)$ satisfying the initial condition $\varphi(t_0) = \xi$.*

Corollary 5 obviously follows from the preceding corollaries. One can also use a rectification that leaves time invariant. For the rectified equation ($\dot{y} = 0$) all the transformations over the time interval from t_0 to t are identities, so that Properties 1)–4) hold.

Let us consider, in particular, the case of an autonomous equation $\dot{x} = v(x)$. In this case the following theorem is obvious.

Theorem. *The transformation over the time interval from t_0 to t for an autonomous equation depends only on the length $t - t_0$ of the time interval and not on the initial instant t_0.*

Proof. A translation of the extended phase space of an autonomous equation along the t-axis maps the direction field onto itself and hence maps integral curves onto one another. Under a translation by s the solution φ with initial condition $\varphi(t_0) = x_0$ maps into the solution ψ with the initial condition $\psi(t_0 + s) = x_0$. For any t we have $\psi(t+s) = \varphi(t)$. Consequently $g_{t_0}^t \equiv g_{t_0+s}^{t+s}$, as asserted. □

For brevity we shall write the mapping $g_{t_0}^{t_0+\tau}$ as g^τ. The mappings g^τ have the following properties:

1) they are defined for sufficiently small $|\tau|$ in a neighborhood of a selected point of the phase space;

2) they are diffeomorphisms of this neighborhood into the phase space and depend smoothly on τ;

3) for all sufficiently small $|s|$ and $|t|$ and for all x in some neighborhood of the selected point the group property $g^s g^t x = g^{s+t} x$ holds;

4) for fixed ξ the function $\varphi(t) = g^t \xi$ is a solution of the equation $\dot{x} = \boldsymbol{v}(x)$ with initial condition $\varphi(0) = \xi$.

The family $\{g^t\}$ is called the *local phase flow* of the vector field $\boldsymbol{v}$.

Problem 1. Assume that the equation $\dot{x} = \boldsymbol{v}(t,x)$ has T-periodic coefficients ($\boldsymbol{v}(t+T,x) \equiv \boldsymbol{v}(t,x)$) and that all the mappings over the time interval from t_0 to t for it are defined everywhere. Prove that the transformations over the times that are multiples of T form a group: $g_0^{kT} = A^k$ for any integer k. Which of the two following relations is true: $g_0^{kT+s} = A^k g_0^s$, $g_0^{kT+s} = g_0^s A^k$?

Answer. The second.

5. Theorems on Continuous and Differentiable Dependence on a Parameter

Assume that the right-hand side of a given equation $\dot{x} = \boldsymbol{v}(t,x;\alpha)$ depends smoothly on a parameter α ranging over some domain A of the space $\boldsymbol{R}^a$.

The following corollary is a consequence of the fundamental Theorem 1 on rectification.

Corollary 6. *The value of the solution with initial condition $\varphi(t_0) = x_0$ at an instant t depends smoothly on the initial condition, the time, and the parameter α.*

We denote this value by $\Phi(t_0, x_0; \alpha; t)$. The corollary asserts that *the function Φ (with values in the phase space) is defined, continuous, and smooth in a neighborhood of each point $(t_0, x_0; \alpha_0, t_0)$ of the product of the extended phase* *space, the time axis, and the domain of variation of the parameter (it is of class C^r if the right-hand side is of class C^r).*

Proof. A small trick is useful here. Consider the "extended equation" $\dot{x} = \boldsymbol{v}(t, x; \alpha)$, $\dot{\alpha} = 0$ with phase space of dimension $m + a$ (where $m = \dim\{x\}$). The solution of this equation with initial condition $(t_0, x_0; \alpha_0)$ is a pair $(x = \varphi(t),\ \alpha = \alpha_0)$ whose first component φ is the solution of the original equation for $\alpha = \alpha_0$ satisfying the initial condition $\varphi(t_0) = x_0$. By Corollary 4 this pair depends smoothly on $(t_0, x_0; t; \alpha_0)$. Consequently the first component also depends smoothly on these arguments, which was to be proved. □

Remark. This extension trick reduces the theorem on smooth dependence on the parameter to smooth dependence on the initial conditions. Conversely, given smooth dependence on the parameter (for a fixed initial condition), it is easy to deduce smooth dependence on the initial condition. It suffices to translate the equation so that the initial condition becomes the parameter $\boldsymbol{v}_\alpha(t, x) = \boldsymbol{v}(t, x - \alpha)$.

The theorem on differentiable dependence on the parameter provides a quite efficient method of approximately solving equations near "unperturbed" equations for which the solution is known. It suffices to represent the solution of the perturbed equation in the form of a Taylor series in powers of the perturbation, substitute this series into the perturbed equation, and equate the coefficients of identical powers of the perturbation. The free term of the series for the solution will be the known solution of the unperturbed equation. Recursively solvable equations will then be obtained to determine the subsequent terms. The most important of them, the equation for the first-degree terms in the perturbation, is an inhomogeneous equation of variations (compare § 3).

The method just described is used in all applications of the theory of differential equations under the name of *perturbation theory* or the *small parameter method.* It is one of the variants of Newton's series method.

Problem 1. Find the derivative of the solution of the logistic equation $\dot{x} = x(a - x)$ with initial condition $x(0) = 1$ with respect to the parameter a at $a = 1$.

Solution. Let $a = 1 + \varepsilon$, and let the perturbed solution be $x = \varphi_0 + \varepsilon\varphi_1 + O(\varepsilon^2)$. Upon substituting into the perturbed equation we obtain the equation

$$\dot{\varphi}_0 + \varepsilon\dot{\varphi}_1 + \cdots = (\varphi_0 + \varepsilon\varphi_1 + \cdots)(1 + \varepsilon - \varphi_0 - \varepsilon\varphi_1 - \cdots).$$

The unperturbed equation $\dot{x} = x(1-x)$ has solution $\varphi_0 \equiv 1$. Equating the coefficients of ε, we obtain the equation of variations $\dot{\varphi}_1 = 1 - \varphi_1$ with the initial condition $\varphi_1(0) = 0$ (why?). Therefore the solution is $1 - e^{-t}$.

Remark. A physicist would state this computation as follows. It is clear that for $a = 1 + \varepsilon$ the solution $x = 1 + y$ differs only a little from 1. Let us ignore the difference between the x in front of the parenthesis in the equation and 1. We then obtain the approximate equation $\dot{x} \approx a - x$, $\dot{y} \approx \varepsilon - y$, whence $y = \varepsilon(1 - e^{-t})$.

Traditional mathematical "rigor" forbids us to neglect the difference between the first x in the equation and 1 while not neglecting this difference for the second

x. Actually this "physical" reasoning is correct – it is simply a convenient shorthand for the calculations given above.

Problem 2. Find the derivative of the solution of the pendulum equation with constant torque, $\ddot{\theta} = a - \sin\theta$ at the instant a, for $a = 0$. At the initial instant the pendulum is at rest ($\theta = \dot{\theta} = 0$).

Solution. $0 = ay + \cdots$, $a\ddot{y} = a - ay$, $\ddot{y} = 1 - y$, $y - 1 = z$, $\ddot{z} = -z$, $z(0) = -1$, $\dot{z}(0) = 0$, $z = -\cos t$, $y = 1 - \cos t$. In first approximation the effect of a small torque is to shift the equilibrium position to the point a while the pendulum undergoes small oscillations with frequency 1 about this point; hence the derivative of the solution with respect to a is $1 - \cos t$.

Caution. Strictly speaking all our approximate solutions are based on the theorem on differentiability only for small $|t|$. In reality it is not difficult to justify them for any finite time interval $|t| \leq T$ provided the size of the perturbation ε does not exceed a certain quantity depending on T. In this time interval the error in the first approximation of perturbation theory is bounded above by the quantity $O(\varepsilon^2)$, but the constant in the O increases with T.

It is extremely risky to extend the conclusions reached in this way to an infinite time interval: the limiting passages $t \to \infty$ and $\varepsilon \to 0$ cannot be interchanged.

Example. Consider a bucket of water at the bottom of which there is a small hole of radius ε (Fig. 71). For every T there exists ε so small that for a long time ($t < T$) the bucket is nearly full. But for any fixed $\varepsilon > 0$ the bucket becomes empty as time tends to infinity.

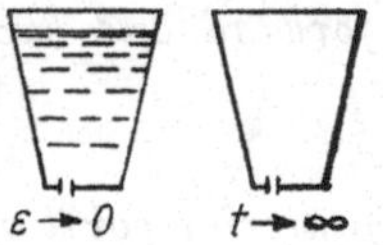

Fig. 71. The asymptotic behavior of the solutions of the perturbed equation

Problem 3. Find the derivative of the solution of the equation of small oscillations of a pendulum $\ddot{\theta} = -\omega^2\theta$ with initial condition $\theta(0) = 1$, $\dot{\theta}(0) = 0$ with respect to the parameter ω for $\omega = 1$.

Solution. The exact solution is given by the formula $\theta = \cos\omega t$. Consequently the derivative is $-t\sin t$.

If we knew the exact solution only for $\omega = 1$ and tried to find a solution for $\omega = 1 + \varepsilon$ by the small parameter method, we would obtain $\theta \approx \cos t - \varepsilon t \sin t$. We might think that the true solution is unbounded if we forget that the approximation can be used only for small εt.

6. Extension Theorems

Consider the differential equation $\dot{x} = v(t, x)$ defined by a smooth direction field in a domain U of the extended phase space. Let Γ be a subset of the domain U.

Definition. A solution φ with the initial condition $\varphi(t_0) = x_0$ *can be extended forward* (resp. *backward*) *to* Γ if there exists a solution with the same initial condition whose graph intersects Γ in a point where $t \geq t_0$ (resp. $t \leq t_0$).

A solution *can be extended forward* (resp. *backward*) *indefinitely* if there exists a solution with the same initial condition defined for all $t \geq t_0$ (resp. for all $t \leq t_0$).

Example. No solution of the equation $\dot{x} = x^2 + 1$ can be extended forward or backward indefinitely.

Definition. A set is called *compact* if from every covering of the set by open sets one can choose a finite subcovering.

The compact subsets of a Euclidean space are the closed and bounded sets.

The *boundary* of a set is the set of points in every neighborhood of which there are both points belonging to the set and points not belonging to the set.

The following corollary is an obvious consequence of the fundamental theorem on rectification.

Corollary 7. *A solution with initial condition in a compact set in the extended phase space can be extended forward and backward to the boundary of the compact set.*

In other words *through any interior point of a compact set there passes an integral curve that intersects the boundary of the compact set in both directions from the initial point* (Fig. 72).

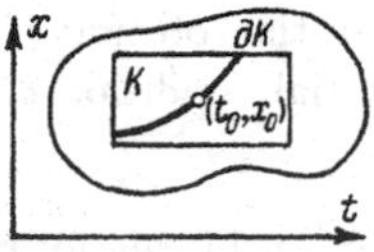

Fig. 72. Extension of a solution to the boundary of a compact set

The extension is unique in the sense that any two solutions with the same initial condition coincide wherever they are both defined.

Problem 1. Is it true that an integral curve of any smooth direction field in a domain of Euclidean space passing through a point of a compact set K can be extended to the boundary of K?

Answer. No. For example, consider the direction field of the phase curves of the pendulum in the domain $x_1^2 + x_2^2 > 0$, K being the annulus $1 \le x_1^2 + x_2^2 \le 2$.

Thus it is essential for the validity of the theorem that the direction field in the extended phase space be "nonvertical."

Proof of Corollary 7. We begin by proving uniqueness. Consider the least upper bound of the time values at which two solutions with the same initial condition coincide. The solutions coincide to the left of this point. If both are defined at this point, then they coincide there also, since they are continuous. But then they must also coincide to the right of the point (by the local uniqueness theorem). Thus the point in question must be an endpoint of one of the intervals of definition. This proves uniqueness of the forward extension of the solution (for the backward extension the reasoning is similar). We now construct the extension.

By the local existence theorem, at each point of the extended phase space there is a neighborhood such that the solution with initial condition at any point of the neighborhood can be continued forward and backward to an interval of time that is the same for all points of this neighborhood. From the covering of the compact set by such neighborhoods we choose a finite subcovering. From the finite collection of time intervals corresponding to these neighborhoods we choose the shortest and denote it by ε.

The solution with initial condition at the original point can be extended forward by ε (since this point belongs to the compact set and hence is covered by one of our neighborhoods). We choose the value of this solution at time $\varepsilon/2$ after the initial instant. If the point of the integral curve corresponding to this value is still inside the compact set, the solution with initial condition at that point can be continued forward again by ε (altogether $3\varepsilon/2$ from the original instant). We again move forward in time by $\varepsilon/2$ (i.e., we consider the value of the extended solution at an instant ε after the original instant) and again extend the solution by ε, etc. After a finite number of steps the integral curve will exit from the compact set (since its projection on the t-axis cannot be unbounded, and t increases by $\varepsilon/2$ at each step). Consequently there will arrive an instant at which the integral curve will intersect the boundary of the compact set, which was to be proved. □

Problem 2. Prove that any solution of an equation $\dot{x} = \boldsymbol{v}(t, x)$ defined by a direction field in $\boldsymbol{R} \times \boldsymbol{R}^n$ can be extended indefinitely if $\boldsymbol{v}$ grows no faster than the first power of x at infinity, i.e., if $|\boldsymbol{v}(t, x)| \le k|x|$ for all t and all $|x| \ge r$, where r and k are constants.

Hint. Comparing with a motion in the field $\dot{x} = kx$, construct compact sets whose boundaries require arbitrarily long times to reach.

We now assume that the domain of definition of the right-hand side of the equation $\dot{x} = \boldsymbol{v}(t, x)$ contains the cylinder $\boldsymbol{R} \times K$, where K is a compact set in the *phase* space.

Definition. The solution φ with the initial condition $\varphi(t_0) = x_0$ *can be extended forward* (resp. *backward*) *to the boundary* of the compact set K if there exists a solution with the same initial condition that assumes values on the boundary of the compact set K for some $t \geq t_0$ (resp. $t \leq t_0$).

The following result is an obvious consequence of Corollary 7.

Corollary 8. *A solution with initial condition in a given compact set K in the phase space can be extended forward (resp. backward) either indefinitely or to the boundary of the compact set K.*

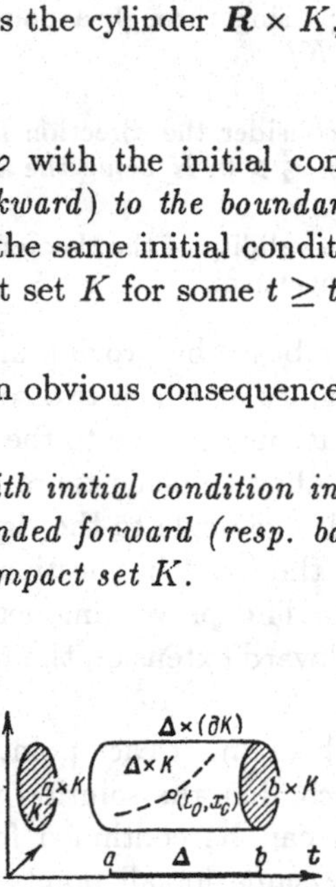

Fig. 73. Extension of a solution to the boundary of a compact set

Example. The solution of the pendulum equation $\dot{x}_1 = x_2$, $\dot{x}_2 = -x_1$ with initial condition $x_1 = 1$, $x_2 = 0$ cannot be extended to the boundary of the compact set $x_1^2 + x_2^2 \leq 2$.

Proof of Corollary 8. Consider a closed interval $\Delta = [a, b]$ of the t-axis containing t_0. The cylinder $\Delta \times K$ in the extended phase space (cf. Fig. 73) is compact. By the preceding theorem the solution can be extended to its boundary. This boundary consists of a "bottom and top" ($a \times K$ and $b \times K$) and a "lateral surface" $\Delta \times (\partial K)$ (by Leibniz' formula $\partial(\Delta \times K) = (\partial\Delta) \times K + \Delta \times (\partial K)$.). If for every $b > t_0$ the integral curve intersects the top $b \times K$, then the solution can be extended forward indefinitely, while if it intersects the lateral surface for some b, it can be extended to the boundary of the compact set. □

Corollary 9. *A solution of the autonomous equation $\dot{x} = \boldsymbol{v}(x)$ with initial value in any compact set of the phase space can be continued forward (resp. backward) either infinitely far or to the boundary of the compact set.*

For the cylinder $\boldsymbol{R} \times K$ belongs to the extended phase space of the autonomous equation for any compact set K in the phase space.

Problem 3. Prove that the vector field $\boldsymbol{v}$ determines a phase flow if all the solutions of the equation $\dot{x} = \boldsymbol{v}(x)$ can be continued infinitely far.

7. Rectification of a Vector Field

Consider a smooth vector field $\boldsymbol{v}$ in the domain U.

A *rectification* of the field is a diffeomorphism that transforms it into a field of parallel vectors of identical length in Euclidean space (Fig. 74).

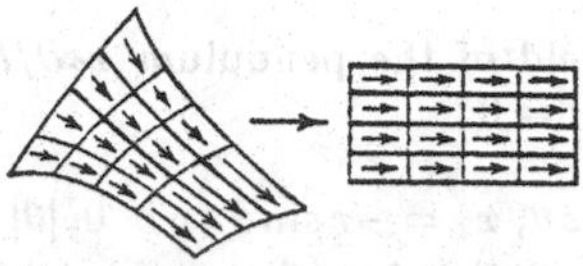

Fig. 74. Rectification of a vector field

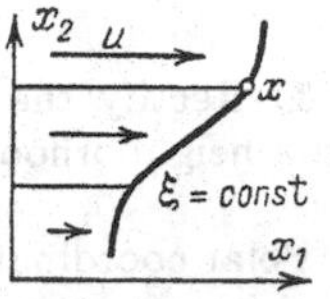

Fig. 75. Construction of rectifying coordinates

The following result is a consequence of the fundamental theorem on rectification.

Corollary 10. *Every smooth vector field is locally rectifiable in a neighborhood of each nonsingular point (a point where the vector field is nonzero).*

Proof. The vector fields in a neighborhood of a nonsingular point are nonzero and hence determine a direction field in this domain of the phase space. By the fundamental theorem this field is rectifiable. Let us perform a rectifying diffeomorphism. We shall thereby get parallel vectors, but their lengths will in general depend on the point where they are attached. In the rectifying coordinates the equation given by our field will assume the form

$$\dot{x}_1 = u(x), \quad \dot{x}_2 = \cdots = \dot{x}_n = 0, \quad \text{with } u(0) \neq 0.$$

In place of x_1 we introduce a new coordinate ξ, defining $\xi(x)$ as the time required to go from the plane $x_1 = 0$ to the point x (Fig. 75). Solving the equation, we find this time by Barrow's formula: $\xi(x) = \displaystyle\int_0^{x_1} \frac{d\eta}{u(\eta; x_2, \ldots, x_n)}$. In the coordinates $(\xi, x_2, \ldots, x_n)$ the equation assumes the form $\dot{\xi} = 1$, $\dot{x}_2 = \cdots = \dot{x}_n = 0$, i.e., the field is rectified. □

Remark. The theorem on rectification of a vector field is yet another reformulation of the theorem on the rectification of a direction field (to deduce the second from the first it suffices to choose one vector (depending smoothly on the point where it is attached) on each line of the given direction field, which is always easy to do locally).

Here are two more obvious reformulations of Corollary 10.

Corollary 11. *Any two smooth vector fields in domains of the same dimension can be transformed into each other by diffeomorphisms in sufficiently small neighborhoods of any nonsingular points.*

Corollary 12. *Every differential equation $\dot{x} = v(x)$ can be written in the normal form $\dot{x}_1 = 1$, $\dot{x}_2 = \cdots = \dot{x}_n = 0$ for a suitable choice of coordinates in a sufficiently small neighborhood of any nonsingular point of the field.*

In other words *every equation $\dot{x} = v(x)$ is locally equivalent to the simplest equation $\dot{x} = v$ ($v \neq 0$ independent of x) in a neighborhood of any nonsingular point.*

Problem 1. Rectify the phase velocity vector field of the pendulum $x_2\partial/\partial x_1 - x_1\partial/\partial x_2$ in a neighborhood of the point $x_1 = 1$, $x_2 = 0$.

Solution. Polar coordinates will do. Let $x_1 = r\cos\theta$, $x_2 = -r\sin\theta$ ($r > 0, |\theta| < \pi$). In these coordinates the equation has the form $\dot{r} = 0$, $\dot{\theta} = 1$, and therefore the field is rectified: it has the form $\partial/\partial\theta$.

Problem 2. Rectify the following fields:

1) $s_1\partial/\partial x_1 + 2x_2\partial/\partial x_2$ for $x_1 > 0$;
2) $\partial/\partial x_1 + \sin x_1\partial/\partial x_2$;
3) $x_1\partial/\partial x_1 + (1 - x_1^2)\partial/\partial x_2$ for $x_1^2 < 1$.

§ 8. Applications to Equations of Higher Order than First

The fundamental theorems on systems of any number of equations of any order will be deduced in this section from the analogous theorems for systems of first-order equations.

1. The Equivalence of an Equation of Order n and a System of n First-order Equations

Definition. A *differential equation of order* n is an equation

$$\frac{d^n x}{dt^n} = F\left(t; x, \frac{dx}{dt}, \ldots, \frac{d^{n-1}x}{dt^{n-1}}\right). \tag{1}$$

where F is a differentiable function (of class C^r, $r \geq 1$) defined in a domain U of a space of dimension $n+1$ (the time t and the derivatives of the unknown function of orders from 0 to $n-1$ inclusive).

A *solution* of Eq. (1) is a C^n-mapping $\varphi : I \to \boldsymbol{R}$ from an interval of the real axis into the real axis for which

1) the point with coordinates $(\tau, \varphi(\tau), \ldots, \varphi^{(n-1)}(\tau))$ belongs to the domain U for any τ in I;

2) for any τ in I

$$\left.\frac{d^n\varphi}{dt^n}\right|_{t=\tau} = F(\tau; \varphi(\tau), \ldots, \varphi^{(n-1)}(\tau)).$$

Example. One solution of the equation of small oscillations of the pendulum $\ddot{x} = -x$ is the function $\varphi(t) = \sin t$; another is $\varphi(t) = \cos t$ (Fig. 76). Consequently ***the graphs of solutions of a second-order equation may intersect*** (in contrast to the graphs of solutions of a first-order equation, i.e., the integral curves, which according to the uniqueness theorem either do not intersect or coincide on the entire interval).

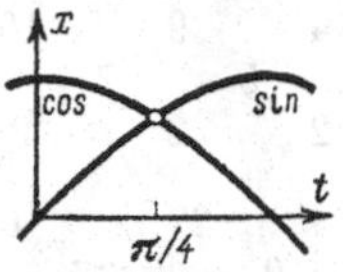

Fig. 76. The graphs of two solutions of a second-order equation

The phase space of the pendulum equation is the plane with coordinates $(x, \dot{x})$: defining these two numbers at the initial instant determines the entire motion of the pendulum. Consider the question of the dimension of the phase space for the general nth-order equation (1): how many numbers must be given at the initial instant in order to determine the solution uniquely at all times?

Theorem. *The nth-order equation* (1) *is equivalent to a system of n first-order equations*

$$\dot{x}_1 = x_2, \ldots, \dot{x}_{n-1} = x_n, \quad \dot{x}_n = F(t; x_1, \ldots, x_{n-1}) \tag{2}$$

in the sense that if φ is a solution of Eq. (1) *then the vector consisting of the derivatives $(\varphi, \dot{\varphi}, \ldots, \varphi^{(n-1)})$ is a solution of the system* (2), *and if $(\varphi_1, \ldots, \varphi_n)$ is a solution of the system* (2), *then φ_1 is a solution of Eq.* (1).

The proof is obvious.

Thus the phase space of the process described by an nth-order differential equation has dimension n; the entire course of the process (φ) is described by giving a set of n numbers at the initial instant of time t_0 – the values of the derivatives of φ of order less than n at the point t_0.

Remark. The reason it is necessary to give n initial conditions at the initial instant in order to determine the solution of an nth-order equation may perhaps become more understandable if we consider a differential equation as a limit of difference equations.

Fix a number $h > 0$ (called the *step size*). The *first difference* of the given function φ with step size h is the function whose value at the point t is $\varphi(t+h) - \varphi(t)$. The first difference is denoted $\Delta\varphi$. The second difference $\Delta^2\varphi$ is defined as $\Delta(\Delta\varphi)$.

Problem 1. Prove that $(\Delta^2\varphi)(t) = \varphi(t+2h) - 2\varphi(t+h) + \varphi(t)$.

The nth difference is defined similarly: $\Delta^n\varphi = \Delta(\Delta^{n-1}\varphi)$.

Problem 2. Prove that $\Delta^n\varphi \equiv 0$ if and only if $\varphi(t+kh)$ is a polynomial of degree less than n in $k \in \mathbf{Z}$.

For example, if we write out successively the values of k^2 and their differences on the line below, then the differences of the differences, the third line will consist of all 2's; if we begin with k^3, then the fourth line will contain nothing but 6's, etc.:

$$\begin{array}{ccccccccc}
1 & & 4 & & 9 & & 16 & & 25 \\
& 3 & & 5 & & 7 & & 9 & \\
& & 2 & & 2 & & 2 & &
\end{array}$$

$$\begin{array}{ccccccccc}
1 & & 8 & & 27 & & 64 & & 125 \\
& 7 & & 19 & & 37 & & 61 & \\
& & 12 & & 18 & & 24 & & \\
& & & 6 & & 6 & & &
\end{array}$$

A *first-order difference equation* is an equation of the form $\dfrac{\Delta\varphi}{\Delta t} = v(t,\varphi)$, i.e., $\dfrac{\varphi(t+h)-\varphi(t)}{h} = v(t,\varphi(t))$. From such an equation, knowing only the number $\varphi(t_0)$, it is possible to find $\varphi(t_0+h)$, and from the latter $\varphi(t_0+2h)$, etc. As $h \to 0$ a difference equation becomes a differential equation. It is therefore not surprising that the solution of a first-order differential equation is also determined by the value of a single number at the initial instant.

A *second-order difference equation* has the form

$$\frac{\Delta^2\varphi}{(\Delta t)^2} = F\left(t;\varphi,\frac{\Delta\varphi}{\Delta t}\right),$$

i.e.,

$$\frac{\varphi(t+2h) - 2\varphi(t+h) + \varphi(t)}{h^2} = F\left(t;\varphi(t),\frac{\varphi(t+h)-\varphi(t)}{h}\right).$$

Knowing the value of φ at two instants separated by a time interval of length h, we can find the value of φ after another interval h from this equation. Thus all the values $\varphi(t_0+kh)$ are determined by the first two of them.

As $h \to 0$ the second-order difference equation becomes a second-order differential equation. It is therefore not surprising that the solution of the differential equation is also determined by giving two numbers at the initial instant (and n numbers for an nth-order equation). The theorem on the preceding page is precisely the justification for passing to the limit as $h \to 0$.

Problem 3. Prove that the equation $d^n x/dt^n = 0$ is satisfied by all polynomials of degree less than n, and only by these functions.

Problem 4. Find the dimension of the manifold of solutions of the Helmholtz equation $\frac{\partial^2 u}{\partial x^2} + \frac{\partial^2 u}{\partial y^2} + u = 0$ in the domain $x^2 + y^2 > 0$, depending only on the distance to the origin.

Solution. The unknown function of r must satisfy a second-order equation; consequently the solutions are determined by two numbers.

2. Existence and Uniqueness Theorems

The following corollary is a consequence of the theorem of Sect. 1 and the existence and uniqueness theorems for systems of first-order equations (§ 7).

Corollary. *Let $\boldsymbol{u} = (u_0; u_1, \ldots, u_n)$ be a point of the domain U in which the right-hand side of Eq. (1) is defined. The solution φ of Eq. (1) with initial condition*

$$\varphi(u_0) = u_1, \quad \dot{\varphi}(u_0) = u_2, \ldots, \varphi^{(n-1)}(u_0) = u_n \tag{3}$$

exists and is unique (in the sense that any two solutions with the same initial condition coincide on the intersection of their intervals of definition).

In writing the initial condition for Eq. (1) it is customary to write x instead of φ.

Example. At $t = \pi/4$ the solutions $\cos t$ and $\sin t$ of the pendulum equation $\ddot{x} = -x$ satisfy respectively the initial conditions $x(\pi/4) = \sqrt{2}/2$, $\dot{x}(\pi/4) = -\sqrt{2}/2$ and $x(\pi/4) = \sqrt{2}/2$, $\dot{x}(\pi/4) = \sqrt{2}/2$ (Fig. 76). These initial conditions are distinct, and so it is not surprising that the graphs of the solutions intersect without coinciding. The uniqueness theorem for a second-order equation forbids only a common tangent at a point of intersection of two noncoincident graphs. The graphs of two solutions of the same third-order equation may be tangent to each other, but then at the point of tangency they must have different curvatures, etc.

Problem 1. Suppose it is known that Eq. (1) has as solutions the functions t and $\sin t$. Find the order n of the equation.

Solution. The functions t and $\sin t$ have the same derivatives of orders 0, 1, and 2 at the point 0. If they satisfied the same third-order equation, they would coincide by virtue of the uniqueness theorem. An equation of order $n \geq 4$ satisfied by both functions is easy to invent, for example, $x^{(n)} + x^{(n-2)} = 0$. Hence $n \geq 4$.

Problem 2. Can the graphs of two solutions of the equation $\ddot{x} + p(t)\dot{x} + q(t)x = 0$ have the form depicted in Fig. 77?

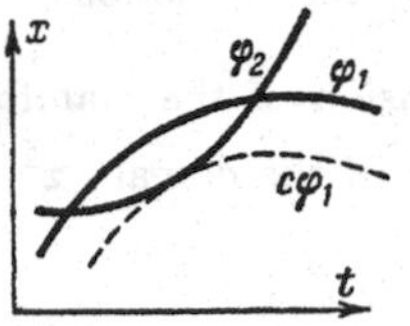

Fig. 77. An impossible configuration of graphs

Answer. No, since the solutions $c\varphi_1$ and φ_2 have the same initial condition and do not coincide.

Problem 3. Consider the equation $2x = t^2\ddot{x}$. The solutions $x \equiv 0$ and $x = t^2$ both satisfy the initial condition $x = \dot{x} = 0$ for $t = 0$. Why don't they coincide?

Answer. The uniqueness theorem applies to equations of the form (1), i.e., to equations that can be solved for the highest-order derivative; but the present equation cannot be written in this form (in a neighborhood of zero).

Problem 4. Solve the difference equation $\Delta^3\varphi = 0$ with the initial condition $\varphi(0) = 0$, $(\Delta\varphi)(0) = 0$, $(\Delta^2\varphi)(0) = 2$ for t a multiple of the step size $h = 1$.

Solution. $\varphi = a + bt + ct^2$, $\Delta\varphi = b + 2ct + c$, $\Delta^2\varphi = 2c$. By the initial conditions $c = 1$, $b = -1$, $a = 0$. Hence $\varphi = t^2 - t$.

3. Differentiability and Extension Theorems

Since the equivalence of an nth-order equation to a system of first-order equations has already been established, we conclude that the solution of an nth-order equation depends smoothly on the initial conditions and the parameters (if the right-hand side depends smoothly on parameters); the reader can easily state an extension theorem as well.

Problem 1. Find in first approximation in ε the influence of a small resistance in the medium $\varepsilon F(x, \dot{x})$ on the motion of a body falling from height h.

Solution. The question involves the equation $\ddot{x} = -g + \varepsilon F(x, \dot{x})$ and the initial conditions $x(0) = h$, $\dot{x}(0) = 0$.

By the theorem on differentiable dependence on the parameter the solution has the form $\varphi = \varphi_0 + \varepsilon\varphi_1 + \cdots$, where $\varphi_0(t) = h - gt^2/2$. Substituting $x = \varphi(t)$ into the equation and equating the terms of the series in ε, we find $\ddot{\varphi}_1 = F(\varphi_0, \dot{\varphi}_0)$, whence $\varphi_1(t) = \int_0^t \int_0^s F(\varphi_0(\tau), \dot{\varphi}_0(\tau))\,d\tau\,ds$. For example, if $F = -\dot{x}$, then $\varphi_1 = gt^3/6$. Hence the retardation during the fall is in first approximation proportional to the height: $-\varepsilon\varphi_1/\dot{\varphi}_0 = \varepsilon t^2/6 = \varepsilon h/3g$.

Problem 2. Prove that all solutions of the pendulum equation $\ddot{\theta} = -\sin\theta$ can be extended indefinitely.

Problem 3. For which natural numbers k can all the solutions of the equation $\ddot{x} = x^k$ be extended infinitely far?

Answer. Only for $k = 1$.

4. Systems of Equations

By a *system of differential equations* we shall mean a system of equations in n unknown functions

$$\frac{d^{n_i} x_i}{dt^{n_i}} = F_i(t; x, \ldots), \quad i = 1, \ldots, n, \tag{4}$$

where the arguments of each function F_i are the independent variable t, the dependent variables x_j, and their derivatives of orders less than n_j $(j = 1, \ldots, n)$ respectively.

A solution of the system is defined as in Sect. 1. It should be emphasized that a solution of the system is a vector-valued function defined on an interval. Thus $(\varphi_1, \ldots, \varphi_n)$ is not n solutions, but only one solution of a system of n equations – a remark that applies equally to systems of both algebraic and differential equations.

First of all we determine the phase space that corresponds to the system (4).

Theorem. *The system* (4) *is equivalent to a system of* $N = \sum_{i=1}^{n} n_i$ *first-order equations.*

In other words *the dimension of the phase space of the system* (4) *is* N.

For the proof we must introduce as coordinates in the phase space the derivatives of x_j of order less than n_j.

For example, let $n = n_1 = n_2 = 2$. Then the system has the form

$$\ddot{x}_1 = F_1(t; x_1, \dot{x}_1, x_2, \dot{x}_2), \quad \ddot{x}_2 = F_2(t; x_1, \dot{x}_1, x_2, \dot{x}_2)$$

and is equivalent to the system of four equations

$$\dot{x}_1 = x_3, \;\; \dot{x}_2 = x_4, \;\; \dot{x}_3 = F_1(t; x), \;\; \dot{x}_4 = F_2(t; x),$$

where $x = (x_1, x_3, x_2, x_4)$.

Example. The system of n second-order differential equations of Newtonian mechanics

$$m_i \ddot{q}_i = -\frac{\partial U}{\partial q_i}, \quad i = 1, \ldots, n, \tag{5}$$

where U is the potential energy and $m_i > 0$ are masses, is equivalent to the Hamiltonian system of $2n$ equations

$$\dot{q}_i = \frac{\partial H}{\partial p_i}, \quad \dot{p}_i = -\frac{\partial H}{\partial q_i}, \quad i = 1, \ldots, n,$$

where $p_i = m_i \dot{q}_i$ and $H = T + U$ is the total energy ($T = \sum m_i \dot{q}_i^2/2 = \sum p_i^2/(2m_i)$ is the kinetic energy). Thus the dimension of the phase space of the system (5) is $2n$.

Theorems on existence, uniqueness, and differentiability with respect to the initial conditions and parameters, as well as extension theorems carry over automatically to systems of the form (4): to determine the solution uniquely it suffices to prescribe the derivatives of x_i of order less than n_i at the initial instant. For example, for the system of Newtonian equations (5) it suffices to prescribe n coordinates and n velocities at the initial instant.

Problem 1. In coordinates fixed on the Earth the Coriolis force $\boldsymbol{F} = 2m[\boldsymbol{v}, \boldsymbol{\Omega}]$, where $\boldsymbol{\Omega}$ is the angular velocity vector of the Earth, acts on a material point of mass m moving with velocity $\boldsymbol{v}$ with respect to the Earth. A stone is thrown (with initial velocity 0) into a mine shaft of depth 10 m at the latitude of Leningrad ($\lambda = 60°$). By how much will the Coriolis force cause it to deviate from the vertical?

Solution. By hypothesis $\ddot{\boldsymbol{x}} = \boldsymbol{g} + 2[\dot{\boldsymbol{x}}, \boldsymbol{\Omega}]$. The magnitude of the angular velocity of the Earth, $\Omega \approx 7.3 \cdot 10^{-5}\text{sec}^{-1}$, will be considered a small parameter. By the theorem on differentiability $\boldsymbol{x} = \boldsymbol{x}_0 + \Omega \boldsymbol{y} + O(\Omega^2)$ and $\boldsymbol{x}_0 = \boldsymbol{g}t^2$. Substituting $\boldsymbol{x}$ into the equation, we obtain $\Omega \ddot{\boldsymbol{y}} = 2[\boldsymbol{g}t, \boldsymbol{\Omega}]$, $\boldsymbol{y}(0) = \dot{\boldsymbol{y}}(0) = \mathbf{o}$. Hence $\Omega \boldsymbol{y} = [\boldsymbol{g}, \boldsymbol{\Omega}]t^3/3$, and consequently $|\Omega \boldsymbol{y}| = \frac{2t}{3}|h||\boldsymbol{\Omega}| \cos \lambda$. Therefore the stone deviates eastward by 0.3 mm.

Remark. The problem of the deviation of the stone played a prominent role in the history of physics. The phenomenon of eastward deviation (rather than westward, as one might expect at first glance) was predicted by Newton in a letter to Hooke of 28 November 1679; Newton asked Hooke to carry out an experiment with a stone to prove the rotation of the earth, which at the time was not universally accepted.

In his reply (of 6 January 1680) Hooke stated the law of universal gravitation. At the time Newton had an inaccurate idea of the stone's orbit. The resulting discussion caused Newton to give up his plan to abandon the study of science and led him to write *The Mathematical Principles of Natural Philosophy*, his famous *Principia*, which was the beginning of modern physics.

In his letter Hooke gave the correct exponent (-2) in the law of gravity (in the *Principia* Newton writes that Wren, Hooke, and Halley had independently discovered that Kepler's third law corresponds to just this exponent). Besides Kepler's law, Hooke refers to Halley's observations on the retardation of a pendulum clock carried up Mount St. Elena. In the letter Hooke says explicitly that the stone is moved by the same force that causes the planets to move in Keplerian ellipses; criticizing the spiral Newton had sketched, Hooke asserted that the orbit of the stone in the absence of air resistance would be an "eccentric elliptoid."

Newton interpreted an elliptoid to be an ellipse and was interested in knowing how Hooke had found the orbit. After great labors he succeeded in proving that the orbit is indeed an ellipse (for falling both onto the earth and down a mine). The proof was (and remains) so difficult mathematically that Newton arrived at the conclusion that Hooke "was asserting more than he knew." He never afterwards referred to Hooke's letter. In a letter to Halley on his discussion with Hooke Newton gave a description of the difference between the approaches of a mathematician and a physicist to natural science, which remains current even today: "Mathematicians, that find out, settle & do all the business, must content themselves with being nothing but dry calculators and drudges, & another, that does nothing but pretend and grasps things, must carry away all the inventions as well of those that were to follow him as of those that went before."

Hooke dropped steel balls from a height of 10 m and asserted that he observed a systematic south-eastward deviation (which is practically impossible because of the extreme smallness of this deviation in comparison with aerodynamic effects). In the absence of resistance the stone inside the mine shaft in a homogeneous Earth would be subject to Hooke's law (the attractive force directly proportional to the distance from the center of the Earth), but Hooke himself could hardly have known this. The orbit of the stone in this case is an ellipse (in a system of coordinates not rotating with the Earth) with center at the center of the Earth and minor semi-axis about 400 km (why?); the orbit would be traversed in the same time that a low-orbit satellite would circle the earth, i.e., in an hour and a half (why?).

Problem 2. It is known from the newspapers that the cosmonaut Leonov, going for a walk in space, threw the lens cap of his movie camera toward the earth. Where did it go?

Solution. This is a problem involving the influence of a *small* perturbation in the initial conditions on the solution. The equation of motion, by the law of universal gravitation, can be written in the form $\ddot{\boldsymbol{r}} = -\gamma \boldsymbol{r}/r^3$. The motion of both the cosmonaut and the lens cap occurs in the plane of the circular orbit, so that we may assume $\boldsymbol{r} \in \boldsymbol{R}^2$. Let us write the equation of motion in polar coordinates. To do this we introduce the unit vector $\boldsymbol{e}_r = \boldsymbol{r}/r$ and the unit vector $\boldsymbol{e}_\varphi$ perpendicular to it and directed forward along the circular orbit. It is clear that $\dot{\boldsymbol{e}}_r = \dot{\varphi}\boldsymbol{e}_\varphi$ and $\dot{\boldsymbol{e}}_\varphi = -\dot{\varphi}\boldsymbol{e}_r$. Differentiating the quantity $\boldsymbol{r} = r\boldsymbol{e}_r$, we find $\boldsymbol{r} = \dot{r}\boldsymbol{e}_r + r\dot{\varphi}\boldsymbol{e}_\varphi$, $\ddot{\boldsymbol{r}} = \ddot{r}\boldsymbol{e}_r + 2\dot{r}\dot{\varphi}\boldsymbol{e}_\varphi + r\ddot{\varphi}\boldsymbol{e}_\varphi - r\dot{\varphi}^2\boldsymbol{e}_r$. Consequently *Newton's equation in polar coordinates* assumes the form of a system of two second-order equations

$$\ddot{r} - r\dot{\varphi}^2 = -\gamma r^2, \quad r\ddot{\varphi} + 2\dot{r}\dot{\varphi} = 0.$$

We take as a unit of length the radius of the circular orbit of the space station (≈ 6400 km). We choose the unit of time so that the angular velocity of the motion in the orbit is 1. Then the motion over the orbit is described by the equations $r = 1$, $\varphi = t$, and so $\gamma = 1$. The initial conditions for the space station (and the cosmonaut) are $r(0) = 1$, $\dot{r}(0) = 0$, $\varphi(0) = 0$, $\dot{\varphi}(0) = 1$. The initial conditions for the lens cap differ only in that $\dot{r}(0) = -v$ is the velocity of the throw, i.e., the initial velocity of the lens cap relative to the cosmonaut. Assume that the velocity of the throw is 10 m/sec. Then $v \approx 1/800$ (since our unit of velocity is nearly the first cosmic velocity, i.e., about 8 km/sec).

The quantity 1/800 is small compared to 1, and therefore we must study the influence of *a small deviation in the initial condition* on the unperturbed solution $r = 1$, $\varphi = t$. By the theorem on differentiability with respect to the initial condition we seek a solution close to the unperturbed solution in the form $r = 1 + r_1 + \cdots$, $\varphi = t + \varphi_1 + \cdots$, where the dots indicate infinitesimals of order v^2. Substituting these expressions in Newton's equation with $\gamma = 1$ and rejecting infinitesimals of order v^2, we obtain the *equations of variations*

$$\ddot{r}_1 = 3r_1 + 2\dot{\varphi}_1, \quad \ddot{\varphi}_1 + 2\dot{r}_1 = 0.$$

The solution of the equations of variations with the initial conditions of the lens cap ($r_1(0) = \varphi_1(0) = \dot{\varphi}_1(0) = 0$, $\dot{r}_1(0) = -v$) can be easily found by observing that $\dot{\varphi}_1 + 2r_1 \equiv 0$, so that $\ddot{r}_1 = -r_1$. This solution has the form $r_1 = -v \sin t$, $\varphi_1 = 2v(1 - \cos t)$. By the theorem on differentiability the true solution of Newton's equations differs from the one we have found by infinitesimals of order v^2 (for t not too large). Consequently the lens cap describes an ellipse relative to the cosmonaut (Fig. 78) with semiaxes v and $2v$. Our unit of length is the radius of the orbit, and $v \approx 1/800$. Thus the lengths of the semiaxes of the ellipse are about 8 and 16 km.

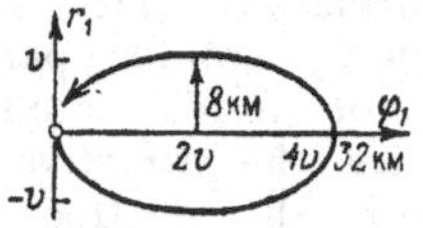

Fig. 78. The motion of the lens cap with respect to the space station

At first the lens cap moves downward (toward the Earth), but then begins to overtake the cosmonaut and moves 32 km ahead in the orbit; finally it returns upward, describing approximately a 100-kilometer ellipse in exactly the same time required for the space station to complete one revolution in its orbit.

Of course in this computation we have neglected quantities of order v^2 and the actual motion of the lens cap relative to the cosmonaut will not be periodic (the coil does not close, and the error will be of order 1/800 the size of the ellipse, i.e., the lens cap will orbit at a distance of about 10 m from the space station). We have also neglected many effects (the pressure of light rays, the deviation of the direction of the throw from the vertical, the deviation of the orbit of the space station from a circle, etc.) which give larger errors.

V. V. Beletskii, from whose charming book *Essays on the Motion of the Celestial Bodies* (Nauka, Moscow, 1972) the lens cap problem was taken, remarks that the lens cap would hardly be visible at a distance greater than one kilometer, and the first kilometer of the ellipse is very nearly a straight line. It was for that reason that Leonov saw the lens cap he had thrown *fly straight toward the Earth.*

5. Remarks on Terminology

The equations and systems of equations considered above are called *normal* or *solved with respect to the highest-order derivatives.* No other equations and systems are considered in this course, so that the term equation or system denotes a normal system or a system equivalent to a normal system (such as, for example, the system of Newton's equations (5)).

The functions occurring on the right-hand side of a system can be defined in various ways: explicitly, implicitly, parametrically, and the like.

Example. The notation $\dot{x}^2 = x$ is an abbreviation for the two distinct differential equations $\dot{x} = \sqrt{x}$ and $\dot{x} = -\sqrt{x}$, each of which has the half-line $x \geq 0$ as phase space. These equations are defined by two distinct vector fields, which are smooth for $x > 0$ (Fig. 79).

When the right-hand side is given implicitly, special care must be taken to establish its domain of definition and avoid ambiguous notation.

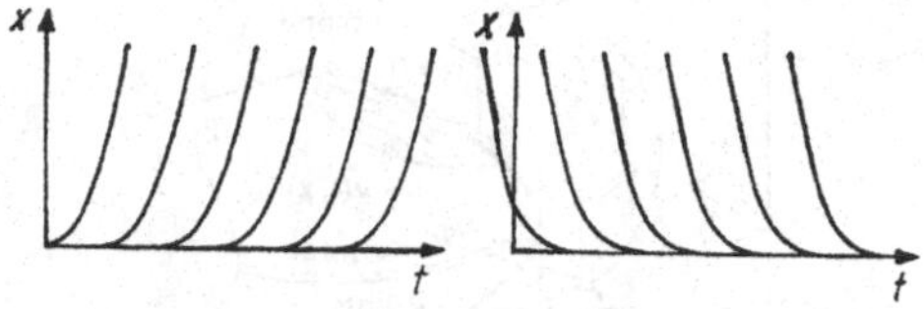

Fig. 79. The integral curves of the two equations combined in the notation $\dot{x}^2 = x$

Example. A *Clairaut equation* is an equation $x = \dot{x}t - f(\dot{x})$. The Clairaut equation

$$x = \dot{x}t - \dot{x}^2/2 \tag{6}$$

is an abbreviation for two distinct differential equations defined for $x \le t^2/2$. Each of them satisfies an existence and uniqueness theorem in the domain below a parabola: $x < t^2/2$ (Fig. 80). Through each point of this domain pass two tangents to the parabola. Each tangent consists of two half-lines. Each half-tangent is an integral curve of one of the two equations combined by formula (6).

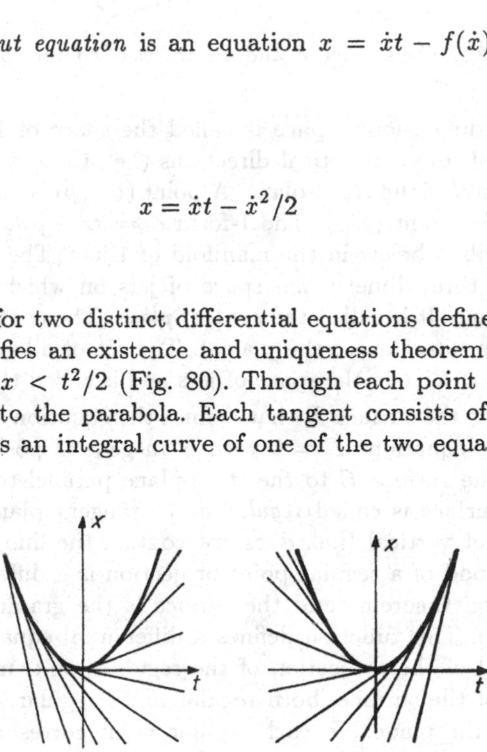

Fig. 80. The integral curves of two equations written together in the form of a Clairaut equation

Problem 1. Study the Clairaut equation $x = \dot{x}t - \dot{x}^3$.

Remark. In studying equations whose right-hand sides are given implicitly, i.e., equations of the form $F(t, x, \dot{x}) = 0$, it is often useful to consider the direction field defined by this equation not in the (t, x)-plane, but on the surface E in three-dimensional (t, x, p)-space given by the equation $F(t, x, p) = 0$ (Fig. 81).

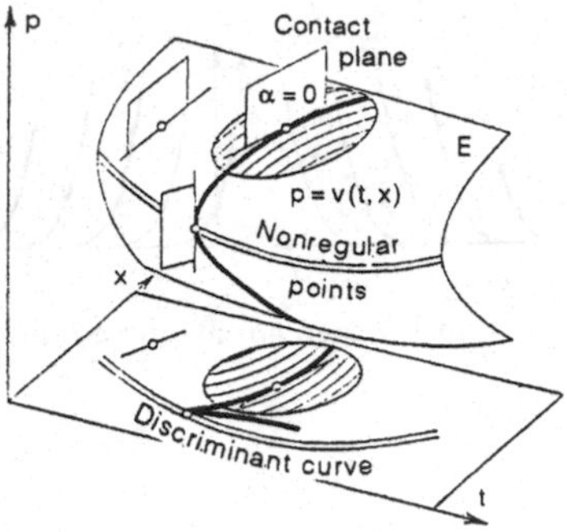

Fig. 81. The surface E and the traces of contact planes on it

This three-dimensional space is called the space of 1-*jets*[2] of functions. Its points are all the nonvertical directions (i.e., those not parallel to the x-axis) at all points of the (t, x)-plane. A point (t, x, p) is the direction of a line $dx = p\,dt$ at the point (t, x). The 1-form $\alpha = dx - p\,dt$ defines the contact structure described below in the manifold of 1-jets. The vectors attached at a point of the three-dimensional space of jets on which this form vanishes constitute a plane. It is called the *contact plane*. The contact plane is vertical (it contains the direction of the p-axis). The set of all contact planes forms the *contact plane field* in the space of jets and is called the *contact structure*.

Assume that the surface E that defines the equation is smooth (this condition holds for equations $F = 0$ with F in general position). Consider the projection of the surface E to the (t, x)-plane parallel to the p-direction. A point on the surface is called *regular* if the tangent plane to the surface at that point is not vertical (i.e., does not contain the line in the p-direction). In a neighborhood of a regular point projection is a diffeomorphism (by the implicit function theorem) and the surface is the graph of a smooth function $p = v(t, x)$. This function defines a differential equation $\dot{x} = v(t, x)$ (in a neighborhood of the projection of the regular point under consideration). Other points of the surface, both regular and irregular, may project to the same point of the plane. To each regular point corresponds its own direction field on the plane and its own differential equation; all these different differential equations are combined in the equation $F = 0$.

Consider the contact plane at a regular point of the surface E. It intersects the tangent plane in a line. Thus in a neighborhood of a regular point on E there arises a smooth direction field – the field of traces of contact planes. The following theorem is obvious.

Theorem. *Under a projection of the surface E defined by the equation $p = v(t, x)$ onto the (t, x)-plane along the p-axis the field of traces of contact planes on E maps into the direction field of the equation $dx/dt = v(t, x)$ on the plane.*

[2] The k-*jet* of a function is its Taylor polynomial of degree k.

Corollary. *This projection maps integral curves of the trace field on E into integral curves of the equation in the plane.*

The tangent plane of the surface E at irregular points is vertical. But it may nevertheless intersect the contact plane in a line (for a surface E in general position the tangent plane will coincide completely with the contact plane only at individual exceptional points).

In a neighborhood of a nonexceptional irregular point on the surface E the traces of contact planes define a smooth direction field. Thus the field of traces of contact planes on the surface E can be extended to nonexceptional irregular points. The extended field is called the *direction field of the equation* $F = 0$ *on* E and its integral curves are called *integral curves of the equation* $F = 0$ *on* E.

The projections of the pieces of these curves between irregular points to the (t, x)-plane are locally the integral curves of the corresponding equations $dx/dt = v(t, x)$ (this does not hold globally, even when there are no irregular points!).

The transition from the plane to the surface E is often useful for both studying and solving the equation.

Problem 2. Find the integral curves of the equation $\dot{x}^2 = t$ on the surface $p^2 = t$ and their projections on the (t, x)-plane.

Solution. We take p and x as coordinates on E. In these coordinates the equation of the traces of contact planes ($dx = p\,dt$) assumes the form $dx = 2p^2\,dp$. The integral curves are $x + C = 2p^3/3$. Their projections are the semicubical parabolas $(x + C)^2 = 4t^3/9$ (Fig. 82). The irregular points form the line $p = 0$. They are all nonexceptional.

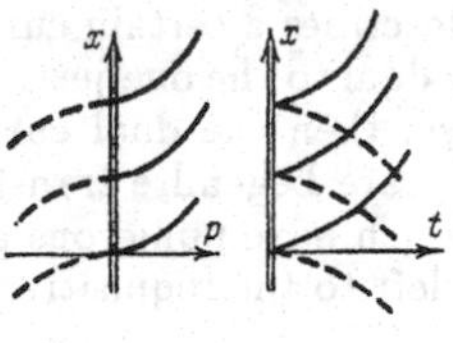

Fig. 82. Projections of integral curves

The projection of the line of irregular points on the (t, x)-plane is called the *discriminant curve*. In the present case the discriminant curve is the x-axis.

The cusp divides the semicubical parabola into two parts. Each of them is an integral curve of one of the two equations $\dot{x} = \sqrt{t}$ (or $-\sqrt{t}$) in the half-plane $t > 0$. It can be shown that the projections of integral curves in E onto the plane for an equation in general position has a cusp at the general point of the discriminant curve (moreover in the neighborhood of such a point the equation reduces to the form $\dot{x}^2 = t$ by a diffeomorphism of the (t, x)-plane). However this is not the case for all equations.

Problem 3. Find the integral curves of the Clairaut equation $x = t\dot{x} - f(\dot{x})$ on the surface $x = pt - f(p)$, their projections on the (t, x)-plane, and the discriminant curve.

Solution. We take p and t as coordinates on E. The equation of the traces of contact planes ($dx = p\,dt$) assumes the form ($t\,dp + p\,dt - f'\,dp = p\,dt$, or $(t - f')\,dp = 0$. The irregular points are those at which $t = f'$. They are all exceptional. The integral curves on E are $p =$ const (in the domain where $t \neq f'$). These are lines. Their projections on the (t, x)-plane are also lines: $x = tC - f(C)$. A Clairaut equation is simply *the equation of a family of lines parametrized by the slope.*

The discriminant curve is given parametrically by the equations $t = f'(C)$, $x = tC - f(C)$. In a neighborhood of a point where $f'' \neq 0$ these formulas define a smooth curve that is the graph of a function $x = g(t)$. Indeed, near a point where $f'' \neq 0$ we can express C in terms of t and then x in terms of t. The line $x = tC - f(C)$ is tangent to the discriminant curve at such a point (why?). Thus *the discriminant curve of a Clairaut equation is the envelope of the family of lines described by the equation.*

The transition from the function f to the function g is called *Legendre's transformation.* The Legendre transformation of the function g will again be f (prove this!). Therefore the functions f and g are called *duals of each other.*

Problem 4. Calculate the Legendre transformation of the function $|p|^\alpha/\alpha$ ($\alpha > 1$).

Answer. $|t|^\beta/\beta$, where $\alpha^{-1} + \beta^{-1} = 1$.

The geometric meaning of Legendre's transformation is as follows. Consider the set of all nonvertical lines (those not parallel to the x-axis) in the (t, x)-plane. A line is defined by its equation $x = at - b$. Thus nonvertical lines can be regarded as points in the (a, b)-plane. This plane is called the *dual* of the original plane. The coordinates a and b are called the *tangential coordinates* of the line.

The plane dual to the (a, b)-plane is the original (t, x)-plane because of the complete symmetry of the equation $x + b = at$ under the change of variable $(t, x) \mapsto (a, b)$: a line in the plane of lines is a point in the original plane.

Consider a smooth curve $x = g(t)$ in the (t, x)-plane. The tangent to this curve varies under motion along the curve. When this happens, the point of the dual plane corresponding to the tangent describes a certain curve. This curve is called the *dual* of the original curve. The curve dual to the one just constructed is the original curve. If $g'' \neq 0$ for the original curve, then the dual curve is the graph of the function $b = f(a)$. The functions f and g are Legendre transformations of each other.

The proof of these facts (which have numerous generalizations and applications in all areas of mathematics) is left to the inquisitive reader as an exercise.

§ 9. The Phase Curves of an Autonomous System

In this section we study the simplest geometric properties of phase curves of autonomous systems, i.e., systems whose right-hand sides are independent of time.

1. Autonomous Systems

Definition. A system of differential equations is called *autonomous* if it maps into itself under arbitrary translations along the time axis.

In other words a system is autonomous if its right-hand side is independent of time. For example, an autonomous nth-order equation is

$$x^{(n)} = F(x, \ldots, x^{(n-1)}).$$

Remark. In the description of evolutionary processes by differential equations it is usually autonomous systems that arise: the independence of the right-hand side of t reflects the time-independence of the laws of nature (without which scientific study of nature would be impossible). The term "autonomous" means "independent" and reflects the independence of the evolution of the state of the system under consideration from all others. Nonautonomous systems arise in the description of nature most often in the following way. Assume that we are studying part I of a physical system $I + II$. Then, although the law of evolution of the entire system does not change with time, the influence of part II on part I may lead to a law of evolution for part I that does change with time.

For example, the influence of the Moon on the Earth causes the tides. Mathematically this influence is expressed by the fact that the magnitude of the gravitational acceleration occurring in the equation of motion of terrestrial objects varies with time.

In such situations we say that the distinguished part I is nonautonomous. Therefore all systems whose right-hand side depends explicitly on time are called nonautonomous. Of course nonautonomous systems may occur in other cases also, for example, in the transformation process when solving autonomous systems. As an example: the transition to the nonautonomous equation with separable variables in integrating the Lotka-Volterra system (Sect. 7 of § 2).

Problem 1. Is the equation of variations for the small perturbations of a solution of an autonomous system under small variations in initial conditions an autonomous equation?

Answer. If the unperturbed solution is a state of equilibrium, it is autonomous; in general it isn't.

2. Translation over Time

Let us begin with an example. Consider the autonomous nth-order equation

$$x^{(n)} = F(x, \dot{x}, \ldots x^{(n-1)}). \tag{1}$$

Theorem. *Assume that* $x = \sin t$ *is a solution of Eq.* (1). *Then the function* $x = \cos t$ *is also a solution.*

This follows immediately from the following proposition.

Theorem. *Let* $\varphi : \boldsymbol{R} \to U$ *be a solution of the autonomous equation* $\dot{x} = \boldsymbol{v}(x)$ *defined by a vector field* $\boldsymbol{v}$ *in the phase space* U, *and let* $h^s : \boldsymbol{R} \to \boldsymbol{R}$ *be a translation along the time axis,* $h^s(t) = s + t$. *Then* $\varphi \circ h^s$ *is also a solution for any* s.

In other words, if $x = \varphi(t)$ is a solution, so is $x = \varphi(t + s)$.

Proof. This is obvious: the direction field of an autonomous equation maps into itself under translations along the time axis, and consequently the integral curves map into integral curves under such translations.

Corollary. *Through each point of the phase space of an autonomous system there passes one and only one phase curve.*

Remark. Here and throughout the following the discussion concerns maximal phase curves that are the images of solutions that cannot be extended to any larger interval (it may be impossible to extend a solution $\varphi : I \to U$ either because the interval I is already the entire line or because $\varphi(t)$ approaches the boundary of the domain U when t approaches an endpoint of the interval).

Proof of the corollary: Assume that two phase curves pass through a point – the images of solutions φ and ψ, defined on the entire line (the case when the solutions cannot be extended infinitely far is left to the reader). Then there exist instants of time a and b such that $\varphi(a) = \psi(b)$ (since both curves pass through one point). Translating one of the curves along the time axis, we obtain a new solution $\varphi \circ h^{a-b}$. This solution has a common initial condition with the solution ψ at $t = b$. Hence they coincide. Consequently ψ is obtained from φ by translation along the time axis. Thus the images of the mappings φ and ψ coincide, as was to be proved. □

Remark. The phase curves of a *nonautonomous system* (the images of solutions in the phase space) may intersect without coinciding. Therefore it is better to trace the solutions of nonautonomous systems through integral curves.

Problem 1. Suppose one and only one phase curve passes through each point of the phase space of the system $\dot{x} = \boldsymbol{v}(t, x)$. Does it follow from this that the system is autonomous?

Answer. No, for example, $\dot{x} = 1 + t^2$.

3. Closed Phase Curves

We know already that distinct phase curves of an autonomous system do not intersect. Let us see if a single phase curve can intersect itself. In other words, can a solution of a first-order autonomous system take on the same value more than once?

Theorem. *A maximal phase curve of an autonomous system either has no self-intersections or reduces to a single point, or is a closed phase curve (diffeomorphic to a circle).*

We have already encountered examples of closed phase curves (for example, limit cycles, cf. § 2).

The proof of the theorem is based on the following four lemmas.

Lemma 1. *A solution φ of a first-order autonomous system that takes on the same value twice ($\varphi(a) = \varphi(b)$, $b > a$) can be extended to the entire time axis as a periodic mapping Φ with period $T = b - a$.*

Proof. Every s is uniquely representable as $s = nT + \sigma$, where $0 \leq \sigma < T$. Set $\Phi(a+s) = \varphi(a+\sigma)$. Then Φ is a solution of period T, coinciding with φ on the interval $[a, b]$. Indeed Φ coincides with a translate of the solution φ in a neighborhood of each point, and so is itself a solution (by the theorem of Sect. 2). □

The solution obtained can have periods other than T. We shall study the set of all periods of a mapping of the line.

Lemma 2. *The set of all periods of any mapping is a subgroup of the group* $\boldsymbol{R}$.

Proof. The number T is a period of the mapping f if and only if a translation of the line by T maps f into itself. The translations that map f into itself form a subgroup of the group of all translations. For if two translations map f into itself, then their composition and their inverses do also.

Remark. This reasoning also shows that if any group acts on any set, then the set of all transformations of the group that leave fixed a given element of the set forms a subgroup of the original group. This subgroup is called the *stationary group* of the given element.

Lemma 3. *The set of all periods of a continuous mapping of the line is closed.*

Proof. Suppose a sequence of periods T_i of the mapping f converges to the number T. Then $f(t+T) = \lim f(t+T_i) = \lim f(t) = f(t)$ for any t. □

Thus *the set of all periods of a continuous mapping of the line is a closed subgroup of the line.*

Lemma 4. *Every closed subgroup G of the group of real numbers $\boldsymbol{R}$ is either $\boldsymbol{R}$ or an arithmetic progression formed by integer multiples of some number, or $\{0\}$.*

Proof. If $G \neq \{0\}$, then G has positive elements (since $-t$ belongs to G if t does).

Two cases are possible:

1) G contains positive elements arbitrarily close to zero;

2) the distance from 0 to all positive elements of the group is larger than some positive number.

In the first case G contains arithmetic progressions with arbitrarily small differences, and hence there are elements of G in every neighborhood of every point of the line. Since G is closed, $G = \boldsymbol{R}$. In the second case, consider the positive element T of G closest to 0 (such an element exists, since the group is closed). The arithmetic progression of integer multiples of the element T belongs to the group. We shall prove that there are no other elements in the group. Indeed, any other number t is representable in the form $nT + \tau$, where $0 < \tau < T$. If $t \in G$, then $t - nT = \tau < T$ is a positive element of the group, contradicting the minimality of the element T. □

Problem 1. Find all closed subgroups 1) of the plane $\boldsymbol{R}^2$; 2) of the space $\boldsymbol{R}^n$; 3) of the circle $\boldsymbol{S}^1 = \{z \in \boldsymbol{C} : |z| = 1\}$.

Solution. The subgroups of 1) and 2) are direct sums of closed subgroups of the line (Fig. 83); those of 3) are regular n-gons formed by the nth roots of unity and $\boldsymbol{S}^1$ itself.

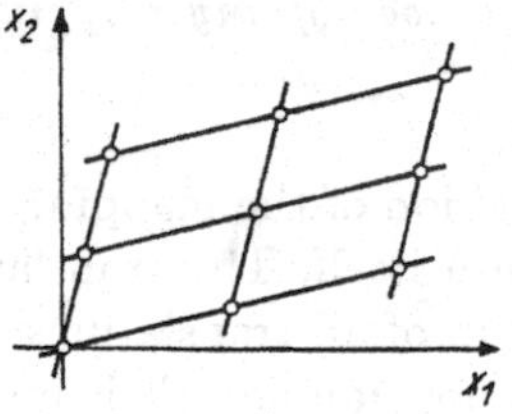

Fig. 83. A closed subgroup of the plane

Combining Lemmas 2, 3, and 4, we conclude that *the set of all periods of a continuous periodic mapping of the line either consists of all integer multiples of one smallest period or is the entire line* (in the latter case the mapping is constant).

In particular the solution Φ of Lemma 1 is either constant (and then the corresponding phase curve is an equilibrium position) or it has a smallest period θ. We define a mapping A of the circle into the phase curve by the formula $A : (\cos\alpha, \sin\alpha) \mapsto \Phi(\alpha\theta/2\pi)$. This mapping A is well-defined, since Φ has period θ. A is differentiable, since Φ is a solution. The mapping A is a

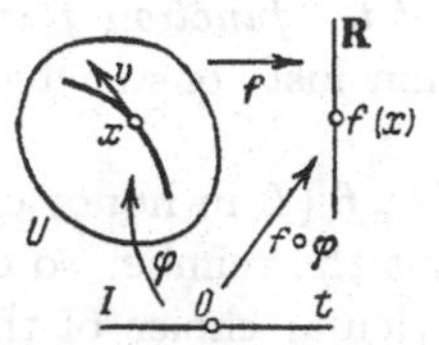

Fig. 84. The derivative of the function f in the direction of the vector $\boldsymbol{v}$

one-to-one mapping of the circle onto a phase curve, since Φ cannot take on the same value twice within its smallest period (by Lemma 1).

The derivative of A with respect to α is nonzero, since otherwise the solution would take on a value that is an equilibrium position, and then by the uniqueness theorem it would be a constant. By the implicit function theorem A is a local diffeomorphic mapping of the α-axis onto the image of Φ in the phase space, i.e., onto a phase curve. Hence the mapping inverse to A is differentiable, i.e., A is a diffeomorphism.

The theorem is now proved. □

Nonclosed phase curves, though they cannot intersect themselves, may wind about themselves in a very complicated manner.

Problem 2. Find the closures of the phase curves of the double pendulum $\ddot{x}_1 = -x_1$, $\ddot{x}_2 = -2x_2$.

Answer. A point, circles, and tori. Cf. § 24 and § 25, Sect. 6.

§ 10. The Derivative in the Direction of a Vector Field and First Integrals

Many geometric concepts can be described in two ways: in the language of *points* of a space or using *functions* defined on the space. Such a duality often turns out to be useful in the most varied areas of mathematics.

In particular vector fields can be described not only using velocities of motions, but also as *differentiations* of functions, and the basic theorems of differential equations can be stated in terms of *first integrals*.

1. The Derivative in the Direction of a Vector

Let $\boldsymbol{v}$ be a vector attached at the point x of a domain U, and let $f : U \to \boldsymbol{R}$ be a differentiable function. Let $\varphi : I \to U$ be some parametrized curve leaving x with velocity $\boldsymbol{v}$, so that $\varphi(0) = x$ and $\dot{\varphi}(0) = \boldsymbol{v}$. A composite mapping of the interval I of the real axis into the real axis then arises, $f \circ \varphi : I \to \boldsymbol{R}$, given by $(f \circ \varphi)(t) = f(\varphi(t))$, i.e., a real-valued function of a real variable t (Fig. 84).

Definition. The *derivative of the function f in the direction of the vector $\boldsymbol{v}$* is the derivative of the function just constructed at the point 0.

This number is denoted $L_{\boldsymbol{v}}f$ (L in honor of Sophus Lie). To justify this definition, we must verify that the number so obtained depends only on the vector $\boldsymbol{v}$ and not on the particular choice of the curve φ. This can be seen, for example, from the expression for the directional derivative in terms of coordinates: by the rule for differentiating a composite function

$$L_{\boldsymbol{v}}f = \frac{d}{dt}\bigg|_{t=0} f \circ \varphi = \sum_{i=1}^{n} \frac{\partial f}{\partial x_i} v_i,$$

where the derivatives are taken at the point where the vector is attached: here x_i are the coordinates in a neighborhood of this point and v_i are the components of the velocity vector in this coordinate system.

The same thing can be expressed in another way by saying that $L_{\boldsymbol{v}}f$ is the value of the 1-form df on the vector $\boldsymbol{v}$.

Problem 1. Calculate the derivative of the function H in the direction of the vector

$$\sum \left(\frac{\partial H}{\partial p_i} \frac{\partial}{\partial q_i} - \frac{\partial H}{\partial q_i} \frac{\partial}{\partial p_i} \right).$$

Answer. 0.

2. The Derivative in the Direction of a Vector Field

Now let $\boldsymbol{v}$ be a vector field in a domain U.

Definition. The *derivative of the function $f : U \to \boldsymbol{R}$ in the direction of the field $\boldsymbol{v}$* is a new function $L_{\boldsymbol{v}}f : U \to \boldsymbol{R}$ whose value at each point x is the derivative of the function f in the direction of the vector of the field attached at x: $(L_{\boldsymbol{v}}f)(x) = L_{\boldsymbol{v}(x)}f$. The function $L_{\boldsymbol{v}}f$ is called the *Lie derivative* of the function f.

Example. Let $\boldsymbol{v} = \partial/\partial x_1$ be the basic vector field whose components in the coordinate system $(x_1, \ldots, x_n)$ are $(1, 0, \ldots, 0)$. Then $L_{\boldsymbol{v}}f = \partial f/\partial x_1$ is the partial derivative of the function f.

Caution. In working with partial derivatives one must keep firmly in mind that a danger lurks in the very notation used for them: the partial derivative of a function f with respect to x_1 depends not only on the function taken as the x_1-coordinate in the domain under consideration but even more on the choice of the other coordinates. For example, in the (x, y)-plane the partial derivative $\partial f/\partial x$ of a function $f(y)$ is zero, but the partial derivative $\partial f/\partial x$

of the same function of a point in the plane on the same variable x in the coordinate system (x, z), where $z = x + y$, is -1. One really should write $\partial f/\partial x\big|_{y=\text{const}}$ and $\partial f/\partial x\big|_{z=\text{const}}$.

The derivative of a function in the direction of a vector field does not have the defect of the partial derivatives just indicated: it is a geometric object independent of the coordinate system by its very definition. If a smooth function f and a smooth vector field $\boldsymbol{v}$ are given, then $L_{\boldsymbol{v}} f$ is a well-defined function (of class C^{r-1} if f and $\boldsymbol{v}$ are of class C^r). In other words if a diffeomorphism maps the vector field and the function to a new place, then the derivative of the translated function in the direction of the translated field coincides with the translation of the derivative of the original function in the direction of the original field. This property of the operation of differentiation in a direction is called *naturalness*. Other examples of natural operations are addition and multiplication of functions, addition of fields, and multiplication of fields by functions.

3. Properties of the Directional Derivative

In this section we undertake the formalization of some obvious facts. We denote by F the set of all infinitely differentiable functions $f : U \to \boldsymbol{R}$. This set has the natural structure of a real vector space (since addition of functions preserves differentiability), and even that of a ring (since the product of infinitely differentiable functions is differentiable), or as one should say, an $\boldsymbol{R}$-algebra (a ring for whose elements multiplication by scalars is defined satisfying the usual requirements).

Let $\boldsymbol{v}$ be an infinitely differentiable vector field in U. The derivative of a function of F in the direction of the field $\boldsymbol{v}$ again belongs to F (here infinite differentiability is essential). Thus differentiation in the direction of the field $\boldsymbol{v}$ is a mapping $L_{\boldsymbol{v}} : F \to F$ of the algebra of infinitely differentiable functions into itself. Let us consider several properties of this mapping:

1. $L_{\boldsymbol{v}}(f+g) = L_{\boldsymbol{v}} f + L_{\boldsymbol{v}} g$; 2. $L_{\boldsymbol{v}}(fg) = f L_{\boldsymbol{v}} g + g L_{\boldsymbol{v}} f$;

3. $L_{\boldsymbol{u}+\boldsymbol{v}} = L_{\boldsymbol{u}} + L_{\boldsymbol{v}}$; 4. $L_{f\boldsymbol{u}} = f L_{\boldsymbol{u}}$; 5. $L_{\boldsymbol{u}} L_{\boldsymbol{v}} = L_{\boldsymbol{v}} L_{\boldsymbol{u}}$

(f and g are smooth functions and $\boldsymbol{u}$ and $\boldsymbol{v}$ are smooth vector fields).

Problem 1. Prove properties 1)–5), except for the one that is not true.

Remark on terminology. Algebraists call a mapping of a (commutative) ring into itself a *derivation* if it possesses Properties 1 and 2 of the mapping $L_{\boldsymbol{v}}$. The set of all derivations of a ring forms a *module* over the ring (a module over a ring is a generalization of a vector space over $\boldsymbol{R}$; the elements of a module can be added to one another and multiplied by elements of the ring).

The vector fields in U form a module over the $\boldsymbol{R}$-algebra F of functions in U. Properties 3 and 4 say that the operation L that maps the field $\boldsymbol{v}$ into the derivation $L_{\boldsymbol{v}}$ is a homomorphism of the F-module of fields into the F-module of derivations of the algebra F. Property 5), when it holds, says that the derivations $L_{\boldsymbol{u}}$ and $L_{\boldsymbol{v}}$ commute.

Problem * 2. Is the homomorphism L an isomorphism?

Analysts call the mapping $L_{\boldsymbol{v}}$ a *first-order homogeneous linear differential operator.* This name is given because, according to 1 and 2, the operator $L_{\boldsymbol{v}} : F \to F$ is $\boldsymbol{R}$-linear. In coordinates this operator is written as follows: $L_{\boldsymbol{v}} = v_1\partial/\partial x_1 + \cdots + v_n\partial/\partial x_n$. Previously (p. 71) we have denoted the vector field itself by the same symbol: the field is frequently identified with the operation of differentiating along it.

A Lie derivative operator along the vector field $\boldsymbol{v}$ analogous to $L_{\boldsymbol{v}}$ can be defined not only for functions, but for arbitrary differential-geometric objects (vector fields, forms, tensors) that can be transformed by diffeomorphisms. The derivative of each object is an object of the same type. The French call the operator $L_{\boldsymbol{v}}$ the *fisherman's derivative*: the fisherman sits still and differentiates the objects carried past him by the phase flow.

4. The Lie Algebra of Vector Fields

Property 5) does not always hold for vector fields $\boldsymbol{u}$ and $\boldsymbol{v}$. For example, for the fields $\boldsymbol{u} = \partial/\partial x$ and $\boldsymbol{v} = x\partial/\partial x$ on the x-axis we have

$$L_{\boldsymbol{u}}L_{\boldsymbol{v}} = \partial/\partial x + x\partial^2/\partial x^2, \quad L_{\boldsymbol{v}}L_{\boldsymbol{u}} = x\partial^2/\partial x^2.$$

Problem 1. Prove that the differential operator $L_{\boldsymbol{a}}L_{\boldsymbol{b}} - L_{\boldsymbol{b}}L_{\boldsymbol{a}}$ is of first order, not of second order, as it appears to be at first glance: $L_{\boldsymbol{a}}L_{\boldsymbol{b}} - L_{\boldsymbol{b}}L_{\boldsymbol{a}} = L_{\boldsymbol{c}}$, where $\boldsymbol{c}$ is a vector field depending on the fields $\boldsymbol{a}$ and $\boldsymbol{b}$.

Definition. The field $\boldsymbol{c}$ is called the *commutator* or the *Poisson bracket* of the fields $\boldsymbol{a}$ and $\boldsymbol{b}$ and is denoted $[\boldsymbol{a}, \boldsymbol{b}]$.

Problem 2. Prove the following three properties of the commutator:

1. $[\boldsymbol{a}, \boldsymbol{b} + \lambda\boldsymbol{c}] = [\boldsymbol{a}, \boldsymbol{b}] + \lambda[\boldsymbol{a}, \boldsymbol{c}]$, for $\lambda \in \boldsymbol{R}$ (linearity);
2. $[\boldsymbol{a}, \boldsymbol{b}] + [\boldsymbol{b}, \boldsymbol{a}] = 0$ (skew-symmetry);
3. $\big[[\boldsymbol{a}, \boldsymbol{b}], \boldsymbol{c}\big] + \big[[\boldsymbol{b}, \boldsymbol{c}], \boldsymbol{a}\big] + \big[[\boldsymbol{c}, \boldsymbol{a}], \boldsymbol{b}\big] = 0$ (the Jacobi identity).

Definition. A vector space with a binary operation possessing properties 1, 2, and 3 is called a *Lie algebra.*

Thus *vector fields with the operation of commutation form a Lie algebra.* This operation is just as fundamental for all of mathematics as addition and multiplication.

Problem 3. Prove that the oriented three-dimensional Euclidean space becomes a Lie algebra if the operation is defined as the vector cross product.

Problem 4. Prove that the space of square matrices of order n becomes a Lie algebra if the operation is defined as $AB - BA$.

Problem 5. Do the symmetric matrices form a Lie algebra with the same operation? What about the skew-symmetric matrices?

Problem 6. Knowing the components of the fields $\boldsymbol{a}$ and $\boldsymbol{b}$ in some coordinate system, find the components of their commutator.

Answer. $[\boldsymbol{a}, \boldsymbol{b}]_i = \sum a_j \partial b_i / \partial x_j - b_j \partial a_i / \partial x_j = L_{\boldsymbol{a}} b_i - L_{\boldsymbol{b}} a_i$.

Problem 7. Let $\{g^t\}$ be the phase flow of the field $\boldsymbol{a}$ and $\{h^s\}$ the phase flow of the field $\boldsymbol{b}$. Prove that the flows commute ($g^t h^s \equiv h^s g^t$) if and only if the commutator of the fields is zero.

Problem 8. Let $\boldsymbol{a}_{\boldsymbol{\omega}}$ be the velocity field of the points of a body rotating with angular velocity $\boldsymbol{\omega}$ about the point $\mathbf{o} \in \boldsymbol{R}^3$. Find the commutator of the fields $\boldsymbol{a}_{\boldsymbol{\alpha}}$ and $\boldsymbol{a}_{\boldsymbol{\beta}}$.

Answer. $[\boldsymbol{a}_{\boldsymbol{\alpha}}, \boldsymbol{a}_{\boldsymbol{\beta}}] = \boldsymbol{a}_{\boldsymbol{\gamma}}$, where $\boldsymbol{\gamma}$ is the cross product of $\boldsymbol{\alpha}$ and $\boldsymbol{\beta}$.

5. First Integrals

Let $\boldsymbol{v}$ be a vector field in a domain U and $f : U \to \boldsymbol{R}$ a differentiable function.

Definition. The function f is called a *first integral* of the equation $\dot{x} = \boldsymbol{v}(x)$ if its derivative in the direction of the field $\boldsymbol{v}$ is zero: $L_{\boldsymbol{v}} f \equiv 0$.

The strange-sounding name *first integral* is a relic of the times when mathematicians tried to solve all differential equations by integration. In those days the name *integral* (or a *partial integral*) was given to what we now call a *solution*.

The following two properties of a first integral are obviously equivalent to the relation $L_{\boldsymbol{v}} f \equiv 0$ and could be taken as the definition of it.

1. *The function f is constant along each solution $\varphi : I \to U$, i.e., each function $f \circ \varphi$ is constant.*

2. *Each phase curve of the field $\boldsymbol{v}$ belongs to one and only one of the level sets of the function f* (Fig. 85).

Example. Consider the system $\dot{x}_1 = x_1$, $\dot{x}_2 = x_2$, whose phase space is the whole plane. The phase curves (rays) are depicted in Fig. 86. We shall show that this system has no first integral except a constant. Indeed a first integral is a function continuous in the entire plane and constant on each ray emanating from the origin; hence it is constant.

Problem 1. Prove that in a neighborhood of a limit cycle every first integral is constant.

Fig. 85. A phase curve lies entirely on one level surface of an integral

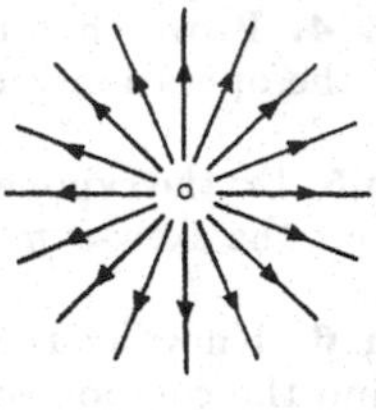

Fig. 86. A system having no first integrals

Problem 2. For which k does the system of equations $\dot{x}_1 = x_1$, $\dot{x}_2 = kx_2$ on the whole plane have a non-constant first integral?

Answer. For $k \leq 0$ (cf. Fig. 30 on p. 40).

Problem 3. Prove that the set of all first integrals of a given field forms an algebra: the sum and product of first integrals are also first integrals.

Nonconstant first integrals are rarely encountered. Nevertheless in the cases where they exist and can be found the reward is quite significant.

Example. Let H be a function of the $2n$ variables $(p_1, \dots, q_n)$ that is differentiable (r times, $r \geq 2$). The system of $2n$ equations $\dot{p}_i = -\partial H/\partial q_i$, $\dot{q}_i = \partial H/\partial p_i$ is called the *system of canonical Hamiltonian equations.* (Hamilton showed that the differential equations of a large number of problems of mechanics, optics, calculus of variations, and other areas of science can be written in this form). The function H is called the *Hamiltonian* (in mechanics it is usually the total energy of the system).

Theorem (Law of Conservation of Energy). *The Hamiltonian is a first integral of the system of canonical Hamiltonian equations.*

Proof.

$$L_v H = \sum_{i=1}^{n} \left(\frac{\partial H}{\partial p_i} \dot{p}_i + \frac{\partial H}{\partial q_i} \dot{q}_i \right) = \sum_{i=1}^{n} \left[\frac{\partial H}{\partial p_i} \left(-\frac{\partial H}{\partial q_i} \right) + \frac{\partial H}{\partial q_i} \frac{\partial H}{\partial p_i} \right] = 0,$$

which was to be shown. □

6. Local First Integrals

The absence of nonconstant first integrals is connected with the topological structure of the phase curves. In the general situation the phase curves cannot be packed globally on a level surface of any function, and so there is no nonconstant first integral. Locally, however, in a neighborhood of a nonsingular point, the phase curves have a simple structure and nonconstant first integrals do exist.

Let U be a domain in n-dimensional space, $\boldsymbol{v}$ a differentiable vector field on U, and x_0 a nonsingular point of the field ($\boldsymbol{v}(x_0) \neq 0$).

Theorem. *There exists a neighborhood V of the point x_0 such that in V the equation $\dot{x} = \boldsymbol{v}(x)$ has $n-1$ functionally independent first integrals $f_1, \ldots, f_{n-1}$, and any first integral of the equation is a function of $f_1, \ldots, f_{n-1}$ in V.*

[A set of m functions is *functionally independent* in a neighborhood of a point x_0 if the rank of the derivative of the mapping $f : U \to \boldsymbol{R}^m$ defined by these functions is m at the point x_0 (cf., for example, G. M. Fikhtengol'ts, *The Fundamentals of Mathematical Analysis*, Pergamon Press, New York, 1965, Vol. 1, Chapt. 6).]

Proof. For the standard equation in $\boldsymbol{R}^n$ $\dot{y}_1 = 1$, $\dot{y}_2 = \cdots = \dot{y}_n = 0$, this is obvious: the first integrals are any smooth functions of $y_2, \ldots, y_n$. The same is true for this equation in any convex domain (a domain is called *convex* if it contains the line segment joining any two of its points). In a convex domain any integral of the standard equation reduces to functions of $y_2, \ldots, y_n$. Every equation in a suitable neighborhood of a nonsingular point can be written in standard form in suitable coordinates y. This neighborhood can be considered convex in the y-coordinates. (If it is not convex, it can be replaced by a smaller convex domain.)

It remains only to remark that both the property of being a first integral and functional independence are independent of the system of coordinates. □

Problem 1. Give an example of a domain in which the standard equation has a first integral that does not reduce to a function of $y_2, \ldots, y_n$.

7. Time-Dependent First Integrals

Let f be a differentiable function on the extended phase space of the equation $\dot{\boldsymbol{x}} = \boldsymbol{v}(t, \boldsymbol{x})$, in general nonautonomous.

Let us form the autonomous system whose phase curves are the integral curves of the original equation. To do this we enlarge the equation by adding the trivial equation $\dot{t} = 1$:

$$\dot{\boldsymbol{X}} = \boldsymbol{V}(\boldsymbol{X}), \quad \boldsymbol{X} = (t, \boldsymbol{x}), \quad \boldsymbol{V}(t, \boldsymbol{x}) = (1, \boldsymbol{v}).$$

Definition. A function f is said to be a *time-dependent first integral* of the equation $\dot{\boldsymbol{x}} = \boldsymbol{v}(t, \boldsymbol{x})$ if it is the first integral of the extended autonomous equation (Fig. 87).

In other words: *each integral curve of the original equation lies in a level set of the function.*

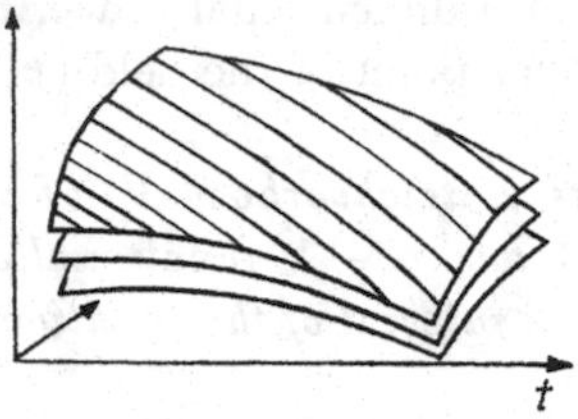

Fig. 87. Integral curves on the level surface of a time-dependent first integral

The vector field $\boldsymbol{V}$ does not vanish anywhere. According to the theorem of Sect. 6 *in some neighborhood of each point of the extended phase space the equation* $\dot{\boldsymbol{x}} = \boldsymbol{v}(t, \boldsymbol{x})$ *has a number of functionally independent (time-dependent) first integrals equal to the dimension of the phase space (the number of components of the vector* $\boldsymbol{x}$*); moreover every (time-dependent) first integral can be expressed in terms of these particular integrals in this neighborhood.*

In particular an autonomous equation with an n-dimensional phase space has in a neighborhood of any (not necessarily nonsingular) point n time-dependent functionally independent first integrals.

A *first integral* of a differential equation (or system) of any order is a first integral of the equivalent first-order system.

Problem 1. Prove that the system of Newtonian equations $\ddot{\boldsymbol{r}} = -\boldsymbol{r}/r^3$ has a first integral that can be written in polar coordinates as $r^2\dot{\varphi}$ ($\boldsymbol{r} \in \boldsymbol{R}^2$).

This integral, called the *sectorial velocity,* was discovered by Kepler from observations of the motion of Mars ("Kepler's Second Law").

Problem 2. Prove that the sectorial velocity is a first integral of the equation $\ddot{\boldsymbol{r}} = \boldsymbol{r}a(r)$ for any form of the function a.

A force field of the form $\boldsymbol{r}a(r)$ is called *central.* The preceding problem shows why the law of universal gravitation cannot be deduced from Kepler's Second Law: the third law is needed.

Problem 3. Prove that each component of the cross product $[\boldsymbol{r}, \dot{\boldsymbol{r}}]$ is a first integral for motion in any central force field in three-dimensional space ("the law of conservation of angular momentum").

Problem 4. Prove that if the Hamiltonian function is independent of q_i, then p_i is a first integral of the Hamilton equations.

Problem 5. Assume that each solution of the equation $\dot{\boldsymbol{x}} = \boldsymbol{v}(t, \boldsymbol{x})$ with an n-dimensional phase space can be extended to the whole t-axis. Prove that such an equation has n functionally independent (time-dependent) first integrals on the entire extended phase space, in terms of which all of its (time-dependent) first integrals can be functionally expressed.

§ 11. First-order Linear and Quasi-linear Partial Differential Equations

Partial differential equations have not been nearly so well studied as ordinary differential equations. The theory of one first-order partial differential equation has been successfully reduced to the study of special ordinary differential equations, the so-called *characteristic equations.* The essence of the connection between a partial differential equation and a characteristic equation is that motion of a solid medium can be described using both the ordinary differential equations of motion of its particles and the partial differential equations for a field. In this section the simplest special cases of linear and so-called quasi-linear partial differential equations are studied in detail and a rule for solving the general equation is given.

1. The Homogeneous Linear Equation

Definition. A *first-order homogeneous linear equation* in a domain U is an equation $L_{\boldsymbol{a}}u = 0$, where $\boldsymbol{a}$ is a known vector field in the domain U and u is an unknown function. In coordinates it has the form $a_1\partial u/\partial x_1 + \cdots + a_n\partial u/\partial x_n = 0$, $a_k = a_k(x_1, \ldots, x_n)$. The phase curves of the vector field $\boldsymbol{a}$ are called the *characteristics* of the equation $L_{\boldsymbol{a}}u = 0$. The equation $\dot{x} = \boldsymbol{a}(x)$ is called the *characteristic equation.*

Remark. The adjective "characteristic" in mathematics always means "connected in an invariant manner" (in the present case, invariant with respect to the choice of the coordinate system). Thus a characteristic subgroup of a group is the subgroup that maps into itself under all automorphisms of the group, the characteristic equation of the matrix of an operator is independent of the choice of basis, the characteristic classes in topology map into themselves under diffeomorphisms, etc.

The characteristics of the equation $L_{\boldsymbol{a}}u = 0$ are connected with the equation invariantly with respect to diffeomorphisms: if a diffeomorphism maps the old equation into a new one, then it maps the characteristics of the old equation into the characteristics of the new one. In addition one can even multiply the field $\boldsymbol{a}$ by a nonvanishing function – this changes neither the solutions nor the characteristics of the equation.

Problem 1. Find the characteristics of the equation $\partial u/\partial x = y\partial u/\partial y$.

Solution. $\dot{x} = 1$, $\dot{y} = -y$; $y = Ce^{-x}$.

Theorem. *A function u is a solution of the equation $L_{\boldsymbol{a}}u = 0$ if and only if it is a first integral of the characteristic equation.*

Proof. This is the definition of a first integral. □

Despite the obviousness of this theorem, it is very useful, since the ordinary characteristic equation is easier to solve than the original partial differential equation.

Problem 2. Solve the equation of Problem 1.

Solution. $u = ye^x$ is a solution, and all solutions can be obtained as functions of this one.

Problem 3. Solve the equation $y\partial u/\partial x = x\partial u/\partial y$ on the entire plane.

Answer. Any solution is a function of $x^2 + y^2$.

Problem 4. Are all solutions of the equation $x\partial u\partial x = y\partial u/\partial y$ on $\boldsymbol{R}^2$ functions of xy?

Answer. No. There exists a solution for which $u(1,1) \neq u(-1,-1)$.

2. The Cauchy Problem

Definition. The *Cauchy problem* for the equation $L_a u = 0$ is the problem of determining a function u from its values on a given hypersurface (a *hypersurface* in $\boldsymbol{R}^n$ is an $(n-1)$-dimensional surface. For example, in the case $n = 2$ a hypersurface is a curve, for $n = 3$ an ordinary surface).

The given hypersurface is called the *initial hypersurface* and prescribing the unknown function on it is called an *initial condition*, $u|_\gamma = \varphi$. The function φ is called the *initial function.* It is defined on the initial hypersurface.

The Cauchy problem does not always have a solution. Indeed, along each characteristic the value of u is constant. But a characteristic can intersect the initial surface more than once (Fig. 88). If the values of the initial function at these points are distinct, then the corresponding Cauchy problem has no solution in any domain containing the characteristic in question.

Fig. 88. An unsolvable Cauchy problem

Definition. A point on the initial hypersurface is called *noncharacteristic* if the characteristic passing through the point is transversal (nontangential) to the initial hypersurface.

Theorem. *Let x be a noncharacteristic point on the initial hypersurface. Then there exists a neighborhood of the point x in which the Cauchy problem has one and only one solution.*

Proof. By the rectification theorem one can choose coordinates in a neighborhood of the point x such that the field a will have components $(1, 0, \ldots, 0)$ and the initial hypersurface will assume the form $x_1 = 0$. In these coordinates the Cauchy problem assumes the form $\partial u/\partial x_1 = 0$, $u|_{x_1=0} = \varphi$. The unique solution in a convex domain is $u(x_1, \ldots, x_n) = \varphi(x_2, \ldots, x_n)$. □

Problem 1. Solve the Cauchy problem $u|_{x=0} = \sin y$ for the equation $\partial u/\partial x = y\partial u/\partial y$.

Solution. On a characteristic $y = Ce^{-x}$; according to the initial condition $u = \sin C$. Hence $u = \sin(e^x y)$.

Problem 2. Which points of the line $x = 1$ are noncharacteristic for the equation $y\partial u/\partial x = x\partial u/\partial y$?

Answer. $y \neq 0$.

Problem 3. Does the Cauchy problem $u|_{x=1} = y^2$ for this equation on $\boldsymbol{R}^2$ have a solution? Is it unique?

Answer. There is a solution, but it is not unique.

Remark. The solutions of an ordinary differential equation form a finite-dimensional manifold: each solution is defined by a finite set of numbers (initial conditions). We see that a homogeneous linear partial differential equation of first order with respect to a function of n variables has "as many solutions as there are functions of $n - 1$ variables." An analogous phenomenon holds also for general first-order partial differential equations.

The reason will become clear if we regard a differential equation as the limit of difference equations. The same considerations suggest that for a second-order partial differential equation it is necessary to prescribe two functions on the initial hypersurface (the values of the function and its derivative in a direction transversal to the initial hypersurface), etc. Of course these considerations do not replace the proofs of the corresponding existence and uniqueness theorems for solutions. These proofs can be found in textbooks on the theory of partial differential equations, for example, in the book of Courant and Hilbert, *Methods of Mathematical Physics*, Interscience, New York, 1953–1962.

3. The Inhomogeneous Linear Equation

Definition. A *first-order inhomogeneous linear equation* in a domain U is an equation $L_a u = b$, where a is a given vector field, b is a given function, and u

is the unknown function in the domain U. In coordinate notation: $a_1\partial u/\partial x_1 + \cdots + a_n\partial u/\partial x_n = b$, where $a_k = b$ are known functions of $x_1, \ldots, x_n$.

The Cauchy problem is posed just as for a homogeneous equation.

Theorem. *In some sufficiently small neighborhood of any noncharacteristic point of the initial surface the solution exists and is unique.*

Proof. The derivative with respect to time of the unknown function along a characteristic is known (equal to b), and hence its increment along a segment of the characteristic equals the integral of b with respect to the time of motion along this segment. For example, if $a_1 \neq 0$ at the point being studied, this increment equals $\int b/a_1\, dx_1$ along a segment of the characteristic. □

Problem 1. Solve the Cauchy problem $u|_{x=0} = \sin y$ for the equation $\partial u/\partial x = y\partial u/\partial y + y$.

Solution. As x changes with velocity 1 the value of u on the characteristic $y = Ce^{-x}$ varies with velocity Ce^{-x}. Consequently the increment of u along this characteristic as x changes from 0 to X is $C(1 - e^{-X})$.

The point (X, Y) lies on the characteristic where $C = e^X Y$. At this point $u = \sin C + C(1 - e^{-X})$. Thus $u = \sin(e^x y) + y(e^x - 1)$.

4. The Quasi-linear Equation

Definition. A *first-order quasi-linear equation* is an equation $L_{\boldsymbol{\alpha}} u = \beta$ with respect to the function u, where $\boldsymbol{\alpha}(x) = \boldsymbol{a}(x, u(x))$ and $\beta(x) = b(x, u(x))$. Here $\boldsymbol{a}$ is a vector field in x-space depending on a point of the u-axis as a parameter and b is a function on x-space also depending on a point of the u-axis as a parameter. In coordinate notation the equation has the form

$$a_1(x, u)\frac{\partial u}{\partial x_1} + \cdots + a_n(x, u)\frac{\partial u}{\partial x_n} = b(x, u).$$

The difference from a linear equation is only that *the coefficients a and b may depend on the value of the unknown function.*

Example. Consider a one-dimensional medium of particles moving along a line by inertia, so that the velocity of each particle remains constant. We denote the velocity of the particle at the point x at time t by $u(t, x)$. We then write Newton's equation: the acceleration of the particle equals zero. If $x = \varphi(t)$ is the motion of a particle, then $\dot{\varphi} = u(t, \varphi(t))$ and $\ddot{\varphi} = \dfrac{\partial u}{\partial t} + \dfrac{\partial u}{\partial x}\dot{\varphi} = \dfrac{\partial u}{\partial t} + u\dfrac{\partial u}{\partial x}$. Thus *the velocity field of a medium consisting of noninteracting particles satisfies the quasi-linear equation $u_t + uu_x = 0$.*

Problem 1. Construct the graph of the solution at the instant t if $u = \arctan x$ for $t = 0$.

Solution. The diffeomorphism of the plane $(x, u) \mapsto (x + ut, u)$ moves each line $u = \text{const}$ along the x-axis by ut and maps the graph of the solution at the instant 0 to the graph of the solution at the instant t (this diffeomorphism is none other than the transformation of the phase flow of the Newton equation for particles: the (x, u)-plane is the phase plane of the particle). For the answer see Fig. 89.

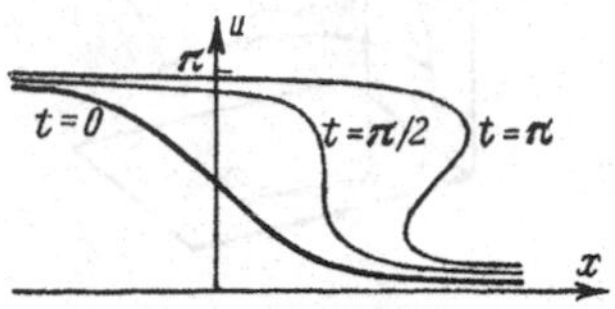

Fig. 89. The graph of a solution is obtained from the graph of the initial condition by the action of the phase flow

Remark. For $t \geq \pi/2$ no smooth solution exists. Starting from this instant the particles collide and the assumption of the absence of interaction among them becomes physically unrealistic. In these circumstances the motion of the medium is described by the so-called shock waves – discontinuous solutions satisfying the equation to the left and right of a discontinuity and satisfying additional conditions derived from physical considerations at the discontinuity (depending on the nature of the interaction when the particles collide).

Problem 2. Construct the equation for the evolution of the velocity field of a medium of noninteracting particles in a force field with force $F(x)$ at the point x.

Answer. $u_t + uu_x = F$.

Problem 3. Solve this equation with the initial condition $u|_{t=0} = 0$ for the force $F(x) = -x$.

Solution. The phase flow consists of rotations, so that the graph of $u(t, \cdot)$ is a straight line with inclination $-t$. Hence $u(t, x) = -x \tan t$, $|t| < \pi/2$.

Problem 4. Find the maximum width of a strip $0 \leq t < C$ in which there exists a solution of the equation $u_1 + uu_x = \sin x$ with the initial condition $u|_{t=0} = 0$.

Answer. $C = \pi/2$.

5. The Characteristics of a Quasi-linear Equation

The example just studied shows the usefulness of passing from the partial differential equations for the velocity field to ordinary differential equations for the motion of particles of the medium. Something analogous can be done also in the case of a general quasi-linear first-order equation.

The equation $L_{a(x,u(x))}u = b(x, u(x))$ says that if the point x leaves x_0 with velocity $a_0 = a(x_0, u_0)$, where $u_0 = u(x_0)$, then the value of $u(x)$ begins to change with velocity $b_0 = b(x_0, u_0)$ (Fig. 90). In other words the vector A_0 of the direct product of the x-space and the u-axis attached at the point (x_0, u_0) and with components a_0 and b_0 is tangent to the graph of the solution. Suppose $A_0 \neq 0$.

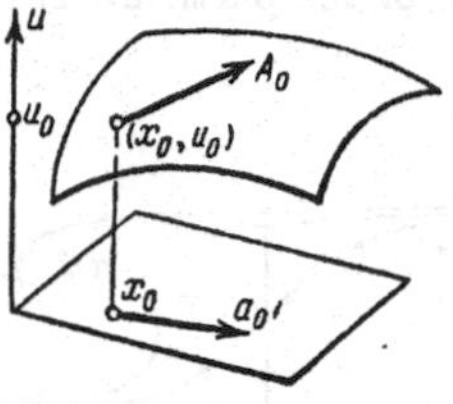

Fig. 90. The geometric meaning of a quasi-linear equation

Definition. The direction line of the vector A_0 is called the *characteristic direction* of the quasi-linear equation at the point (x_0, u_0).

The set of characteristic directions at all points of the domain of definition of the coefficients of the equation forms a direction field. This field is called the *characteristic direction field* of the equation. In coordinate notation the characteristic directions are the directions of the vectors of the field

$$A = \sum a_k(x, u)\frac{\partial}{\partial x_k} + b(x, u)\frac{\partial}{\partial u}.$$

The differential equation defined by the characteristic direction field is called the *characteristic equation*, and its integral curves are called *characteristics*. Thus characteristics are phase curves of the vector field A.

Problem 1. Find the characteristics of the equation of a medium of noninteracting particles $u_t + uu_x = 0$.

Solution. $\dot{x} = u$, $\dot{t} = 1$, $\dot{u} = 0$. The characteristics are the lines $x = x_0 + u_0 t$, $u = u_0$.

Remark 1. A linear equation is a special case of a quasi-linear equation, but the characteristics of a linear equation, when regarded as a quasi-linear equation, differ from its characteristics as a linear equation: the former lie in (x, u)-space, the latter are the projections of the former on the x-space.

Remark 2. Quasilinear equations preserve the quasi-linear form under diffeomorphisms of the x-space and even under diffeomorphisms of the product space in which its coefficients a and b are defined. Characteristics are invariantly connected with the equation: if such a diffeomorphism maps the old equation into a new one, then the characteristics of the old equation map into the characteristics of the new one. Moreover the equation can be multiplied by a nonvanishing function of x and u, and in the process neither the solutions nor the characteristics will change (although the vector field $\boldsymbol{A}$ changes).

Problem 2. Prove that a quasi-linear equation can be reduced by a suitable local diffeomorphism of the product space to the standard form $\partial u/\partial x_1 = 0$ in a neighborhood of any point (x, u) at which the value of $\boldsymbol{a}$ is nonzero.

6. Integration of a Quasi-linear Equation

The characteristic equation for the equation $\sum a_k \partial u/\partial x_k = b$ is customarily written in the so-called *symmetric form*

$$\frac{dx_1}{a_1} = \cdots = \frac{dx_n}{a_n} = \frac{du}{b},$$

which expresses the collinearity of the tangent to the characteristic with the characteristic vector (these relations signify that the 1-forms are equal on vectors tangent to characteristics if the denominators are nonzero).

Definition. A surface is called an *integral surface* of a direction field if the direction field lies in its tangent plane at every point.

Theorem. *A necessary and sufficient condition for a smooth surface to be the integral surface of a smooth direction field is that each integral curve having a point in common with the surface lies entirely on the surface.*

Proof. By the rectification theorem the field can be transformed into a field of parallel lines by a diffeomorphism. For such a field the theorem is obvious. □

The following theorem is a consequence of the definition of a characteristic direction.

Theorem. *A function u is a solution of a quasi-linear equation if and only if its graph is an integral surface of the characteristic direction field.*

The following corollary is a consequence of the last two theorems.

Corollary. *A function u is a solution of a quasi-linear equation if and only if its graph contains an interval of the characteristic passing through each of its points.*

Thus finding the solutions of a quasi-linear equation reduces to finding its characteristics. If the characteristics are known, it remains only to form a surface from them that is the graph of a function: this function will be a solution of the quasi-linear equation and all solutions can be obtained in this way.

Problem 1. Prove that the Cauchy problem for a first-order quasi-linear equation has a solution, and moreover only one in a sufficiently small neighborhood of a point x_0 of the initial hypersurface for an initial condition such that the vector $\boldsymbol{a}(x_0, u(x_0))$ is not tangent to the initial hypersurface.

Remark. In contrast to the linear equation one cannot speak of points of the initial hypersurface themselves as being characteristic for a quasi-linear equation: whether a given point is characteristic or not depends on the initial value as well.

7. The First-order Nonlinear Partial Differential Equation

Like linear or quasi-linear equations, nonlinear equations of the most general form $F(x, \partial u/\partial x, u) = 0$ can be integrated using characteristics. But while the characteristics of a linear equation with respect to a function in $\boldsymbol{R}^n$ lie in $\boldsymbol{R}^n$ and those of a quasi-linear equation lie in the $(n+1)$-dimensional space $\boldsymbol{R}^n \times \boldsymbol{R}$, the characteristics of the general nonlinear equation are curves in the $(2n+1)$-dimensional space of 1-jets of functions on which the function F that defines the equation is defined.

Definition. The space of *1-jets of functions of* $x = (x_1, \ldots, x_n)$ is the $(2n+1)$-dimensional space with coordinates $(x_1, \ldots, x_n; p_1, \ldots, p_n; y)$. The 1-jet of a function u at the point x is the point of this space with coordinates $(x, p = \partial u/\partial x, y = u(x))$. The set of 1-jets of the function u at all points x of its domain of definition is called the *1-graph* of the function.

The equation $F(x, \partial u/\partial x, u) = 0$ defines a hypersurface E in the space of 1-jets, namely the surface on which $F(x, p, y) = 0$. A solution of the equation $F = 0$ is a function whose 1-graph belongs to the hypersurface E.

We shall assume that the vector of derivatives F_p (with components $\partial F/\partial p_i$) is nonzero: without this restriction the equation might not contain $\partial u/\partial x$ and hence would not even be a differential equation. It follows from the condition $F_p \neq 0$ that the hypersurface E is smooth (by the implicit function theorem). The hardest part of the theory of a first-order nonlinear partial differential equation is to invent the following definition.

Definition. The *characteristics* of the equation $F = 0$ are the phase curves of the following difficult-to-remember system of differential equations on the hypersurface E in the space of 1-jets:

$$\dot{x} = F_p, \quad \dot{p} = -F_x - pF_y, \quad \dot{y} = pF_p.$$

Problem 1. Prove that a phase curve of this system originating on the hypersurface E lies entirely in E.

Solution. $\dot{F} = F_x\dot{x} + F_p\dot{p} + F_y\dot{y} = 0.$

Problem 2. Prove that *the 1-graph of each solution of the equation $F = 0$ contains an interval of the characteristic passing through each of its points. Conversely if the 1-graph of a function consists of whole characteristics, then the function is a solution.*

Solution. Along the 1-graph of the solution $dy = p\,dx$ and $dp = (\partial^2 u/\partial x^2)\,dx$. For a characteristic vector the first condition obviously holds and the second follows from the fact that the restriction of dF to the 1-graph is zero: the restriction of $F_x\,dx + F_p\,dp + F_y\,dy$ to the 1-graph has the form

$$(F_x + pF_y)\,dx + F_p \partial^2 u/\partial x^2\,dx.$$

The proof of the converse (and also the geometric motivation for the strange definition of characteristics) can be found in the book of V. I. Arnol'd, *Geometrical Methods in the Theory of Ordinary Differential Equations*, Springer-Verlag, New York, 1988, § 8, or in the book of V. I. Arnol'd, *Mathematical Methods of Classical Mechanics*, Springer-Verlag, 1989, pp. 369–370: these proofs are based on the geometry of the field of contact planes in the space of jets.

The result of Problem 2 reduces the integration of a first-order nonlinear equation (for example, finding a solution of the Cauchy problem) to integrating a system of ordinary differential equations, the characteristic equations. From the initial condition one constructs a submanifold of the space of 1-jets, and the characteristics passing through it form the 1-graph of the desired solution.

Problem 3. Prove that the characteristics of a nonlinear equation that is quasi-linear project to characteristics of this quasi-linear equation under the mapping $(x, p, y) \mapsto (x, y)$.

Problem * 4. Prove that the characteristics of the nonlinear equation $F = 0$ are invariantly connected with the equation: under diffeomorphisms of the $\boldsymbol{x}$-space or even the product of the $\boldsymbol{x}$-space with the axis of values of the function the derivatives transform so that the characteristics of the old equation map into characteristics of the new one; under multiplication of F by a nonvanishing function the characteristics do not change.

Remark. In reality the connection between the hypersurface E and the characteristics on it is invariant with respect to the even larger group of diffeomorphisms of the space of jets, which permutes the arguments not only with the values but also with the derivatives: all that matters is that the diffeomorphism of the space of jets preserve the field of contact planes (defined by the equation $dy = p\,dx$). Such diffeomorphisms are called *contact diffeomorphisms* and form the *contact group*, which is fundamental for the theory of first-order partial differential equations and for geometric optics.

Definition. A *Hamilton-Jacobi equation* is a first-order partial differential equation in which the value of the unknown function does not occur explicitly, i.e., an equation of the form $H(x, \partial u/\partial x) = 0$.

Problem 5. Prove that the distance from a point of the plane to a smooth curve in the plane (Fig. 91) satisfies the Hamilton-Jacobi equation $\sum(\partial u/\partial x_i)^2 = 1$ in a neighborhood of this curve (excluding the curve itself).

Problem 6. Prove that the distance from a point in Euclidean space to a smooth submanifold (of any dimension) in the space satisfies the Hamilton-Jacobi equation

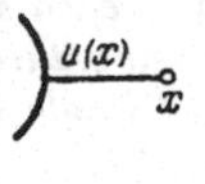

Fig. 91. The solution of a Hamilton-Jacobi equation

$\sum(\partial u/\partial x_i)^2 = 1$ in a neighborhood of the submanifold (excluding the submanifold itself).

Problem 7. Prove that in a sufficiently small neighborhood of any point of a Euclidean space every solution of the Hamilton-Jacobi equation $\sum(\partial u/\partial x_i)^2 = 1$ is the distance to some smooth hypersurface plus a constant.

Problem 8. Prove that the characteristics of a Hamilton-Jacobi equation $H = 0$ project to (x,p)-space as phase curves of the Hamilton equations $\dot{x} = H_p$, $\dot{p} = -H_x$ lying on the zero level surface of the Hamiltonian function.

§ 12. The Conservative System with one Degree of Freedom

As an example of the application of the first integral to the study of a differential equation we shall consider here a mechanical system with one degree of freedom, without friction.

1. Definitions

A *conservative system with one degree of freedom* is a system described by a differential equation

$$\ddot{x} = F(x), \tag{1}$$

where F is a function that is differentiable on some interval I of the x-axis.

Eq. (1) is equivalent to the system

$$\dot{x}_1 = x_2, \quad \dot{x}_2 = F(x_1), \quad (x_1, x_2) \in I \times \boldsymbol{R}. \tag{2}$$

In mechanics the following terminology is adopted:

I is the configuration space;
$x_1 = x$ is the coordinate;
$x_2 = \dot{x}$ is the velocity;
$\ddot{x}$ is the acceleration;
$I \times \boldsymbol{R}$ is the phase space;
(1) is Newton's equation;
F is the force field;
$F(x)$ is the force.

Consider also the following functions on the phase space:

$$T = \frac{\dot{x}^2}{2} = \frac{x_2^2}{2} \quad \textit{the kinetic energy;}$$

$$U = -\int_{x_0}^{x} F(\xi)\,d\xi \quad \textit{the potential energy;}$$

$$E = T + U \quad \textit{the total mechanical energy.}$$

It is obvious that $F(x) = -\dfrac{dU}{dx}$ so that *the potential energy determines the system.*

Example 1. For the pendulum of § 1 (Fig. 92) we have $\ddot{x} = -\sin x$, where x is the angular displacement, $F(x) = -\sin x$, $U(x) = -\cos x$. For the equation of small oscillations of the pendulum $\ddot{x} = -x$

$$F(x) = -x, \quad U(x) = \frac{x^2}{2}.$$

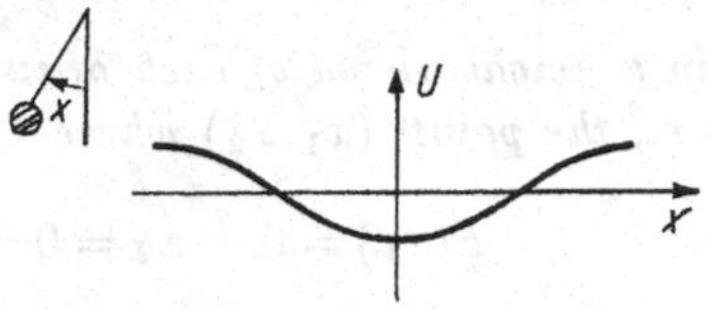

Fig. 92. The potential energy of a pendulum

For the equation of small oscillations of the inverted pendulum $\ddot{x} = x$

$$F(x) = x, \quad U(x) = -\frac{x^2}{2}$$

(cf. Fig. 93).

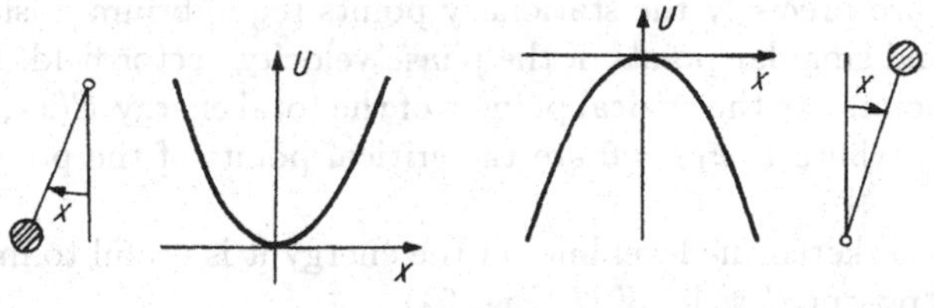

Fig. 93. The potential energy of a pendulum near the lower and upper equilibrium positions

2. The Law of Conservation of Energy

Theorem. *The total energy E is a first integral of the system* (2).

Proof. We have

$$\frac{d}{dt}\left(\frac{x_2(t)^2}{2} + U(x_1(t))\right) = x_2\dot{x}_2 + U'\dot{x}_1 = x_2F(x_1) - F(x_1)x_2 = 0,$$

which was to be proved. □

The theorem just proved makes it possible to study and solve explicitly "in quadratures" equations of the form (1), for example the pendulum equation.

3. The Level Lines of the Energy

Let us study the phase curves of the system (2). Each of these lies entirely on one level set of the energy. We shall study these level sets.

Theorem. *The level set of the energy*

$$\left\{(x_1, x_2) : \frac{x_2^2}{2} + U(x_1) = E\right\}$$

is a smooth curve in a neighborhood of each of its points, except for the equilibrium positions, i.e., the points (x_1, x_2) *where*

$$F(x_1) = 0, \quad x_2 = 0.$$

Proof. Using the implicit function theorem, we have

$$\frac{\partial E}{\partial x_1} = -F(x_1), \quad \frac{\partial E}{\partial x_2} = x_2.$$

If one of the derivatives is nonzero, then the level set of E in a neighborhood of that point is the graph of a differentiable function of the form $x_1 = x_1(x_2)$ or $x_2 = x_2(x_1)$. The theorem is now proved. □

We remark that the points (x_1, x_2) excluded above, where $F(x_1) = 0$ and $x_2 = 0$, are precisely the stationary points (equilibrium positions) of the system (2) and singular points of the phase velocity vector field. Furthermore these same points are the critical points[3] of the total energy $E(x_1, x_2)$. Finally, the points x_1 where $F(x_1) = 0$ are the critical points of the potential energy U.

In order to sketch the level lines of the energy it is useful to imagine a ball rolling in a "potential well" of U (Fig. 94).

Fix the value of the total energy E. We remark that the kinetic energy is nonnegative. Therefore the potential energy cannot exceed the total energy, and so a level line of the energy E projects to the configuration space (the x_1-axis) into the set of values of the potential energy not exceeding E $\{x_1 \in I : U(x_1) \leq E\}$ (the ball cannot rise above the level E in the potential well).

---

[3] A *critical point* of a function is a point at which the total differential of the function vanishes. The value of the function at such a point is also called a *critical value*.

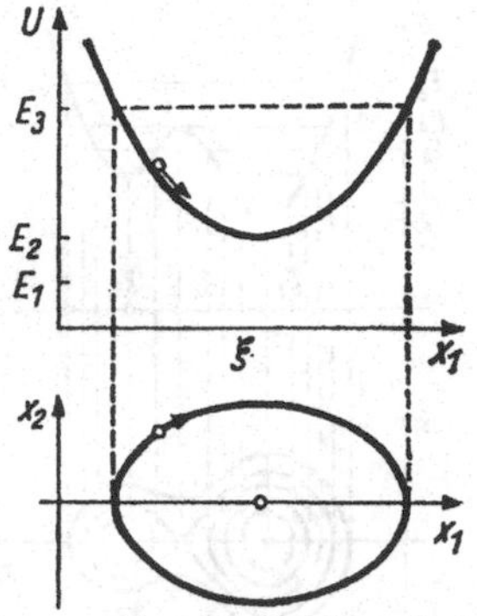

Fig. 94. A ball in a potential well and the phase curve

Furthermore the smaller the potential energy the larger the velocity (in absolute value): $|x_2| = \sqrt{2(E - U(x_1))}$ (when rolling into the well the ball picks up speed, which it loses when ascending). At the "turning points" where $U(x_1) = E$ the velocity is zero.

It follows from the fact that the energy is an even function of x_2 that a level line of the energy is symmetric with respect to the x_1-axis (the ball traverses each point with the same velocity in both directions).

These simple considerations suffice to sketch the level lines of the energy of systems with various potentials U. We begin by considering the simplest case (an infinitely deep potential well with one center of attraction ξ), where $F(x)$ is monotonically decreasing, $F(\xi) = 0$, and $I = \boldsymbol{R}$ (Fig. 94).

If the value of the total energy E_1 is less than the minimum of the potential E_2, then the level set $E = E_1$ is empty (the motion of the ball is physically impossible). The level set $E = E_2$ consists of a single point $(\xi, 0)$ (the ball rests at the bottom of the well).

If the value E_3 of the total energy is larger than the critical value $E_2 = U(\xi)$, then the level set $E = E_3$ is a smooth closed symmetric curve enclosing the equilibrium position $(\xi, 0)$ on the phase plane (the ball rolls back and forth in the well; it rises to the height E_3, and at that instant its velocity becomes zero, and it rolls back down the well and passes through ξ, at which instant its velocity is a maximum; it then rises on the other side, etc.).

In studying more complicated cases one must proceed similarly, successively increasing the value of the total energy E and stopping at values of E equal to the critical values of the potential energy $U(\xi)$ (where $U'(\xi) = 0$), each time examining the curves with values of E slightly smaller and slightly larger than the critical values.

Example 1. Suppose the potential energy U has three critical points: ξ_1 (a minimum), ξ_2 (a local maximum), and ξ_3 (a local minimum). Figure 95 shows the level lines $E_1 = U(\xi_1)$, $U(\xi_1) < E_2 < U(\xi_3)$, $E_3 = U(\xi_3)$, $U(\xi_3) < E_4 < U(\xi_2)$, $E_5 = U(\xi_2)$, and $E_6 > U(\xi_2)$.

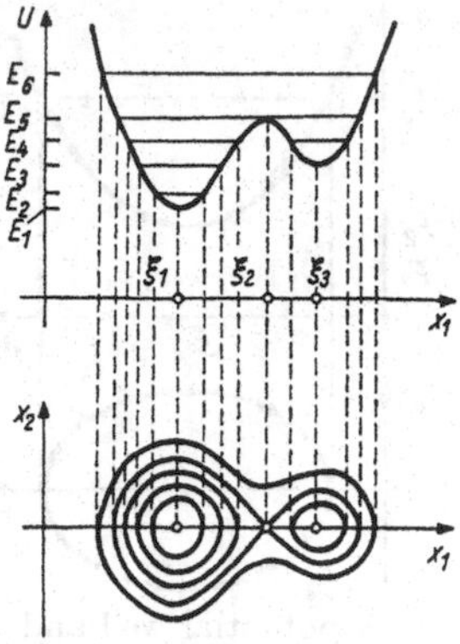

Fig. 95. The level lines of the energy

Problem 1. Sketch the level lines of the energy for the pendulum equation $\ddot{x} = -\sin x$ and for the equations of the pendulum near the lower and upper equilibrium positions ($\ddot{x} = -x$ and $\ddot{x} = x$).

Problem 2. Sketch the level lines of the energy for the *Kepler problem*[4] $U = -\dfrac{1}{x} + \dfrac{C}{x^2}$ and for the potentials depicted in Fig. 96.

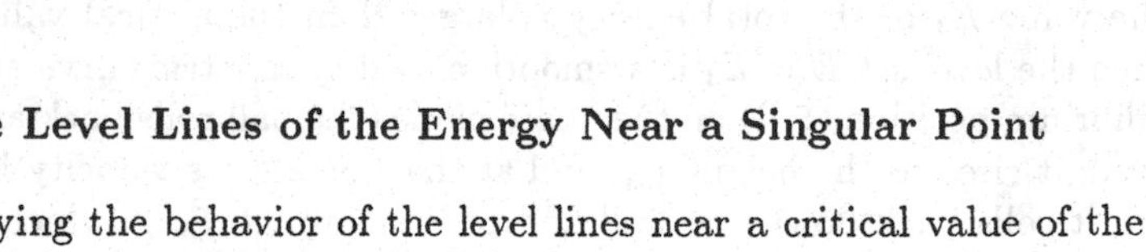

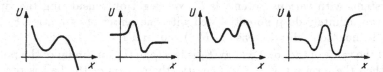

Fig. 96. Sketch the level lines of the energy

4. The Level Lines of the Energy Near a Singular Point

In studying the behavior of the level lines near a critical value of the energy it is useful to keep in mind the following circumstances.

Remark 1. If the potential energy is a quadratic form $U = kx^2/2$, then the level lines of the energy are second-order curves $2E = x_2^2 + kx_1^2$.

In the case of an attraction we have $k > 0$ (i.e., the critical point 0 is a minimum of the potential energy (Fig. 97)). In this case the level lines of the energy are similar ellipses with center at 0.

In the case of repulsion we have $k < 0$ (i.e., the critical point 0 is a maximum of the potential energy (Fig. 97)). In this case the level lines of the

[4] The variation in the distance of planets and comets from the Sun is described by Newton's equation with this potential.

energy are similar hyperbolas with center at 0 and asymptotes given by $x_2 = \pm\sqrt{k}x_1$. These asymptotes are also called *separatrices*, since they separate hyperbolas of different types.

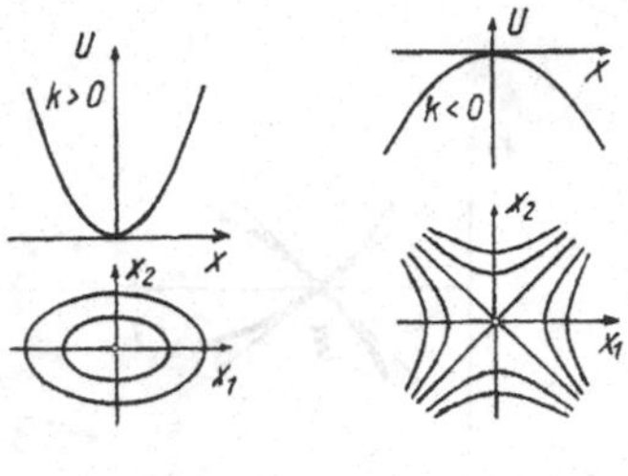

Fig. 97. The level lines of the energy for attractive and repellent quadratic potentials

Remark 2. In a neighborhood of a nondegenerate critical point the increment of the function is a quadratic form provided the coordinates are suitably chosen.

The point 0 is a critical point of the differentiable function f if $f'(0) = 0$. The critical point 0 is *nondegenerate* if $f''(0) \neq 0$. Let us assume that $f(0) = 0$.

Lemma (Morse[5]). *In a neighborhood of the critical point* 0 *the* y*-coordinate can be chosen so that* $f = Cy^2$, $C = \operatorname{sgn} f''(0)$.

Such a coordinate will of course be $y = \operatorname{sgn} x\sqrt{|f(x)|}$. The assertion is that the correspondence $x \mapsto y$ is diffeomorphic in a neighborhood of the point 0.

For the proof it is convenient to use the following proposition:

Lemma (Hadamard). *Let* f *be a differentiable function (of class* C^r*) that vanishes at the point* $x = 0$. *Then* $f(x) = xg(x)$, *where* g *is a differentiable function (of class* C^{r-1} *in a neighborhood of the point* $x = 0$*).*

Proof. We have

$$f(x) = \int_0^1 \frac{df(tx)}{dt} = \int_0^1 f'(tx)x\,dt = x\int_0^1 f'(tx)\,dt;$$

the function $g(x) = \int_0^1 f'(tx)\,dt$ is of class C^{r-1}, and the lemma is proved. □

We now apply Hadamard's Lemma twice to the function f of Morse's Lemma. We find that $f = x^2\varphi(x)$, where $2\varphi(0) = f''(0) \neq 0$. Thus $y = x\sqrt{|\varphi(x)|}$. Morse's Lemma is now proved, since the function $\sqrt{|\varphi(x)|}$ is differentiable in a neighborhood of the point $x = 0$ ($r - 2$ times if f is of class C^r).

Thus in a neighborhood of a nondegenerate critical point the level lines of the energy become either ellipses or hyperbolas under a diffeomorphic change of the coordinates (x_1, x_2).

[5] Both this lemma and the one that follows can be extended to functions of several variables.

Problem 1. Find the tangents to the separatrices of a repelling singular point ($U''(\xi) < 0$).

Answer. $x_2 = \pm\sqrt{|U''(\xi)|}(x_1 - \xi)$ (Fig. 98).

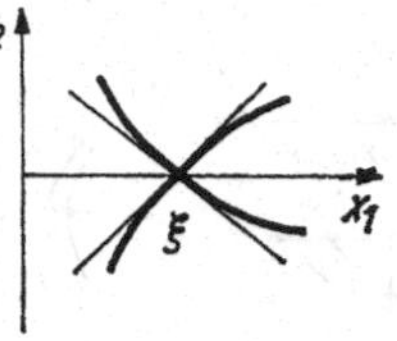

Fig. 98. Tangents to the separatrices of a repelling singular point

5. Extension of the Solutions of Newton's Equation

Suppose the potential energy is defined on the entire x-axis. The following result is an immediate consequence of the law of conservation of energy.

Theorem. *If the potential energy U is everywhere positive*[6], *then every solution of the equation*

$$\ddot{x} = -\frac{dU}{dx} \tag{1_1}$$

can be extended indefinitely.

Example 1. Let $U = -x^2/2$. The solution $x = 1/(t-1)$ cannot be extended to $t = 1$.

We begin by establishing the following proposition, known as an *a priori estimate.*

Lemma. *If there exists a solution for $|t| < \tau$, it satisfies the inequalities $|\dot{x}(t)| \leq \sqrt{2E_0}$, $|x(t) - x(0)| < \sqrt{2E_0}|t|$, where $E_0 = \dfrac{\dot{x}(0)^2}{2} + U(x(0))$ is the initial value of the energy.*

Proof. According to the law of conservation of energy

$$\frac{\dot{x}^2(t)}{2} + U(x(t)) = E_0,$$

and since $U > 0$, the first inequality is proved. The second inequality follows from the first, since $x(t) - x(0) = \int_0^t \dot{x}(\theta)\, d\theta$. The lemma is now proved. □

[6] Of course changing the potential energy U by a constant does not change Eq. (1_1). All that matters is that U be bounded below.

Proof of the theorem. Let T be an arbitrary positive number.

Consider the rectangle Π (Fig. 99) on the phase plane given by

$$|x_1 - x_1(0)| \leq 2\sqrt{2E_0}T, \quad |x_2| \leq 2\sqrt{2E_0}.$$

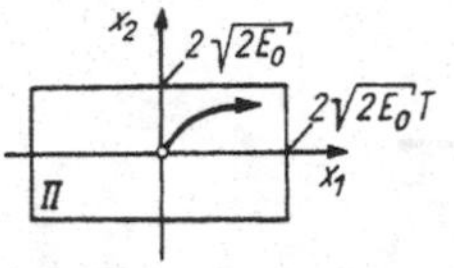

Fig. 99. A rectangle from which the phase point does not exit in time T

Consider the parallelepiped $|t| \leq T$, $(x_1, x_2) \in \Pi$ in the extended phase space of (x_1, x_2, t). By the extension theorem the solution can be extended to the boundary of the parallelepiped. It follows from the lemma that the solution can exit only through a face of the parallelepiped where $|t| = T$. Thus the solution can be extended to any $t = \pm T$, and hence indefinitely. □

Problem 1. Prove that the solutions of the system of Newton equations $m_i \ddot{x}_i = -\frac{\partial U}{\partial x_i}$, $i = 1, \ldots, N$, $m_i > 0$, $x \in \boldsymbol{R}^N$ can be extended indefinitely in the case of a positive potential energy ($U > 0$).

6. Noncritical Level Lines of the Energy

Assume that the potential energy U is defined on the whole x-axis. Let E be a noncritical value of the energy, i.e., E does not equal the value of the function U at any of its critical points.

Consider the set of points where the value of U is less than E, $\{x : U(x) < E\}$. This set (Fig. 100) consists of a finite or countable number of intervals, since the function U is continuous (two of these intervals may extend to infinity). At the endpoints of the intervals we have $U(x) = E$, so that $U'(x) \neq 0$ (because E is a noncritical value).

Each point of the set $\{x : U(x) \leq E\}$ is for this reason an endpoint of an interval of smaller values. Therefore the whole set $\{x : U(x) \leq E\}$ is the union of an at most countable number of pairwise disjoint segments and, perhaps, one or two rays extending to infinity, or else it coincides with the entire x-axis.

Consider (Fig. 101) one such interval $a \leq x \leq b$, where

$$U(a) = U(b) = E, \quad U(x) < E \text{ for } a < x < b.$$

Theorem. *The equation* $\frac{x_2^2}{2} + U(x_1) = E$ *for* $a \leq x_1 \leq b$ *defines a smooth curve diffeomorphic to a circle in the* (x_1, x_2)*-plane. This curve is a phase curve of the system* (2).

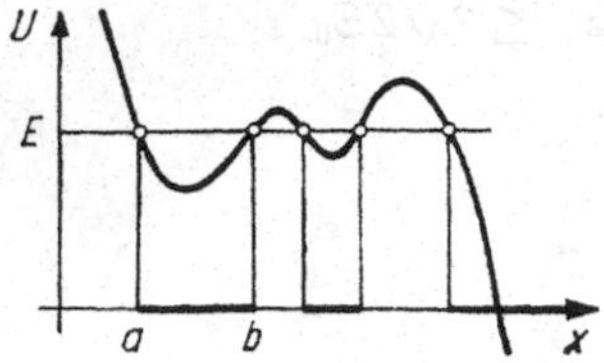

Fig. 100. The set of points x where $U(x) < E$

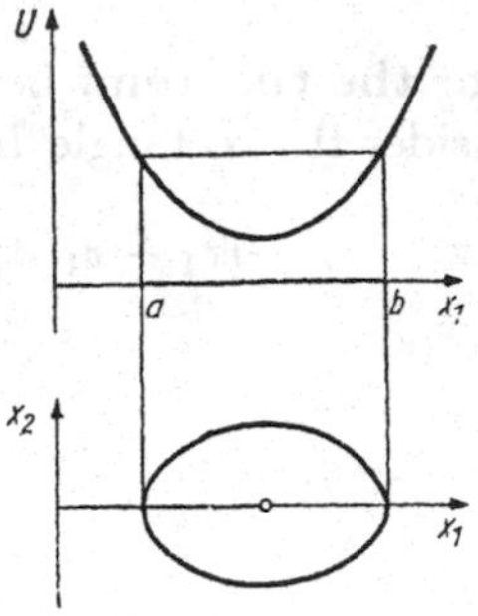

Fig. 101. A phase curve diffeomorphic to a circle

Similarly a ray $a \leq x < \infty$ *(or* $-\infty < x \leq b$*) on which* $U(x) \leq E$ *is the projection to the* x_1*-axis of a phase curve diffeomorphic to a straight line (Fig. 102). Finally in the case when* $U(x) < E$ *on the whole line the level set corresponding to* E *consists of two phase curves*

$$x_2 = \pm\sqrt{2(E - U(x_1))}.$$

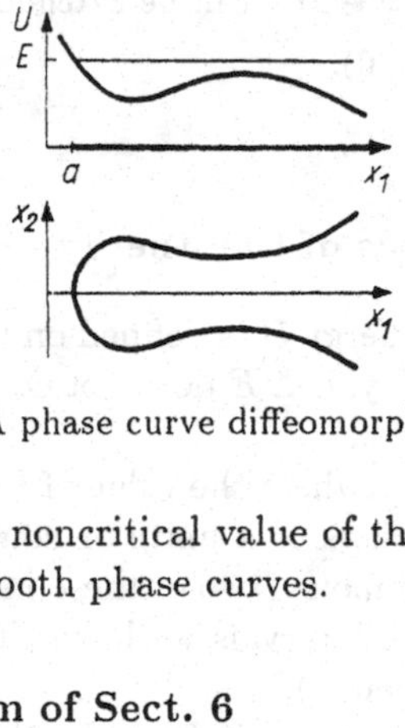

Fig. 102. A phase curve diffeomorphic to a line

Thus the level set for a noncritical value of the energy consists of a finite or countable number of smooth phase curves.

7. Proof of the Theorem of Sect. 6

The law of conservation of energy makes it possible to solve Newton's equation explicitly. Indeed, for a fixed value of the total energy E the magnitude (but not the sign) of the velocity $\dot{x}$ is defined by the position x:

$$\dot{x} = \pm\sqrt{2(E - U(x))}. \tag{3}$$

and this is an equation with a *one-dimensional* phase space that we already know how to solve.

Let (x_1, x_2) be a point of our level set with $x_2 > 0$ (Fig. 103). We seek the solution φ of Eq. (1) with initial condition $\varphi(t_0) = x_1$, $\dot{\varphi}(t_0) = x_2$, from relation (3):

$$t - t_0 = \int_{x_1}^{\varphi(t)} \frac{d\xi}{\sqrt{2(E - U(\xi))}} \tag{4}$$

for t near t_0.

We now remark that the integral $\frac{T}{2} = \int_a^b \frac{d\xi}{\sqrt{2(E - U(\xi))}}$ converges, since $U'(a) \neq 0$ and $U'(b) \neq 0$. It follows from this that formula (4) defines a function φ continuous on some closed interval $t_1 \leq t \leq t_2$, and moreover $\varphi(t_1) = a$ and $\varphi(t_2) = b$. This function satisfies Newton's equation everywhere (Fig. 104).

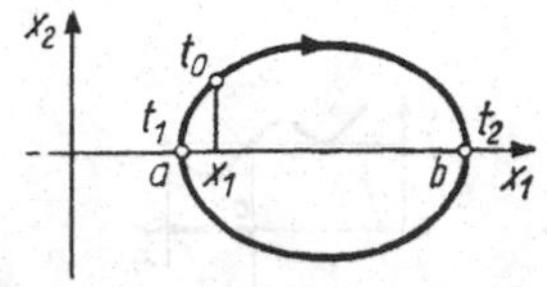

Fig. 103. A phase point traverses half of the phase curve (from a to b) in the finite time $T/2 = t_2 - t_1$

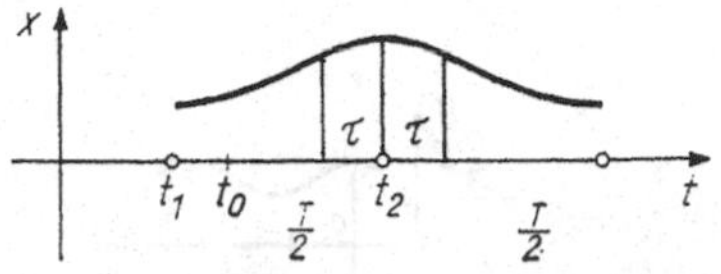

Fig. 104. Extension of a solution of Newton's equation using reflections

The interval (t_1, t_2) has length $T/2$. We extend φ to the next interval of length $T/2$ by symmetry considerations: $\varphi(t_2 + \tau) = \varphi(t_2 - \tau)$, $0 \leq \tau \leq T/2$, and then periodically: $\varphi(t + T) \equiv \varphi(t)$. The function φ now constructed on the entire line satisfies Newton's equation everywhere. In addition $\varphi(t_0) = x_1$, $\dot{\varphi}(t_0) = x_2$.

Thus we have constructed a solution of the system (2) with the initial condition (x_1, x_2). It turned out to be periodic, with period T. The corresponding closed phase curve is precisely the part of the level set corresponding to E over the interval $a \leq x \leq b$. This curve is diffeomorphic to a circle, like every closed phase curve (cf. § 9).

The case when the interval extends to infinity (in one direction or both) is simpler than the one we have considered and is left to the reader.

8. Critical Level Lines

Critical level lines can have a more complicated arrangement. We remark that such a line contains fixed points (x_1, x_2) (where $U'(x_1) = 0$ and $x_2 = 0$), each of which is a phase curve. If $U(x) < E$ everywhere on a phase curve except at $U(a) = U(b) = E$ and both ends are critical points ($U'(a) = U'(b) = 0$), then both of the open arcs $x_2 = \pm\sqrt{2(E - U(x_1))}$, $a < x_1 < b$, (Fig. 105) are phase curves. The time spent by a phase point in traversing such an arc is infinite (by the extension theorem of Sect. 5 and uniqueness).

If $U'(a) = 0$ and $U'(b) \neq 0$ (Fig. 105), then the equation

$$\frac{x_2^2}{2} + U(x_1) = E, \quad a < x_1 \leq b,$$

determines a nonclosed phase curve. Finally, if $U'(a) \neq 0$ and $U'(b) \neq 0$ (Fig. 105), the part of the critical level set over the interval $a \leq x_1 \leq b$ is a closed phase curve, as in the case of a noncritical level E.

9. An Example

Let us apply all that has just been said to the pendulum equation

$$\ddot{x} = -\sin x.$$

The potential energy is $U(x) = -\cos x$ (Fig. 106). The critical points are $x_1 = k\pi$, $k = 0, \pm 1, \ldots$.

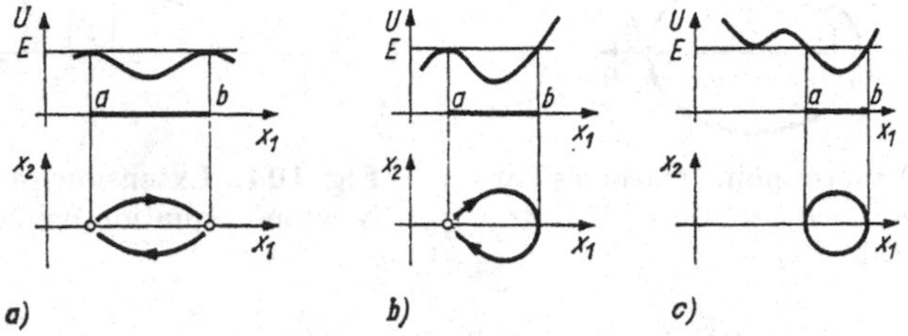

Fig. 105. Partition of a critical level line of the energy into phase curves

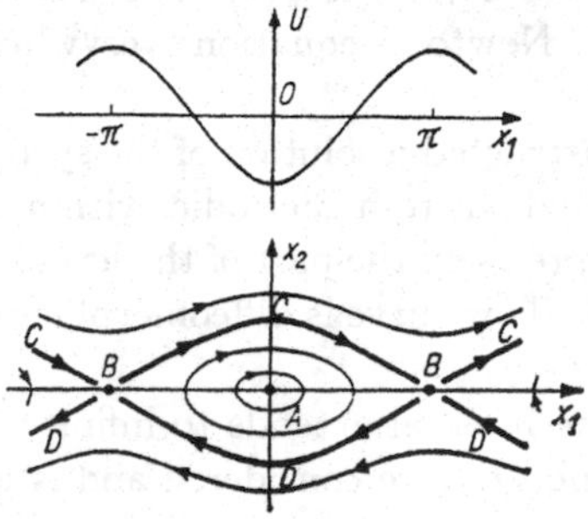

Fig. 106. The phase curves of the pendulum equation

The closed phase curves near $x_1 = 0$, $x_2 = 0$ resemble ellipses. Small oscillations of the pendulum correspond to these phase curves. Their period T is nearly independent of the amplitude as long as the latter is small. For larger values of the constant energy larger closed curves are obtained, until the energy reaches the critical value equal to the potential energy of the pendulum turned upside down. As this happens the period of oscillation grows (since the time of motion along the separatrices forming the critical level set is infinite).

To still larger values of the energy there correspond nonclosed curves on which x_2 is of constant sign, i.e., the pendulum rotates rather than oscillates. Its velocity attains its greatest value at the lower equilibrium position and its smallest value at the higher equilibrium position.

We remark that values of x_1 differing by $2k\pi$ correspond to the same position of the pendulum. Therefore it is natural to consider the phase space of the pendulum to be a cylinder $\{x_1 \mod 2\pi, x_2\}$ rather than a plane (Fig. 107).

Wrapping the picture already sketched in the plane onto the cylinder, we obtain the phase curves of the pendulum on the surface of the cylinder. They are all closed smooth curves except the two stationary points A and B (the lower and upper equilibrium positions) and the two separatrices C and D.

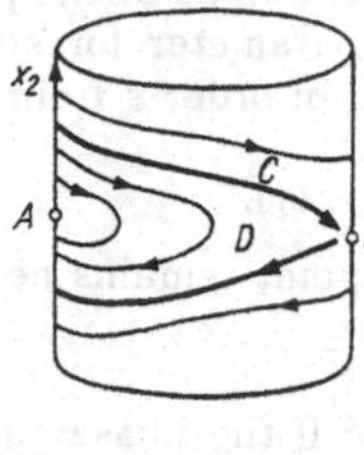

Fig. 107. The cylindrical phase space of the pendulum

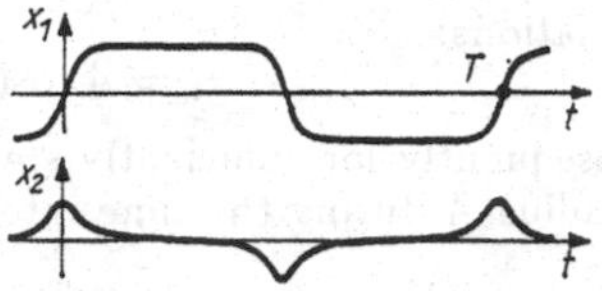

Fig. 108. The angle of deviation of the pendulum and its rate of change for amplitude near π

Problem 1. Sketch the graphs of the functions $x_1(t)$ and $x_2(t)$ for a solution with energy close to the critical energy in the upper position, but slightly smaller.

Solution. See Fig. 108. The functions $x_1(t)$ and $x_2(t)$ can be expressed in terms of the elliptic sine sn and the elliptic cosine cn. As E tends to the smaller critical value, the oscillations of the pendulum approach harmonic oscillations, and sn and cn become sin and cos.

Problem 2. With what velocity does the period of oscillation T of the pendulum tend to infinity when the energy E tends to the critical value E_1?

Answer. With logarithmic velocity ($\sim C \ln(E_1 - E)$).

Hint. See formula (4).

Problem 3. Sketch the phase curves of the systems with potential energies $U(x) = \pm x \sin x$, $U(x) = \pm \dfrac{\sin x}{x}$, $U(x) = \pm \sin x^2$.

Problem 4. Sketch the phase curves of Newton's equation with the force fields $F(x) = \pm x \sin x$, $F(x) = \pm \dfrac{\sin x}{x}$, $F(x) = \pm \sin x^2$.

10. Small Perturbations of a Conservative System

By studying the motion of a conservative system we can study nearby systems of a general type using the theorem on differentiability with respect to a parameter (cf. § 7, Sect. 5). In doing this we encounter qualitatively new phenomena that are quite important for applications – the so-called *self-oscillations*.

Problem 1. Study the phase curves of a system near the system of equations of small oscillations of the pendulum:

$$\begin{cases} \dot{x}_1 = x_2 + \varepsilon f_1(x_1, x_2), \\ \dot{x}_2 = -x_1 + \varepsilon f_2(x_1, x_2), \end{cases} \qquad \varepsilon \ll 1, \quad x_1^2 + x_2^2 \le R^2.$$

Solution. For $\varepsilon = 0$ we obtain the equations of small oscillations of the pendulum. By the theorem on differentiability with respect to a parameter for small ε the solution (on a finite time interval) differs by a correction of order ε from harmonic oscillations:

$$x_1 = A\cos(t - t_0), \quad x_2 = -A\sin(t - t_0).$$

Consequently for sufficiently small $\varepsilon < \varepsilon_0(T)$ the phase point remains near a circle of radius A during the time interval T.

In contrast to the conservative case ($\varepsilon = 0$) for $\varepsilon \neq 0$ the phase curve is not necessarily closed: it may have the form of a spiral (Fig. 109) in which the distance between adjacent coils is small (of order ε). To determine whether the phase curve approaches the origin or recedes from it, we consider the increment in the energy $E = \frac{x_1^2}{2} + \frac{x_2^2}{2}$ over one revolution about the origin. We shall be particularly interested in the sign of this increment: the increment is positive on an expanding spiral, negative on a contracting spiral, and zero on a cycle. We shall deduce an approximate formula (5) for the increment in the energy.

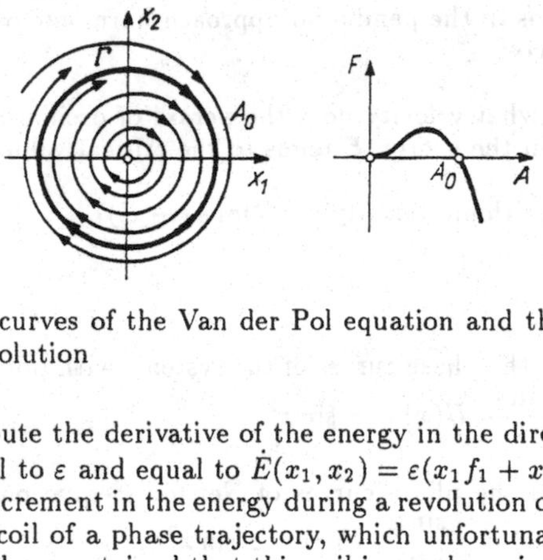

Fig. 109. The phase curves of the Van der Pol equation and the increment in the energy during one revolution

It is easy to compute the derivative of the energy in the direction of our vector field: it is proportional to ε and equal to $\dot{E}(x_1, x_2) = \varepsilon(x_1 f_1 + x_2 f_2)$.

To compute the increment in the energy during a revolution one should integrate this function along a coil of a phase trajectory, which unfortunately is unknown to us. But we have already ascertained that this coil is nearly a circle. Therefore up to order $O(\varepsilon^2)$ the integral can be taken over a circle S of radius A:

$$\Delta E = \varepsilon \int_0^{2\pi} \dot{E}(A\cos t, -A\sin t)\, dt + O(\varepsilon^2).$$

Substituting the computed value of $\dot{E}$, we find[7]

$$\Delta E = \varepsilon F(A) + O(\varepsilon^2), \tag{5}$$

where $F(A) = \oint f_1\, dx_2 - f_2\, dx_1$ (the integral is taken counterclockwise over the circle of radius A).

[7] We are using the fact that $dx_1 = x_2\, dt$ and $dx_2 = -x_1\, dt$ along S.

Having computed the function F, we can study the behavior of the phase curves. If the function F is positive, the increment of the energy ΔE over a revolution is also positive (for small positive ε). In this case the phase curve is an expanding spiral; the system undergoes increasing oscillations. If $F < 0$, then $\Delta E < 0$ and the phase spiral contracts. In this case the oscillations die out.

It can happen that the function F changes sign (Fig. 109). Let A_0 be a simple zero of the function F. Then for small ε the equation $\Delta E(x_1, x_2) = 0$ is satisfied by a closed curve Γ on the phase plane close to a circle of radius A_0 (this follows from the implicit function theorem).

It is obvious that the curve Γ is a closed phase curve – a limit cycle of our system.

Whether the nearby phase curves wind onto the cycle or unwind from it is determined by the sign of the derivative $F' = \left.\frac{dF}{dA}\right|_{A=A_0}$. If $\varepsilon F' > 0$, the cycle is unstable, and if $\varepsilon F' < 0$ it is stable. Indeed, in the former case the increment of the energy during a revolution is positive if the phase curve is outside the cycle and less than zero if it is inside. Therefore the phase curve always recedes from the cycle. In the latter case, however, the phase curves approach the cycle both from within and from without, as in Fig. 109.

Example 1. Consider the equation $\ddot{x} = -x + \varepsilon\dot{x}(1 - x^2)$ (called the *Van der Pol equation*). Computing the integral (5) for $f_1 = 0$, $f_2 = x_2(1 - x_1^2)$, we obtain $F(A) = \pi\left(A^2 - \frac{A^4}{4}\right)$.

This function has a simple root $A_0 = 2$ (Fig. 109); for smaller A it is positive and for large A it is negative. Therefore the Van der Pol equation has a stable limit cycle close to the circle $x^2 + \dot{x}^2 = 4$ in the phase plane.

Let us compare the motion of the original conservative system ($\varepsilon = 0$) with what happens for $\varepsilon \neq 0$. In a conservative system oscillations with any amplitude are possible (all phase curves are closed). The amplitude is determined by the initial conditions in this case.

In the nonconservative system qualitatively different phenomena are possible, for example a stable limit cycle. In this case a periodic oscillation of one and the same completely definite amplitude is established for quite different initial conditions. This established mode is called a *self-oscillating mode.*

Problem * 2. Study the self-oscillating modes of the motion of a pendulum with small friction under the action of a constant torque M:

$$\ddot{x} + \sin x + \varepsilon\dot{x} = M.$$

Hint: This problem is studied in detail for any ε and M in the book of A. A. Andronov, A. A. Vitt, and S. E. Khaikin, *Theory of Oscillators*, (Pergamon Press, New York, distributed by Addison-Wesley Publishing Co., Reading, Mass., 1966, Chapter 7). In the first edition of this classical book the name of the second author was omitted, due to "a tragic mistake," as the later editions explain it. (The author had been shot!)

Chapter 3. Linear Systems

Linear equations are about the only large class of differential equations for which a fairly complete theory exists. This theory, which is really a branch of linear algebra, makes it possible to solve completely all autonomous linear equations.

The theory of linear equations is also useful as a first approximation in the study of nonlinear problems. For example, it makes it possible to study the stability of equilibrium positions and the topological type of singular points of generic vector fields.

§ 13. Linear Problems

We begin by considering two examples of situations where linear equations arise.

1. Example: Linearization

Consider a differential equation defined by a vector field $\boldsymbol{v}$ in the phase space. We already know that in a neighborhood of a nonsingular point ($\boldsymbol{v} \neq \mathbf{o}$) the field has a simple structure: it can be rectified by a diffeomorphism. Now let us consider the structure of the field in the neighborhood of a singular point, i.e., a point where the vector field vanishes. Such a point x_0 is a stationary solution of our equation. If the equation describes a physical process, then x_0 is a stationary state of the process, an "equilibrium position" of it. Therefore the study of a neighborhood of a singular point is the study of the way in which the process develops under a small deviation of the initial conditions from equilibrium (for example: the upper and lower equilibrium positions of a pendulum).

In studying a vector field in a neighborhood of a point x_0 where the vector field is 0 it is natural to expand the field in a neighborhood of this point in a Taylor series. The first term of the Taylor series is the linear term. Discarding all the other terms is called *linearization*. The linearized vector field can be regarded as an example of a vector field with a singular point x_0. On the other hand, one may hope that the behavior of the solutions of the original and the linearized equations are closely related (since the infinitesimals discarded in linearization are of higher order). Of course the question of the connection between the solutions of the original and the linearized equations requires special investigation. This investigation is based on a detailed analysis of the linear equation, which we take up first.

Problem 1. Show that linearization is an invariant operation, i.e., independent of the system of coordinates.

More precisely, suppose the field $\boldsymbol{v}$ in the domain U is given by components $v_i(x)$ in the system of coordinates x_i. Suppose the singular point has coordinates $x_i = 0$, (so that $v_i(0) = 0$, $i = 1, \ldots, n$). Then the original equation can be written in the form of the system

$$\dot{x}_i = v_i(x), \quad i = 1, \ldots, n.$$

Definition. The *linearized equation* is the equation

$$\dot{\xi}_i = \sum_{j=1}^{n} a_{i,j}\xi_j, \quad i = 1, \ldots, n, \quad a_{i,j} = \left.\frac{\partial v_i}{\partial x_j}\right|_{x=0}.$$

Consider a tangent vector $\boldsymbol{\xi} \in T_0U$ with components ξ_i $(i = 1, \ldots, n)$. The linearized equation can be written in the form

$$\dot{\boldsymbol{\xi}} = A\boldsymbol{\xi},$$

where A is the linear mapping, $A : T_0U \to T_0U$ given by the matrix $a_{i,j}$.

We claim that *the mapping A is independent of the system of coordinates x_i occurring in its definition.*

Problem 2. Linearize the equation of the pendulum $\ddot{x} = -\sin x$ near the equilibrium position $x_0 = k\pi$, $\dot{x}_0 = 0$.

2. Example: One-parameter Groups of Linear Transformations of $\boldsymbol{R}^n$

Another problem that reduces immediately to linear differential equations is the problem of describing the one-parameter groups of linear transformations[1] of the vector space $\boldsymbol{R}^n$.

We remark that *the tangent space to the vector space $\boldsymbol{R}^n$ at any point can be naturally identified with the vector space itself.* To be specific, we identify the element $\dot{\varphi}$ of the tangent space $T_x\boldsymbol{R}^n$ represented by the curve $\varphi : I \to \boldsymbol{R}^n$, $\varphi(0) = x$ with the vector

$$v = \lim_{t\to 0} \frac{\varphi(t) - x}{t} \in \boldsymbol{R}^n$$

[1] We recall that part of the definition of a one-parameter group $\{g^t\}$ is the requirement that $g^t x$ be differentiable with respect to x and t.

of the space $\boldsymbol{R}^n$ itself (the correspondence $\boldsymbol{v} \leftrightarrow \dot{\varphi}$ is one-to-one and onto).

This identification depends on the *vector space* structure of $\boldsymbol{R}^n$ and *is not* preserved by diffeomorphisms. Nevertheless in the linear problems that we shall now be studying (for example, in the problem of one-parameter groups of linear transformations), the vector space structure of $\boldsymbol{R}^n$ is fixed once and for all. For that reason *henceforth until we return to nonlinear problems we make the identification* $T_x\boldsymbol{R}^n \equiv \boldsymbol{R}^n$.

Let $\{g^t,\ t \in \boldsymbol{R}\}$ be a one-parameter group of linear transformations. Consider the motion $\varphi : \boldsymbol{R} \to \boldsymbol{R}^n$ of a point $x_0 \in \boldsymbol{R}^n$.

Problem 1. Prove that $\varphi(t)$ is a solution of the equation

$$\dot{\boldsymbol{x}} = A\boldsymbol{x} \tag{1}$$

with the initial condition $\varphi(0) = x_0$, where $A : \boldsymbol{R}^n \to \boldsymbol{R}^n$ is the linear operator ($\equiv \boldsymbol{R}$-endomorphism) defined by the relation $A\boldsymbol{x} = \left.\frac{d}{dt}\right|_{t=0} (g^t\boldsymbol{x})$ for all $\boldsymbol{x} \in \boldsymbol{R}^n$.

Hint. Cf. § 4, Sect. 4.

Equation (1) is called *linear*. Thus to describe all one-parameter groups of linear transformations it suffices to study the solutions of the linear equations (1).

We shall see below that the correspondence between one-parameter groups $\{g^t\}$ of linear transformations and the linear equations (1) is one-to-one and onto: each operator $A : \boldsymbol{R}^n \to \boldsymbol{R}^n$ defines a one-parameter group $\{g^t\}$.

Example 1. Let $n = 1$, and let A be multiplication by the number k. Then g^t is a dilation by a factor of e^{kt}.

Problem 2. Find the velocity field of the points of a rigid body rotating about an axis passing through the point **o** with angular velocity $\boldsymbol{\omega}$.

Answer. $\boldsymbol{v}(\boldsymbol{x}) = [\boldsymbol{\omega}, \boldsymbol{x}]$.

3. The Linear Equation

Let $A : \boldsymbol{R}^n \to \boldsymbol{R}^n$ be a linear operator on the real n-dimensional space $\boldsymbol{R}^n$.

Definition. A *linear equation* is an equation with the phase space $\boldsymbol{R}^n$ defined by the vector field $\boldsymbol{v}(\boldsymbol{x}) = A\boldsymbol{x}$:

$$\dot{\boldsymbol{x}} = A\boldsymbol{x}. \tag{1}$$

The full name of equation (1) is: *a system of n first-order homogeneous linear ordinary differential equations with constant coefficients.*

If we fix in $\boldsymbol{R}^n$ a (linear) coordinate system x_i, $i = 1, \ldots, n$, then Eq. (1) can be written in the form of a system of n equations: $\dot{x}_i = \sum_{j=1}^{n} a_{ij} x_j$, $i = 1, \ldots, n$, where (a_{ij}) is the matrix of the operator A in the given coordinate system. This matrix is called the *matrix of the system.*

In the case $n = 1$ the solution of Eq. (1) with the initial condition $\varphi(0) = \boldsymbol{x}_0$ is given by the exponential $\varphi(t) = e^{At}\boldsymbol{x}_0$.

It turns out that the solution is given by this same formula in the general case as well: one has only to explain what is meant by the exponential of a linear operator. We now take up this problem.

§ 14. The Exponential Function

The function e^A, $A \in \boldsymbol{R}$, can be defined in either of two equivalent ways:

$$e^A = E + A + \frac{A^2}{2!} + \frac{A^3}{3!} + \cdots, \tag{1}$$

$$e^A = \lim_{n\to\infty} \left(E + \frac{A}{n}\right)^n \tag{2}$$

(where E denotes the number 1).

Now let $A : \boldsymbol{R}^n \to \boldsymbol{R}^n$ be a linear operator. To define e^A we first define the concept of the limit of a sequence of linear operators.

1. The Norm of an Operator

We fix an inner product in $\boldsymbol{R}^n$ and denote by $\|\boldsymbol{x}\| = \sqrt{(\boldsymbol{x}, \boldsymbol{x})}$ ($\boldsymbol{x} \in \boldsymbol{R}^n$) the square root of the inner product of $\boldsymbol{x}$ with itself.

Let $A : \boldsymbol{R}^n \to \boldsymbol{R}^n$ be a linear operator.

Definition. The *norm* of A is the number

$$\|A\| = \sup_{\boldsymbol{x}\neq \boldsymbol{o}} \frac{\|A\boldsymbol{x}\|}{\|\boldsymbol{x}\|}.$$

Geometrically $\|A\|$ denotes the largest "coefficient of dilation" of the transformation A.

Problem 1. Prove that $0 \leq \|A\| < \infty$.

Hint. $\|A\| = \sup_{\|\boldsymbol{x}\|=1} \|A\boldsymbol{x}\|$, the sphere is compact, and the function $\|A\boldsymbol{x}\|$ is continuous.

Problem 2. Prove that $\|\lambda A\| = |\lambda|\,\|A\|$, $\|A + B\| \leq \|A\| + \|B\|$, and $\|AB\| \leq \|A\|\,\|B\|$, where $A : \boldsymbol{R}^n \to \boldsymbol{R}^n$ and $B : \boldsymbol{R}^n \to \boldsymbol{R}^n$ are linear operators, and $\lambda \in \boldsymbol{R}$ is a number.

Problem 3. Let (a_{ij}) be the matrix of the operator A in an orthonormal basis. Show that

$$\max_j \sum_i a_{ij}^2 \leq \|A\|^2 \leq \sum_{i,j} a_{ij}^2.$$

Hint. See G. E. Shilov, *An Introduction to the Theory of Linear Spaces*, Dover, New York, 1974, § 53.

2. The Metric Space of Operators

The set L of all linear operators $A : \boldsymbol{R}^n \to \boldsymbol{R}^n$ is itself a vector space over the field of real numbers (by definition $(A + \lambda B)\boldsymbol{x} = A\boldsymbol{x} + \lambda B\boldsymbol{x}$).

Problem 1. Find the dimension of the vector space L.

Answer. n^2.

Hint. An operator is defined by its matrix.

We define the distance between two operators as the norm of their difference $\rho(A, B) = \|A - B\|$.

Theorem. *The space of linear operators L with the metric ρ is a complete metric space*[2].

Let us verify that ρ is a metric.

By definition $\rho > 0$ if $A \neq B$, $\rho(A, A) = 0$, and $\rho(B, A) = \rho(A, B)$. The triangle inequality $\rho(A, B) + \rho(B, C) \geq \rho(A, C)$ follows from the inequality $\|X + Y\| \leq \|X\| + \|Y\|$ of Problem 2 of Sect. 1 (with $X = A - B$ and $Y = B - C$). Thus the metric ρ makes L into a metric space. It is obviously also a complete space.

3. Proof of Completeness

Let $\{A_i\}$ be a Cauchy sequence, i.e., for every $\varepsilon > 0$ there exists $N(\varepsilon)$ such that $\rho(A_m, A_k) < \varepsilon$ for $m, k > N$. Let $\boldsymbol{x} \in \boldsymbol{R}^n$. Form the sequence of points $\boldsymbol{x}_i \in \boldsymbol{R}^n$

---

[2] A *metric space* is a pair consisting of a set M and a function $\rho : M \times M \to \boldsymbol{R}$, called a *metric*, for which

1) $\rho(x, y) \geq 0$, $(\rho(x, y) = 0) \Leftrightarrow (x = y)$;
2) $\rho(x, y) = \rho(y, x)$ $\forall x, y \in M$;
3) $\rho(x, y) \leq \rho(x, z) + \rho(z, y)$ $\forall x, y, z \in M$.

A sequence of points x_i in a metric space M is called a *Cauchy sequence* if for every $\varepsilon > 0$ there exists N such that $\rho(x_i, x_j) < \varepsilon$ for all $i, j > N$. The sequence x_i *converges to the point* x if for every $\varepsilon > 0$ there exists N such that $\rho(x, x_i) < \varepsilon$ for all $i > N$. A space is called *complete* if every Cauchy sequence converges.

such that $\boldsymbol{x}_i = A_i\boldsymbol{x}$. We shall show that $\{\boldsymbol{x}_i\}$ is a Cauchy sequence in the space $\boldsymbol{R}^n$ endowed with the metric $\rho(\boldsymbol{x},\boldsymbol{y}) = \|\boldsymbol{x}-\boldsymbol{y}\|$. Indeed, by definition of the norm of an operator, for $m, k > N$

$$\|\boldsymbol{x}_m - \boldsymbol{x}_k\| \le \rho(A_m, A_k)\|\boldsymbol{x}\| \le \varepsilon\|\boldsymbol{x}\|.$$

Since $\|\boldsymbol{x}\|$ is a fixed number (independent of m and k), it follows from this that $\{\boldsymbol{x}_i\}$ is a Cauchy sequence. The space $\boldsymbol{R}^n$ is complete. Therefore the following limit exists:

$$\boldsymbol{y} = \lim_{i\to\infty} \boldsymbol{x}_i \in \boldsymbol{R}^n.$$

We remark that $\|\boldsymbol{x}_k - \boldsymbol{y}\| \le \varepsilon\|\boldsymbol{x}\|$ for $k > N(\varepsilon)$, and $N(\varepsilon)$ is the same number as above, independent of $\boldsymbol{x}$.

The point $\boldsymbol{y}$ is a linear function of the point $\boldsymbol{x}$ (the limit of a sum is the sum of the limits). We therefore obtain a linear operator $A : \boldsymbol{R}^n \to \boldsymbol{R}^n$ with $A\boldsymbol{x} = \boldsymbol{y}$, and $A \in L$. We see that for $k > N(\varepsilon)$

$$\rho(A_k, A) = \|A_k - A\| = \sup_{\boldsymbol{x}\neq \boldsymbol{o}} \frac{\|\boldsymbol{x}_k - \boldsymbol{y}\|}{\|\boldsymbol{x}\|} \le \varepsilon.$$

Thus $A = \lim_{k\to\infty} A_k$, and so the space L is complete.

Problem 1. Prove that a sequence of operators A_i converges if and only if the sequence of their matrices in a fixed basis converges. Deduce another proof of completeness from this result.

4. Series

Suppose given a real vector space M made into a complete metric space by a metric ρ such that the distance between any two points of M depends only on their difference, and $\rho(\lambda x, 0) = |\lambda|\rho(x,0)$ ($x \in M$, $\lambda \in \boldsymbol{R}$). Such a space is said to be *normed*, and the function $\rho(x,0)$ is called the *norm* of x and denoted $\|x\|$.

Example 1. The Euclidean space $M = \boldsymbol{R}^n$ with the metric

$$\rho(\boldsymbol{x},\boldsymbol{y}) = \|\boldsymbol{x}-\boldsymbol{y}\| = \sqrt{(\boldsymbol{x}-\boldsymbol{y})(\boldsymbol{x}-\boldsymbol{y})}.$$

Example 2. The space L of linear operators $\boldsymbol{R}^n \to \boldsymbol{R}^n$ with the metric $\rho(A,B) = \|A-B\|$.

We shall denote the distance between the elements A and B of M by $\|A-B\|$.

Since the elements of M can be added and multiplied by scalars and Cauchy sequences in M have limits, the theory of series of the form $A_1 + A_2 + \cdots$, $A_i \in M$, repeats verbatim the theory of numerical series.

The theory of functional series also carries over immediately to functions with values in M.

Problem 1. Prove the following two theorems:

The Weierstrass Criterion[3]**.** *If a series* $\sum_{i=1}^{\infty} f_i$ *of functions* $f_i : X \to M$ *is majorized by a convergent numerical series*

$$\|f_i\| \leq a_i, \quad \sum_{i=1}^{\infty} a_i < \infty, \quad a_i \in \boldsymbol{R},$$

then it converges absolutely and uniformly on X.

Differentiation of a series. *If a series* $\sum f_i$ *of functions* $f_i : \boldsymbol{R} \to M$ *converges and the series of derivatives* $\sum \frac{df_i}{dt}$ *converges uniformly, then the latter converges to the derivative* $\frac{d}{dt} \sum_{i=1}^{\infty} f_i$ *(t is a coordinate on the line* $\boldsymbol{R}$*).*

Hint. The proof for the case $M = \boldsymbol{R}$ can be found in any course of analysis. It carries over verbatim to the general case.

5. Definition of the Exponential e^A

Let $A : \boldsymbol{R}^n \to \boldsymbol{R}^n$ be a linear operator.

Definition. The *exponential* e^A of the operator A is the linear operator from $\boldsymbol{R}^n$ to $\boldsymbol{R}^n$ given by

$$e^A = E + A + \frac{A^2}{2!} + \cdots = \sum_{k=0}^{\infty} \frac{A^k}{k!}$$

(where E is the identity operator, $E\boldsymbol{x} = \boldsymbol{x}$).

Theorem. *The series* e^A *converges for any* A *uniformly on each set* $X = \{A : \|A\| \leq a\}$, $a \in \boldsymbol{R}$.

Proof. Let $\|A\| \leq a$. Then our series is majorized by the numerical series $1 + a + \frac{a^2}{2!} + \cdots$, which converges to e^a. By the Weierstrass criterion the series e^A converges uniformly for $\|A\| \leq a$. □

Problem 1. Compute the matrix e^{At} if the matrix A has the form

[3] Usually referred to in Western literature as the *M-test*, since the constants a_i are usually denoted M_i, *Trans.*

$$1)\ \begin{pmatrix}1 & 0\\ 0 & 2\end{pmatrix}.\quad 2)\ \begin{pmatrix}0 & 1\\ 0 & 0\end{pmatrix}.$$

$$3)\ \begin{pmatrix}0 & 1\\ -1 & 0\end{pmatrix}.\quad 4)\ \begin{pmatrix}0 & 1 & 0\\ 0 & 0 & 1\\ 0 & 0 & 0\end{pmatrix}.$$

Answer.

$$1)\begin{pmatrix}e^t & 0\\ 0 & e^{2t}\end{pmatrix}.\quad 2)\ \begin{pmatrix}1 & t\\ 0 & 1\end{pmatrix},$$

$$3)\ \begin{pmatrix}\cos t & \sin t\\ -\sin t & \cos t\end{pmatrix},\quad 4)\ \begin{pmatrix}1 & t & t^2/2\\ 0 & 1 & t\\ 0 & 0 & 1\end{pmatrix},$$

6. An Example

Consider the set of polynomials of degree less than n in one variable x with real coefficients.

This set has a natural real vector space structure: polynomials can be added and multiplied by numbers.

Problem 1. Find the dimension of the space of polynomials of degree less than n.

Solution. n; a basis, for example, is $1, x, x^2, \ldots, x^{n-1}$.

We shall denote the space of polynomials of degree less than n by $\boldsymbol{R}^n$[4]. The derivative of a polynomial of degree less than n is a polynomial of degree less than n. Therefore a mapping arises

$$A : \boldsymbol{R}^n \to \boldsymbol{R}^n, \quad Ap = \frac{dp}{dx}.$$

Problem 2. Prove that A is a linear operator; find its kernel and image.

Answer. $\operatorname{Ker} A = \boldsymbol{R}^1$, $\operatorname{Im} A = \boldsymbol{R}^{n-1}$.

On the other hand we denote by H^t, $(t \in \boldsymbol{R})$ the operator of translation by t taking the polynomial $p(x)$ into $p(x+t)$.

Problem 3. Prove that $H^t : \boldsymbol{R}^n \to \boldsymbol{R}^n$ is a linear operator. Find its kernel and image.

Answer. $\operatorname{Ker} H^t = 0$, $\operatorname{Im} H^t = \boldsymbol{R}^n$.

---

[4] Thus we are identifying the space of polynomials in which the basis exhibited above was chosen with the coordinate space $\boldsymbol{R}^n$ isomorphic to it.

Finally, we form the operator e^{At}.

Theorem. $e^{At} = H^t$.

Proof. In a course of analysis this theorem is called Taylor's formula for polynomials:

$$p(x+t) = p(x) + \frac{t}{1!}\frac{dp}{dx} + \frac{t^2}{2!}\frac{d^2p}{dx^2} + \cdots .$$

7. The Exponential of a Diagonal Operator

Suppose the matrix of the operator A is diagonal with diagonal elements $\lambda_1, \ldots, \lambda_n$. It is easy to see that the matrix of the operator e^A is also diagonal, with diagonal elements $e^{\lambda_1}, \ldots, e^{\lambda_n}$.

Definition. An operator $A : \boldsymbol{R}^n \to \boldsymbol{R}^n$ is called *diagonalizable* if its matrix in some basis is diagonal. Such a basis is called a *proper basis*.

Problem 1. Give an example of a nondiagonalizable operator.

Problem 2. Prove that the eigenvalues of a diagonalizable operator A are real.

Problem 3. If all n eigenvalues of the operator $A : \boldsymbol{R}^n \to \boldsymbol{R}^n$ are real and distinct, then the operator is diagonalizable.

Let A be a diagonalizable operator. Then the computation of e^A is carried out most simply in a proper basis.

Example 1. Suppose the matrix of the operator A has the form $\begin{pmatrix} 1 & 1 \\ 1 & 1 \end{pmatrix}$ in a basis $\boldsymbol{e}_1, \boldsymbol{e}_2$. Since the eigenvalues $\lambda_1 = 2$, $\lambda_2 = 0$ are real and distinct, the operator A is diagonalizable. A proper basis is $\boldsymbol{f}_1 = \boldsymbol{e}_1 + \boldsymbol{e}_2$, $\boldsymbol{f}_2 = \boldsymbol{e}_1 - \boldsymbol{e}_2$. The matrix of the operator A in the basis of eigenvectors is $\begin{pmatrix} 2 & 0 \\ 0 & 0 \end{pmatrix}$. Therefore the matrix of the operator e^A in the basis of eigenvectors is $\begin{pmatrix} e^2 & 0 \\ 0 & 1 \end{pmatrix}$.

Thus in the original basis the matrix of the operator e^A is $\frac{1}{2}\begin{pmatrix} e^2+1 & e^2-1 \\ e^2-1 & e^2+1 \end{pmatrix}$.

8. The Exponential of a Nilpotent Operator

Definition. The operator $A : \boldsymbol{R}^n \to \boldsymbol{R}^n$ is called *nilpotent* if some power of it equals 0.

Problem 1. Prove that the operator with the matrix $\begin{pmatrix} 0 & 1 \\ 0 & 0 \end{pmatrix}$ is nilpotent. In general if all the elements of a matrix on or below the diagonal are 0, the operator is nilpotent.

Problem 2. Prove that the differentiation operator $\frac{d}{dx}$ in the space of polynomials of degree less than n is nilpotent.

If the operator A is nilpotent, the series for e^A terminates, i.e., reduces to a finite sum.

Problem 3. Calculate e^{tA} ($t \in \boldsymbol{R}$), where $A : \boldsymbol{R}^n \to \boldsymbol{R}^n$ is the operator with the matrix

$$\begin{pmatrix} 0 & 1 & & & 0 \\ & 0 & 1 & & \\ & & 0 & \ddots & \\ & & & \ddots & 1 \\ & & & & 0 \end{pmatrix}$$

(containing 1's only above the main diagonal).

Hint. One of the ways of solving this problem is Taylor's formula for polynomials. The differentiation operator $\frac{d}{dx}$ has a matrix of this form in some basis (which one?). For the solution see § 25.

9. Quasi-polynomials

Let λ be a real number. A *quasi-polynomial with exponent* λ is a product $e^{\lambda x}p(x)$, where p is a polynomial. The degree of the polynomial p is called the degree of the quasi-polynomial. Fix the value of the exponent λ.

Problem 1. Prove that the set of all quasi-polynomials of degree less than n is a vector space. Find its dimension.

Solution. n. A basis is, for example, $e^{\lambda x}, xe^{\lambda x}, \dots, x^{n-1}e^{\lambda x}$.

Remark. The concept of a quasi-polynomial, like the concept of a polynomial, conceals a certain ambiguity. A (quasi-)polynomial can be interpreted as an *expression* composed of signs and letters; in that case the solution of the preceding problem is obvious. On the other hand a (quasi-)polynomial can be interpreted as a *function*, i.e., a mapping $f : \boldsymbol{R} \to \boldsymbol{R}$.

In reality these two interpretations are equivalent (when the coefficients of the polynomials are real or complex numbers; at present we are considering (quasi-)polynomials with real coefficients).

Problem 2. Prove that every function $f : \boldsymbol{R} \to \boldsymbol{R}$ that can be written in the form of a quasi-polynomial has a unique expression of this form.

Hint. It suffices to prove that the relation $e^{\lambda x}p(x) \equiv 0$ implies that all the coefficients of the polynomial $p(x)$ are equal to 0.

We shall denote the n-dimensional vector space of quasi-polynomials of degree less than n with exponent λ by $\boldsymbol{R}^n$.

Theorem. *The differentiation operator* $\frac{d}{dx}$ *is a linear operator from* $\boldsymbol{R}^n$ *into* $\boldsymbol{R}^n$, *and for any* $t \in \boldsymbol{R}$

$$e^{t\frac{d}{dx}} = H^t, \tag{3}$$

where $H^t : \boldsymbol{R}^n \to \boldsymbol{R}^n$ *is the operator of translation by* t *(i.e.,* $(H^t f)(x) = f(x+t)$*).*

Proof. We must first of all prove that the derivative and the translate of a quasi-polynomial of degree less than n with exponent λ are again quasi-polynomials of degree less than n with exponent λ.

Indeed,

$$\frac{d}{dx}(e^{\lambda x}p(x)) = \lambda e^{\lambda x}p(x) + e^{\lambda x}p'(x), \quad e^{\lambda(x+t)}p(x+t) = e^{\lambda x}(e^{\lambda t}p(x+t)).$$

The linearity of differentiation and translation are not in question. It remains only to remark that the Taylor series for a quasi-polynomial converges absolutely on the entire line (since the Taylor series for $e^{\lambda x}$ and for $p(x)$ converge absolutely). This is what formula (3) expresses. □

Problem 3. Compute the matrix of the operator e^{At} if the matrix of A has the form

$$\begin{pmatrix} \lambda & 1 & & 0 \\ & \lambda & \ddots & \\ & & \ddots & 1 \\ 0 & & & \lambda \end{pmatrix}$$

(the diagonal consists of λ's, there are 1's above the diagonal, and 0's everywhere else). For example, compute

$$\exp\begin{pmatrix} 1 & 1 \\ 0 & 1 \end{pmatrix}.$$

Hint. This is precisely the form of the matrix of the differentiation operator in the space of quasi-polynomials (in what basis?). For the solution see § 25.

§ 15. Properties of the Exponential

We now establish a number of properties of the operator $e^A : \boldsymbol{R}^n \to \boldsymbol{R}^n$; these properties enable us to use e^A to solve linear differential equations.

1. The Group Property

Let $A : \boldsymbol{R}^n \to \boldsymbol{R}^n$ be a linear operator.

Theorem. *The family of linear operators $e^{tA} : \boldsymbol{R}^n \to \boldsymbol{R}^n$, $t \in \boldsymbol{R}$, is a one-parameter group of linear transformations of $\boldsymbol{R}^n$.*

Proof. Since we already know that e^{tA} is a linear operator, we need only verify that

$$e^{(t+s)A} = e^{tA}e^{sA} \tag{1}$$

and that e^{tA} depends differentiably on t. We shall prove that

$$\frac{d}{dt}e^{tA} = Ae^{tA}, \tag{2}$$

as an exponential should behave.

To prove the group property (1) we first multiply the power series in A formally:

$$\left(E + tA + \frac{t^2}{2}A^2 + \cdots\right)\left(E + sA + \frac{s^2}{2}A^2 + \cdots\right) =$$
$$= E + (t+s)A + \left(\frac{t^2}{2} + ts + \frac{s^2}{2}\right)A^2 + \cdots.$$

The coefficient of A^k in the product is $(t+s)^k/(k!)$, since the formula (1) holds in the case of numerical series ($A \in \boldsymbol{R}$). All that remains to be done is to justify the termwise multiplication. This can be done in the same way that termwise multiplication of absolutely convergent numerical series is justified (the series for e^{tA} and e^{sA} converge absolutely, since the series $e^{|t|a}$ and $e^{|s|a}$ converge, where $a = \|A\|$). The proof can be directly reduced to the numerical case, as we shall now show.

Lemma. *Let $p \in \boldsymbol{R}[z_1, \ldots, z_n]$ be a polynomial with nonnegative coefficients in the variables $z_1, \ldots, z_n$. Let $A_1, \ldots, A_N : \boldsymbol{R}^n \to \boldsymbol{R}^n$ be linear operators. Then*

$$\|p(A_1, \ldots, A_n)\| \le p(\|A_1\|, \ldots, \|A_n\|).$$

Proof. This follows from the relations

$$\|A + B\| \le \|A\| + \|B\|, \quad \|AB\| \le \|A\|\,\|B\|, \quad \|\lambda A\| = |\lambda|\,\|A\|.$$

The lemma is now proved. □

We denote by $S_m(A)$ the partial sum of the series for e^A:

$$S_m(A) = \sum_{k=0}^{m} \frac{A^k}{k!}.$$

S_m is a polynomial in A with nonnegative coefficients. We must prove that the difference $\Delta_m = S_m(tA)S_m(sA) - S_m((t+s)A)$ tends to zero as $m \to \infty$.

We remark that Δ_m is a polynomial in sA and tA with nonnegative coefficients. Indeed, the terms of degree at most m in A in the product of the series are all obtained by multiplying all the terms of degree not larger than m in the two factor series. Moreover $S_m((s+t)A)$ is the partial sum of the product series. Therefore Δ_m is the sum of all the terms of degree larger than m in A in the product $S_m(tA)S_m(sA)$. But all the coefficients of a product of polynomials with nonnegative coefficients are nonnegative.

By the lemma $\|\Delta_m(tA, sA)\| \leq \Delta_m(\|tA\|, \|sA\|)$. Let us denote the nonnegative numbers $\|tA\|$ and $\|sA\|$ by τ and σ. Then $\Delta_m(\tau, \sigma) = S_m(\tau)S_m(\sigma) - S_m(\tau+\sigma)$. Since $e^\tau e^\sigma = e^{\tau+\sigma}$, the right-hand side tends to 0 as $m \to \infty$. Thus $\lim_{m\to\infty} \Delta_m(tA, sA) = 0$, and relation (1) is proved.

To prove relation (2) we differentiate the series e^{At} formally with respect to t; we then obtain the series of derivatives

$$\sum_{k=0}^{\infty} \frac{d}{dt}\frac{t^k}{k!}A^k = A\sum_{k=0}^{\infty}\frac{t^k}{k!}A^k.$$

This series converges absolutely and uniformly in any domain $\|A\| \leq a$, $|t| \leq T$, just like the original series. Therefore the derivative of the sum of the series is the sum of the series of derivatives. The theorem is now proved. □

Problem 1. Is it true that $e^{A+B} = e^A e^B$?

Answer. No.

Problem 2. Prove that $\det e^A \neq 0$.

Hint. $e^{-A} = (e^A)^{-1}$.

Problem 3. Prove that if the operator A in a Euclidean space is skew-symmetric, then the operator e^A is orthogonal.

2. The Fundamental Theorem of the Theory of Linear Equations with Constant Coefficients

An immediate consequence of the theorem just proved is a formula for solving a linear equation:

$$\dot{\boldsymbol{x}} = A\boldsymbol{x}, \quad \boldsymbol{x} \in \boldsymbol{R}^n. \tag{3}$$

Theorem. *The solution of Eq.* (3) *with initial condition* $\varphi(0) = \boldsymbol{x}_0$ *is*

$$\varphi(t) = e^{tA}\boldsymbol{x}_0, \quad t \in \boldsymbol{R}. \tag{4}$$

Proof. According to the differentiation formula (2)

$$\frac{d\varphi}{dt} = Ae^{tA}\boldsymbol{x}_0 = A\varphi(t).$$

Thus φ is a solution. Since $e^0 = E$, we have $\varphi(0) = \boldsymbol{x}_0$. The theorem is now proved, since by the uniqueness theorem every solution coincides with the solution (4) in its domain of definition. □

3. The General Form of One-parameter Groups of Linear Transformations of the Space $\boldsymbol{R}^n$

Theorem. *Let $\{g^t : \boldsymbol{R}^n \to \boldsymbol{R}^n\}$ be a one-parameter group of linear transformations. Then there exists a linear operator $A : \boldsymbol{R}^n \to \boldsymbol{R}^n$ such that $g^t = e^{At}$.*

Proof. Set $A = \left.\dfrac{dg^t}{dt}\right|_{t=0} = \lim\limits_{t\to 0} \dfrac{g^t - E}{t}$. We have already proved that the motion $\varphi(t) = g^t \boldsymbol{x}_0$ is a solution of Eq. (3) with the initial condition $\varphi(0) = \boldsymbol{x}_0$. According to (4) we have $g^t \boldsymbol{x}_0 = e^{tA}\boldsymbol{x}_0$, which was required. □

The operator A is called the *infinitesimal generator* of the group $\{g^t\}$.

Problem 1. Prove that the infinitesimal generator is uniquely determined by the one-parameter group.

Remark. Thus there is a one-to-one correspondence between the linear differential equations (3) and their phase flows $\{g^t\}$, and the phase flow consists of linear diffeomorphisms.

4. A Second Definition of the Exponential

Theorem. *Let $A : \boldsymbol{R}^n \to \boldsymbol{R}^n$ be a linear operator. Then*

$$e^A = \lim_{m\to\infty} \left(E + \frac{A}{m}\right)^m. \tag{5}$$

Proof. Consider the difference

$$e^A - \left(E + \frac{A}{m}\right)^m = \sum_{k=0}^{\infty} \left(\frac{1}{k!} - \frac{C_m^k}{m^k}\right) A^k.$$

(The series converges since $\left(E + \dfrac{A}{m}\right)^m$ is a polynomial and the series for e^A converges.) We remark that the coefficients of the difference are nonnegative:

$$\frac{1}{k!} \geq \frac{m(m-1)\cdot\ldots\cdot(m-k+1)}{m\cdot m\cdot\cdots\cdot m}\,\frac{1}{k!}.$$

Therefore, setting $\|A\| = a$, we find

$$\left\| e^A - \left(E + \frac{A}{m}\right)^m \right\| \le \sum_{k=0}^{\infty} \left(\frac{1}{k!} - \frac{C_m^k}{m^k}\right) a^k = e^a - \left(1 + \frac{a}{m}\right)^m.$$

This last quantity tends to 0 as $m \to \infty$, and the theorem is proved. □

5. An Example: Euler's Formula for e^z

Let C be the complex line. We can regard it as the real plane $\mathbf{R}^2$, and multiplication by a complex number z as a linear operator $A : \mathbf{R}^2 \to \mathbf{R}^2$. The operator A is a rotation through the angle $\arg z$ together with a dilation by a factor of $|z|$.

Problem 1. Find the matrix of the multiplication by $z = u + iv$ in the basis $\mathbf{e}_1 = 1$, $\mathbf{e}_2 = i$.

Answer. $\begin{pmatrix} u & -v \\ v & u \end{pmatrix}$.

Let us now find e^A. By formula (5) we must first form the operator $E + \dfrac{A}{n}$. This is multiplication by the number $1 + \dfrac{z}{n}$, i.e., a rotation through the angle $\arg\left(1 + \dfrac{z}{n}\right)$ together with dilation by a factor of $\left|1 + \dfrac{z}{n}\right|$ (Fig. 110).

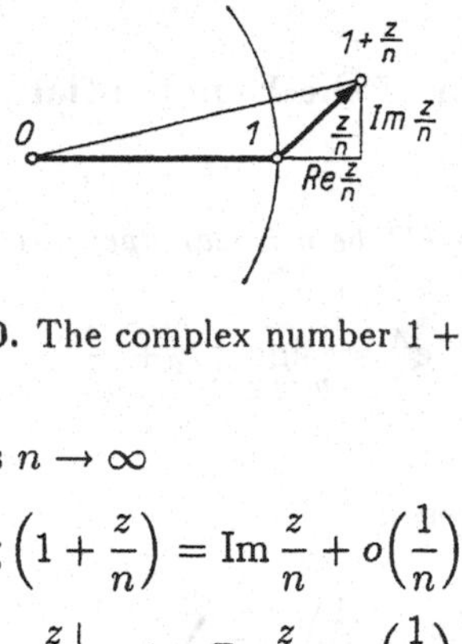

Fig. 110. The complex number $1 + (z/n)$

Problem 2. Prove that as $n \to \infty$

$$\arg\left(1 + \frac{z}{n}\right) = \operatorname{Im}\frac{z}{n} + o\left(\frac{1}{n}\right),$$
$$\left|1 + \frac{z}{n}\right| = 1 + \operatorname{Re}\frac{z}{n} + o\left(\frac{1}{n}\right). \tag{6}$$

The operator $\left(E + \dfrac{A}{n}\right)^n$ is rotation by the angle $n \arg\left(1 + \dfrac{z}{n}\right)$ together with dilation by a factor of $\left|1 + \dfrac{z}{n}\right|^n$. From formula (6) we find the limits of the angle of rotation and the coefficient of dilation:

$$\lim_{n\to\infty} n \arg\left(1 + \frac{z}{n}\right) = \operatorname{Im} z, \quad \lim_{n\to\infty} \left|1 + \frac{z}{n}\right|^n = e^{\operatorname{Re} z}.$$

Thus we have proved the following theorem.

Theorem. *Let $z = u + iv$ be a complex number and $A : \boldsymbol{R}^2 \to \boldsymbol{R}^2$ the operator of multiplication by z. Then e^A is the operator of multiplication by the complex number $e^u(\cos v + i \sin v)$.*

Definition. The complex number

$$e^u(\cos v + i\sin v) = \lim_{n\to\infty}\left(1+\frac{z}{n}\right)^n$$

is called the *exponential* of the complex number $z = u + iv$ and is denoted

$$e^z = e^u(\cos v + i \sin v). \tag{7}$$

Remark. If we do not distinguish a complex number from the operator of multiplication by that complex number, the definition becomes a theorem, since the exponential of an operator is already defined.

Problem 3. Find e^0, e^1, e^i, $e^{\pi i}$, $e^{2\pi i}$.

Problem 4. Prove that $e^{z_1+z_2} = e^{z_1}e^{z_2}$ ($z_1 \in \boldsymbol{C}$, $z_2 \in \boldsymbol{C}$).

Remark. Since the exponential is also defined by a series, we have

$$e^z = 1 + z + \frac{z^2}{2!} + \cdots, \quad z \in \boldsymbol{C} \tag{8}$$

(the series converges absolutely and uniformly in each disk $|z| \le a$).

Problem 5. By comparing this series with Euler's formula (7), deduce the Taylor series for $\sin v$ and $\cos v$.

Remark. Conversely, knowing the Taylor series for $\sin v$, $\cos v$, and e^u, we could have proved formula (7), starting from formula (8) as the definition of e^z.

6. Euler's Broken Lines

Combining formulas (4) and (5), we obtain a method of approximate solution of the differential equation (3) known as *Euler's broken-line method.*

Consider a differential equation defined by a vector field $\boldsymbol{v}$ and whose phase space is the vector space $\boldsymbol{R}^n$. To find the solution φ of the equation $\dot{\boldsymbol{x}} = \boldsymbol{v}(\boldsymbol{x})$, $\boldsymbol{x} \in \boldsymbol{R}^n$, with initial condition $\boldsymbol{x}_0$, we proceed as follows (Fig. 111). The velocity at the point $\boldsymbol{x}_0$ is known: it is $\boldsymbol{v}(\boldsymbol{x}_0)$. We move with constant velocity $\boldsymbol{v}(\boldsymbol{x}_0)$ from $\boldsymbol{x}_0$ during the time interval $\Delta t = t/N$. We arrive at the point $\boldsymbol{x}_1 = \boldsymbol{x}_0 + \boldsymbol{v}(\boldsymbol{x}_0)\Delta t$. During the next time interval of length Δt we move with velocity $\boldsymbol{v}(\boldsymbol{x}_1)$, etc.:

$$\boldsymbol{x}_{k+1} = \boldsymbol{x}_k + \boldsymbol{v}(\boldsymbol{x}_k)\Delta t, \quad k = 0, 1, \ldots, N-1.$$

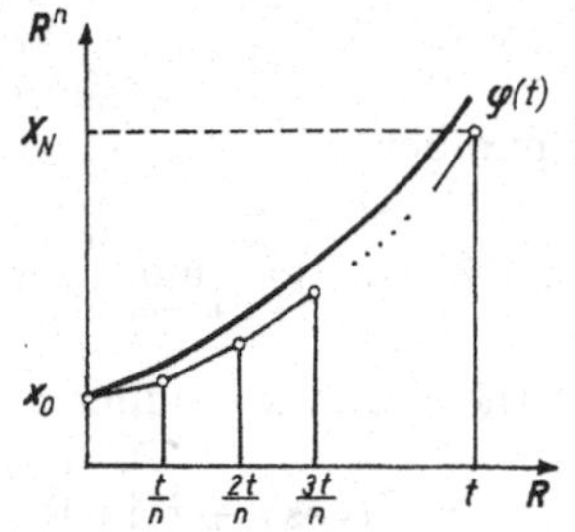

Fig. 111. An Euler broken line

We denote the last point $\boldsymbol{x}_N$ by $\boldsymbol{X}_N(t)$. We remark that the graph describing a motion with piecewise constant velocity is a broken line of N segments in the extended phase space $\boldsymbol{R}\times\boldsymbol{R}^n$. This broken line is called an *Euler broken line.* One would naturally expect that as $N \to \infty$ the sequence of Euler broken lines tends to an integral curve, so that for large N the last point $X_N(t)$ will be near the value of the solution φ with the initial condition $\varphi(0) = \boldsymbol{x}_0$ at the point t.

Theorem. *For a linear equation* (3) *the relation* $\lim_{N\to\infty} \boldsymbol{X}_N(t) = \varphi(t)$ *holds.*

Proof. By definition of an Euler broken line for $\boldsymbol{v}(\boldsymbol{x}) = A\boldsymbol{x}$ we have $\boldsymbol{X}_N = \left(E + \frac{At}{N}\right)^N \boldsymbol{x}_0$. Therefore $\lim_{N\to\infty} \boldsymbol{X}_N = e^{tA}\boldsymbol{x}_0$ (cf. (5)). Thus $\lim_{N\to\infty} \boldsymbol{X}_N = \varphi(t)$ (cf. (4)). □

Problem 1. Prove that not only does the endpoint of the Euler broken line tend to $\varphi(t)$, but also the whole sequence of piecewise-linear functions $\varphi_n : I \to \boldsymbol{R}^n$ whose graphs are the Euler broken lines tends uniformly to the solution φ on the interval $[0, t]$.

Remark. In the general case the Euler broken line (when the vector field $\boldsymbol{v}$ depends nonlinearly on $\boldsymbol{x}$) can also be written in the form $\boldsymbol{X}_N = \left(E + \frac{tA}{N}\right)^n \boldsymbol{x}_0$, where A is the nonlinear operator that maps the point $\boldsymbol{x}$ into the point $\boldsymbol{v}(\boldsymbol{x})$. We shall see below that in this case the sequence of Euler broken lines converges to a solution, at least for sufficiently small $|t|$ (§ 31, Sect. 9). Thus the expression (4), in which the exponential is defined by formula (5), gives the general solution of all differential equations[5].

[5] In practice it is not convenient to solve an equation approximately using Euler broken lines, since one is forced to take a very small step size Δt in order to

The Euler theory of the exponential, which is the same in all its variants from the definition of the number e, the Euler formula for e^z, Taylor's formula, formula (4) for solving linear equations, down to Euler's broken-line method, has many other applications that are beyond the scope of this course.

§ 16. The Determinant of an Exponential

If the operator A is given by its matrix, the computation of the matrix of the operator e^A may require long calculations. However the determinant of the matrix e^A can be computed very easily, as we shall now see.

1. The Determinant of an Operator

Let $A : \boldsymbol{R}^n \to \boldsymbol{R}^n$ be a linear operator.

Definition. The *determinant* of the operator A is the determinant of the matrix of the operator in some basis $\boldsymbol{e}_1, \ldots, \boldsymbol{e}_n$; we denote it by $\det A$.

The determinant of the matrix of the operator A is independent of the basis. Indeed, if (A) is the matrix of the operator A in the basis $\boldsymbol{e}_1, \ldots, \boldsymbol{e}_n$, then the matrix of the operator A in another basis will be $(B)(A)(B^{-1})$, and

$$\det(B)(A)(B^{-1}) = \det(A).$$

The determinant of a matrix is the oriented volume of the parallelepiped[6] *whose edges are the columns of the matrix.* [This definition of a determinant, which makes the algebraic theory of determinants trivial, is kept secret by the authors of most algebra textbooks in order to enhance the authority of their science.]

For example, for $n = 2$ (Fig. 112) the determinant $\begin{vmatrix} x_1 & x_2 \\ y_1 & y_2 \end{vmatrix}$ is the area of the parallelogram spanned by the vectors $\boldsymbol{\xi}_1$ and $\boldsymbol{\xi}_2$ with components (x_1, y_1) and (x_2, y_2) taken with the plus sign if the ordered pair of vectors $(\boldsymbol{\xi}_1, \boldsymbol{\xi}_2)$ gives the same orientation of $\boldsymbol{R}^2$ as the basis pair $(\boldsymbol{e}_1, \boldsymbol{e}_2)$, and taken with the minus sign if not.

The ith column in the matrix of the operator A in the basis $\boldsymbol{e}_1, \ldots, \boldsymbol{e}_n$ is composed of the coordinates of the image $A\boldsymbol{e}_i$ of the basis vector $\boldsymbol{e}_i$. Therefore

---

obtain a prescribed accuracy. More frequently one uses various improvements on this method, in which the integral curve is approximated not by a segment of a line but by a segment of a parabola of one degree or another. The most commonly used methods are those of Adams, Störmer, and Runge. Further information about these methods is contained in textbooks on numerical computation.

[6] The parallelepiped with edges $\boldsymbol{\xi}_1, \ldots, \boldsymbol{\xi}_n \in \boldsymbol{R}^n$ consists of all points of the form $x_1\boldsymbol{\xi}_1 + \cdots + x_n\boldsymbol{\xi}_n$, $0 \le x_i \le 1$. For $n = 2$ a parallelepiped is called a parallelogram. If you know some definition of volume, you will easily be able to prove this assertion. If not, you may take it as the definition of the volume of a parallelepiped.

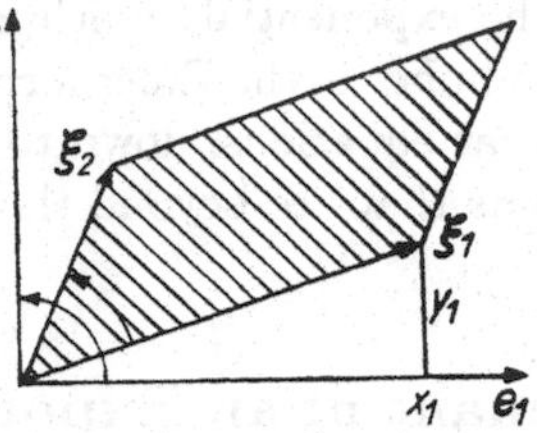

Fig. 112. The determinant of a matrix equals the oriented area of the parallelogram spanned by its columns

the determinant of the operator A is the oriented volume of the image of the unit cube (the parallelepiped with edges $\mathbf{e}_1, \ldots \mathbf{e}_n$*) under the mapping A.*

Problem 1. Let Π be a parallelepiped with linearly independent edges. Prove that the ratio of the (oriented) volume of the image parallelepiped $A\Pi$ to the (oriented) volume of Π is independent of Π and equal to $\det A$.

Remark. The reader familiar with the theory of measurement of volumes in $\boldsymbol{R}^n$ can replace Π by any figure having a volume.

Thus *the determinant of the operator A is the coefficient by which an oriented volume changes: when A is applied, the oriented volume of any figure is altered by a factor of* $\det A$. It is not at all obvious geometrically that the scaling of the volume of all figures is the same (even in the case of the plane); indeed the shape of a figure changes greatly under a linear transformation.

2. The Trace of an Operator

The *trace* of the matrix A is the sum of its diagonal elements. The trace is denoted tr or Sp (from the German *Spur*): $\operatorname{tr} A = \sum_{i=1}^{n} a_{ii}$.

The trace of the matrix of an operator $A : \boldsymbol{R}^n \to \boldsymbol{R}^n$ is independent of the basis and depends only on the operator A itself.

Problem 1. Prove that the trace of a matrix equals the sum of all n of its eigenvalues and the determinant is their product.

Hint. Apply Viète's formula to the polynomial

$$\det |A - \lambda E| = (-\lambda)^n + (-\lambda)^{n-1} \sum_{i=1}^{n} a_{i,i} + \cdots$$

The eigenvalues do not depend on the basis. This enables us to make the following definition.

Definition. The *trace* of an operator A is the trace of its matrix in some (or any) basis.

3. The Connection Between the Determinant and the Trace

Let $A : \boldsymbol{R}^n \to \boldsymbol{R}^n$ be a linear operator, and let $\varepsilon \in \boldsymbol{R}$. The following result is easy to prove.

Theorem. *As* $\varepsilon \to 0$, $\det(E + \varepsilon A) = 1 + \varepsilon \operatorname{tr} A + O(\varepsilon^2)$.

Proof. The determinant of the operator $E+\varepsilon A$ is the product of its eigenvalues. The eigenvalues of the operator $E + \varepsilon A$ (counted according to multiplicity) are equal to $1+\varepsilon\lambda_i$, where λ_i are the eigenvalues of the operator A. Therefore $\det(E+\varepsilon A) = \prod_{i=1}^{n}(1+\varepsilon\lambda_i) = 1+\varepsilon\sum_{i=1}^{n}\lambda_i+O(\varepsilon^2)$, which was to be proved. □

Second proof. It is clear that $\varphi(\varepsilon) = \det(E + \varepsilon A)$ is a polynomial in ε, and $\varphi(0) = 1$. It must be proved that $\varphi'(0) = \operatorname{tr} A$. We shall denote the determinant of the matrix $\|x_{i,j}\|$ by $\Delta(\|x_{i,j}\|)$. By the rule for differentiating a composite function $\left.\frac{d\varphi}{d\varepsilon}\right|_{\varepsilon=0} = \sum_{i,j=1}^{n} \left.\frac{\partial\Delta}{\partial x_{i,j}}\right|_E \frac{dx_{i,j}}{d\varepsilon}$, where $x_{i,j}(\varepsilon)$ are the elements of the matrix $E + \varepsilon A$.

The partial derivative $\left.\frac{\partial\Delta}{\partial x_{i,j}}\right|_E$ is by definition equal to $\left.\frac{d}{dh}\right|_{h=0} \det(E+he_{i,j})$, where $e_{i,j}$ is the matrix whose only nonzero element is a 1 in the ith row and jth column. But $\det(E+he_{i,j}) = 1$ for $i \neq j$ and $1+h$ if $i = j$. Thus $\left.\frac{\partial\Delta}{\partial x_{i,j}}\right|_E = 0$ if $i \neq j$ and 1 if $i = j$. Therefore $\left.\frac{d\varphi}{d\varepsilon}\right|_{\varepsilon=0} = \sum_{i=1}^{n}\frac{dx_{i,i}}{d\varepsilon} = \sum_{i=1}^{n} a_{i,i} = \operatorname{tr} A$, which was to be proved. □

Incidentally we have just given a new proof that the trace is independent of the basis.

Corollary. *Under a small variation in the edges of a parallelepiped only the change in each edge along its own direction influences the change in volume; the contribution to the change in volume by the changes of an edge in the direction of the other edges is a second-order infinitesimal.*

For example, the area of the nearly square parallelogram of Fig. 113 differs from the area of the shaded rectangle by second-order infinitesimals.

This corollary can be proved from elementary geometric considerations; such an approach would lead to a geometric proof of the preceding theorem.

4. The Determinant of the Operator e^A

Theorem. *For any linear operator* $A : \boldsymbol{R}^n \to \boldsymbol{R}^n$

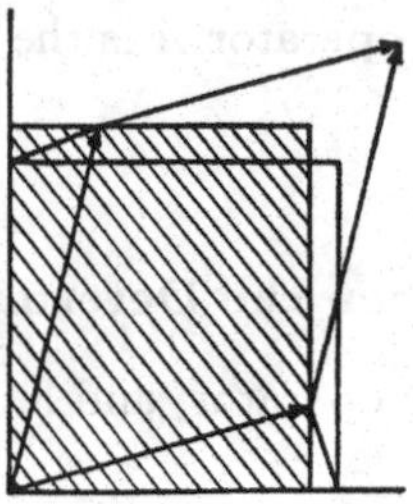

Fig. 113. Approximate determination of the area of a nearly square parallelogram

$$\det e^A = e^{\operatorname{tr} A}.$$

Proof. According to the second definition of the exponential we have $\det e^A = \det\left(\lim_{m\to\infty}\left(E+\frac{A}{m}\right)^m\right) = \lim_{m\to\infty}\left(\det\left(E+\frac{A}{m}\right)^m\right)$, for the determinant of a matrix is a polynomial (and hence a continuous function) of its elements. Furthermore, by the preceding theorem

$$\det\left(E+\frac{A}{m}\right)^m = \left(\det\left(E+\frac{A}{m}\right)\right)^m = \left(1+\frac{1}{m}\operatorname{tr} A + O\left(\frac{1}{m}\right)\right)^m, \quad m\to\infty.$$

It remains only to remark that $\lim_{m\to\infty}\left(1+\frac{a}{m}+O\left(\frac{1}{m^2}\right)\right)^m = e^a$ for any $a \in \boldsymbol{R}$, in particular for $a = \operatorname{tr} A$. □

Corollary 1. *The operator e^A is nondegenerate.*

Corollary 2. *The operator e^A preserves the orientation of $\boldsymbol{R}^n$ (i.e., $\det e^A > 0$).*

Corollary 3 (Liouville's formula). *The phase flow $\{g^t\}$ of the linear equation*

$$\dot{\boldsymbol{x}} = A\boldsymbol{x}, \quad \boldsymbol{x} \in \boldsymbol{R}^n, \tag{1}$$

over time t changes the volume of any figure by a factor of e^{at}, where $a = \operatorname{tr} A$.

Indeed, $\det g^t = \det e^{At} = e^{\operatorname{tr} At} = e^{t \operatorname{tr} A}$.

In particular the following corollary is a consequence of this.

Corollary 4. *If the trace of A is 0, then the phase flow of Eq.* (1) *preserves volumes (i.e., g^t maps any parallelepiped into a parallelepiped of the same volume).*

Indeed $e^0 = 1$.

Example 1. Consider the pendulum equation with friction coefficient $-k$

$$\ddot{x} = -x + k\dot{x},$$

which is equivalent to the system

$$\dot{x}_1 = x_2, \quad \dot{x}_2 = -x_1 + kx_2$$

with the matrix (Fig. 114) $\begin{pmatrix} 0 & 1 \\ -1 & k \end{pmatrix}$.

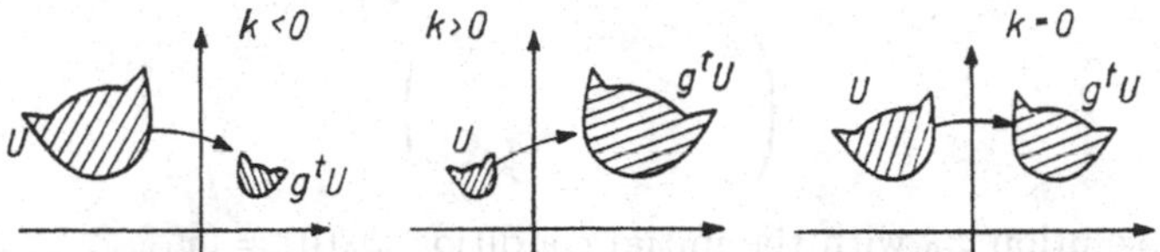

Fig. 114. The behavior of areas under the phase flow transformations of the pendulum equation

The trace of this matrix is k. Hence for $k < 0$ the phase flow transformation g^t $(t > 0)$ maps each domain of the phase plane into a domain of smaller area. In a system with negative friction ($k > 0$), in contrast, the area of the domain g^tU $(t > 0)$ is larger than the area of U. Finally, when there is no friction ($k = 0$), the phase flow g^t preserves area (not surprisingly: in this case, as we already know, g^t is rotation through the angle t).

Problem 1. Suppose the real parts of all the eigenvalues of A are negative. Prove that a phase flow g^t of Eq. (1) decreases volumes $(t > 0)$.

Problem 2. Prove that the eigenvalues of the operator e^A are e^{λ_i}, where λ_i are the eigenvalues of the operator A. Deduce the theorem proved above from this fact.

§ 17. Practical Computation of the Matrix of an Exponential – The Case when the Eigenvalues are Real and Distinct

In the practical solution of differential equations the operator A is defined by its matrix in some basis and an explicit computation of the matrix of the operator e^A in the same basis is required. We begin with the simplest case.

1. The Diagonalizable Operator

Consider the linear differential equation

$$\dot{\boldsymbol{x}} = A\boldsymbol{x}, \quad \boldsymbol{x} \in \boldsymbol{R}^n, \tag{1}$$

where $A : \boldsymbol{R}^n \to \boldsymbol{R}^n$ is a diagonalizable operator. In a basis in which the matrix of the operator A is diagonal it has the form

$$\begin{pmatrix} \lambda_1 & & 0 \\ & \ddots & \\ 0 & & \lambda_n \end{pmatrix},$$

where λ_i are the eigenvalues. The matrix of the operator e^{At} has the diagonal form

$$\begin{pmatrix} e^{\lambda_1 t} & & \\ & \ddots & \\ & & e^{\lambda_n t} \end{pmatrix}.$$

Thus the solution φ_0 with the initial condition $\varphi_0(0) = (x_{1_0}, \dots, x_{n_0})$ has in this basis the form $\varphi_k = e^{\lambda_k t} x_{k_0}$. It is necessary to pass to this basis if the matrix of the operator A is given in a different basis.

If all n of the eigenvalues of the operator A are real and distinct, it is diagonalizable ($\boldsymbol{R}^n$ decomposes into the direct sum of one-dimensional subspaces invariant with respect to A).

Therefore to solve Eq. (1) in the case when the eigenvalues of the operator A are real and distinct, one must proceed as follows to:

1) form the *characteristic* equation

$$\det |A - \lambda E| = 0;$$

2) find its roots $\lambda_1, \dots, \lambda_n$; we are assuming that they are real and distinct;

3) find the eigenvectors $\boldsymbol{\xi}_1, \dots, \boldsymbol{\xi}_n$ from the linear equations $A\boldsymbol{\xi}_k = \lambda_k \boldsymbol{\xi}_k$, $\boldsymbol{\xi}_k \neq \mathbf{o}$;

4) expand the initial condition in eigenvectors

$$\boldsymbol{x}_0 = \sum_{k=1}^{n} C_k \boldsymbol{\xi}_k;$$

5) write the solution $\varphi(t) = \sum_{k=1}^{n} C_k e^{\lambda_k t} \boldsymbol{\xi}_k$.

In particular, we have the following corollary.

Corollary. *Let A be a diagonalizable operator. Then the elements of the matrix e^{At} ($t \in \boldsymbol{R}$) in any basis are linear combinations of the exponentials $e^{\lambda_k t}$, where λ_k are the eigenvalues of the matrix of A.*

2. An Example

Consider the pendulum with friction

$$\dot{x}_1 = x_2, \quad \dot{x}_2 = -x_1 - kx_2.$$

The matrix of the operator A has the form

$$\begin{pmatrix} 0 & 1 \\ -1 & -k \end{pmatrix}, \quad \operatorname{tr} A = -k, \quad \det A = 1.$$

Therefore the characteristic equation has the form $\lambda^2 + k\lambda + 1 = 0$; the roots are real and distinct when the discriminant is positive, i.e., when $|k| > 2$. Thus for a coefficient of friction k that is sufficiently large (in absolute value) the operator A is diagonalizable.

Consider the case $k > 2$. In this case both roots λ_1 and λ_2 are negative. In a proper basis the equation can be written in the form

$$\dot{y}_1 = \lambda_1 y_1, \quad \lambda_1 < 0, \quad \dot{y}_2 = \lambda_2 y_2, \quad \lambda_2 < 0.$$

From this, as in § 2, we obtain the solution $y_1(t) = e^{\lambda_1 t} y_1(0)$, $y_2(t) = e^{\lambda_2 t} y_2(0)$ and the picture (the node of Fig. 115). As $t \to +\infty$ all solutions tend to zero. Almost all the phase curves are tangent to the y_1-axis if $|\lambda_2|$ is larger than $|\lambda_1|$ (in that case y_2 tends to zero faster than y_1). The picture in the (x_1, x_2)-plane is obtained by a linear transformation.

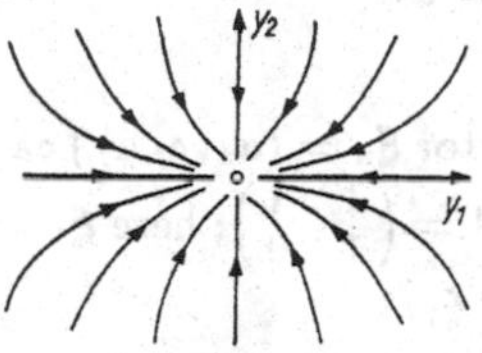

Fig. 115. The phase curves of the pendulum with strong friction in a proper basis

Suppose, for example, $k = 3\frac{1}{3}$, so that $\lambda_1 = -\frac{1}{3}$ and $\lambda_2 = -3$.

We find the eigenvector $\boldsymbol{\xi}_1$ from the condition $x_1 = -3x_2$; we obtain $\boldsymbol{\xi}_1 = \boldsymbol{e}_2 - 3\boldsymbol{e}_1$. Similarly $\boldsymbol{\xi}_2 = \boldsymbol{e}_1 - 3\boldsymbol{e}_2$. Since $|\lambda_1| < |\lambda_2|$, the phase curves have the form depicted in Fig. 116. Examining Fig. 116, we arrive at the following remarkable inference: if the coefficient of friction k is sufficiently large ($k > 2$), the pendulum does not undergo damped oscillations, but instead goes directly to the equilibrium position: its velocity x_2 changes sign at most once.

Problem 1. To which motions of the pendulum do the phase curves I, II, and III of Fig. 116 correspond? Sketch a typical graph $x(t)$.

Problem 2. Study the motion of the inverted pendulum with friction $\ddot{x} = x - k\dot{x}$.

3. The Discrete Case

All that has just been said about the exponential function e^{At} of a continuous argument t applies also to the exponential function A^n of the discrete argu-

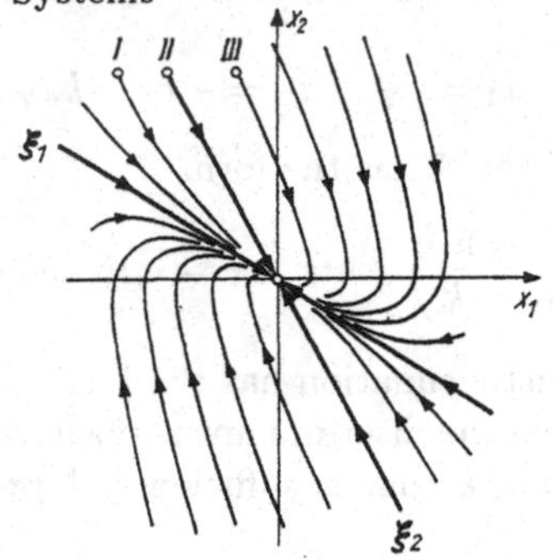

Fig. 116. The phase curves of the pendulum equation with strong friction in the usual basis

ment n. In particular, if A is a diagonalizable operator, then to calculate A^n it is convenient to pass to a diagonal basis.

Example. The Fibonacci sequence $0, 1, 1, 2, 3, 5, 8, 13, \ldots$ is defined by the condition that each successive term is the sum of its two predecessors ($a_n = a_{n-1} + a_{n-2}$) and the two initial terms $a_0 = 0$, $a_1 = 1$.

Problem 1. Find a formula for a_n. Show that a_n increases like a geometrical progression, and find $\lim\limits_{n\to\infty} \dfrac{\ln a_n}{n} = \alpha$.

Hint. We remark that the vector $\boldsymbol{\xi}_n = (a_n, a_{n-1})$ can be expressed linearly in terms of $\boldsymbol{\xi}_{n-1}$: $\boldsymbol{\xi}_n = A\boldsymbol{\xi}_{n-1}$, where $A = \begin{pmatrix} 1 & 1 \\ 1 & 0 \end{pmatrix}$; here $\boldsymbol{\xi}_1 = (1, 0)$. Therefore a_n is the first component of the vector $A^{n-1}\boldsymbol{\xi}_1$.

Answer. $\alpha = \ln((\sqrt{5}+1)/2)$, $a_n = (\lambda_1^n - \lambda_2^n)/\sqrt{5}$, where $\lambda_1 = (1+\sqrt{5})/2$ and $\lambda_2 = (1-\sqrt{5})/2$ are the eigenvalues of A.

The same reasoning reduces the study of any recursive sequence a_n of order k given by a rule

$$a_n = c_1 a_{n-1} + \cdots + c_k a_{n-k}, \quad n = 1, 2, \ldots,$$

and its k initial terms[7], to the study of the exponential function A^n, where $A : \boldsymbol{R}^k \to \boldsymbol{R}^k$ is a linear operator. Therefore, learning how to compute the matrix of an exponential, we are simultaneously studying all recursive sequences.

Returning to the general problem of computing e^{At}, we remark that the roots of the characteristic equation $\det(A - \lambda E) = 0$ may be complex. To study this case, we first consider a linear equation with the complex phase space $\boldsymbol{C}^n$.

[7] The fact that it is necessary to know the first k terms to determine a recursively defined sequence of order k is closely connected with the fact that the phase space of a differential equation of order k has dimension k. This connection becomes comprehensible if the differential equation is written in the form of a limit of difference equations.

§ 18. Complexification and Realification

Before studying complex differential equations we recall what is meant by the complexification of a real space and the realification of a complex space.

1. Realification

We shall denote by $\boldsymbol{C}^n$ an n-dimensional vector space over the field $\boldsymbol{C}$ of complex numbers.

The *realification* of the space $\boldsymbol{C}^n$ is the real vector space that coincides with $\boldsymbol{C}^n$ as a group, and in which multiplication by real scalars is defined as in $\boldsymbol{C}^n$, but multiplication by complex scalars is not defined. (In other words, realifying $\boldsymbol{C}^n$ amounts to forgetting about the $\boldsymbol{C}$-module structure while retaining the $\boldsymbol{R}$-module structure).

It is easy to see that the realification of the space $\boldsymbol{C}^n$ is a $2n$-dimensional real vector space $\boldsymbol{R}^{2n}$. We shall denote the realification by the symbol $\boldsymbol{R}$ above and to the left, for example: ${}^{\boldsymbol{R}}\boldsymbol{C} = \boldsymbol{R}^2$.

If $(\boldsymbol{e}_1, \ldots, \boldsymbol{e}_n)$ is a basis in $\boldsymbol{C}^n$, then $(\boldsymbol{e}_1, \ldots, \boldsymbol{e}_n, i\boldsymbol{e}_1, \ldots i\boldsymbol{e}_n)$ is a basis in ${}^{\boldsymbol{R}}\boldsymbol{C}^n = \boldsymbol{R}^{2n}$.

Let $A : \boldsymbol{C}^m \to \boldsymbol{C}^n$ be a $\boldsymbol{C}$-linear operator. The *realification of the operator* A is the $\boldsymbol{R}$-linear operator ${}^{\boldsymbol{R}}A : {}^{\boldsymbol{R}}\boldsymbol{C}^m \to {}^{\boldsymbol{R}}\boldsymbol{C}^n$ that coincides pointwise with A.

Problem 1. Let $(\boldsymbol{e}_1, \ldots, \boldsymbol{e}_m)$ be a basis of the space $\boldsymbol{C}^m$, $(\boldsymbol{f}_1, \ldots \boldsymbol{f}_n)$ a basis of the space $\boldsymbol{C}^n$, and (A) the matrix of the operator A. Find the matrix of the realified operator ${}^{\boldsymbol{R}}A$.

Answer. $\begin{pmatrix} \alpha & -\beta \\ \beta & \alpha \end{pmatrix}$, where $(A) = (\alpha) + i(\beta)$.

Problem 2. Prove that ${}^{\boldsymbol{R}}(A + B) = {}^{\boldsymbol{R}}A + {}^{\boldsymbol{R}}B$ and ${}^{\boldsymbol{R}}(AB) = {}^{\boldsymbol{R}}A{}^{\boldsymbol{R}}B$.

2. Complexification

Let $\boldsymbol{R}^n$ be a real vector space. The *complexification of the space* $\boldsymbol{R}^n$ is the n-dimensional complex vector space denoted by ${}^{\boldsymbol{C}}\boldsymbol{R}^n$ constructed as follows.

The points of the space ${}^{\boldsymbol{C}}\boldsymbol{R}^n$ are the pairs $(\boldsymbol{\xi}, \boldsymbol{\eta})$, where $\boldsymbol{\xi} \in \boldsymbol{R}^n$ and $\boldsymbol{\eta} \in \boldsymbol{R}^n$. Such a pair is denoted $\boldsymbol{\xi} + i\boldsymbol{\eta}$. The operations of addition and multiplication by complex scalars are defined in the usual manner:

$$(u + iv)(\boldsymbol{\xi} + i\boldsymbol{\eta}) = (u\boldsymbol{\xi} - v\boldsymbol{\eta}) + i(v\boldsymbol{\xi} + u\boldsymbol{\eta}),$$
$$(\boldsymbol{\xi}_1 + i\boldsymbol{\eta}_1) + (\boldsymbol{\xi}_2 + i\boldsymbol{\eta}_2) = (\boldsymbol{\xi}_1 + \boldsymbol{\xi}_2) + i(\boldsymbol{\eta}_1 + \boldsymbol{\eta}_2).$$

It is easy to verify that the $\boldsymbol{C}$-module so obtained is an n-dimensional complex vector space: ${}^{\boldsymbol{C}}\boldsymbol{R}^n = \boldsymbol{C}^n$. If $(\boldsymbol{e}_1, \ldots, \boldsymbol{e}_n)$ is a basis in $\boldsymbol{R}^n$, then the vectors $\boldsymbol{e}_k + i\boldsymbol{o}$ form a $\boldsymbol{C}$-basis in $\boldsymbol{C}^n = {}^{\boldsymbol{C}}\boldsymbol{R}^n$.

The vectors $\boldsymbol{\xi} + i\mathbf{o}$ are denoted more briefly by $\boldsymbol{\xi}$.

Let $A : \boldsymbol{R}^m \to \boldsymbol{R}^n$ be an $\boldsymbol{R}$-linear operator. The *complexification of the operator* A is the $\boldsymbol{C}$-linear operator ${}^{\boldsymbol{C}}A : {}^{\boldsymbol{C}}\boldsymbol{R}^m \to {}^{\boldsymbol{C}}\boldsymbol{R}^n$ defined by the relation ${}^{\boldsymbol{C}}A : (\boldsymbol{\xi} + i\boldsymbol{\eta}) = A\boldsymbol{\xi} + iA\boldsymbol{\eta}$.

Problem 1. Let $(\boldsymbol{e}_1, \ldots, \boldsymbol{e}_m)$ be a basis of $\boldsymbol{R}^m$ and $(\boldsymbol{f}_1, \ldots, \boldsymbol{f}_n)$ a basis of $\boldsymbol{R}^n$. Let (A) be the matrix of an operator A. Find the matrix of the complexified operator $({}^{\boldsymbol{C}}A)$.

Answer. $({}^{\boldsymbol{C}}A) = (A)$.

Problem 2. Prove that ${}^{\boldsymbol{C}}(A + B) = {}^{\boldsymbol{C}}A + {}^{\boldsymbol{C}}B$ and ${}^{\boldsymbol{C}}(AB) = {}^{\boldsymbol{C}}A\,{}^{\boldsymbol{C}}B$.

Remark on terminology. The operations of complexification and realification are defined for both spaces and mappings. Algebraists call operations of this kind *functors*.

3. The Complex Conjugate

Consider the real $2n$-dimensional vector space $\boldsymbol{R}^{2n} = {}^{\boldsymbol{RC}}\boldsymbol{R}^n$ obtained from $\boldsymbol{R}^n$ by a complexification followed by a realification. This space contains the n-dimensional subspace of vectors of the form $\boldsymbol{\xi} + i\mathbf{o}$, $\boldsymbol{\xi} \in \boldsymbol{R}^n$, called the *real plane* $\boldsymbol{R}^n \subset \boldsymbol{R}^{2n}$.

The subspace of vectors of the form $\mathbf{o} + i\boldsymbol{\xi}$, $\boldsymbol{\xi} \in \boldsymbol{R}^n$, is called the *imaginary plane* $i\boldsymbol{R}^n \subset \boldsymbol{R}^{2n}$. The whole space $\boldsymbol{R}^{2n}$ is the direct sum of these two n-dimensional subspaces.

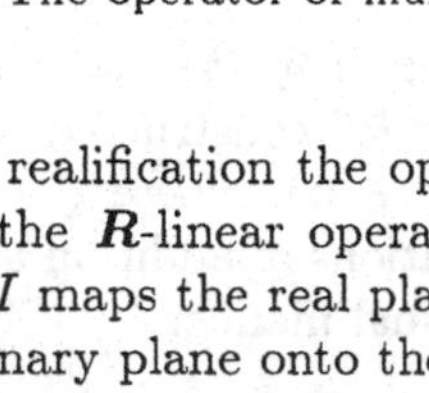

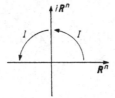

Fig. 117. The operator of multiplication by i

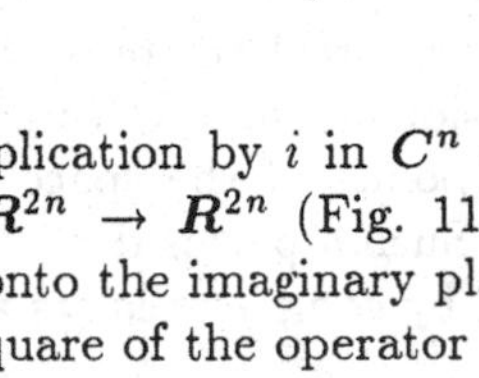

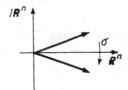

Fig. 118. The complex conjugate

After realification the operator iE of multiplication by i in $\boldsymbol{C}^n = {}^{\boldsymbol{C}}\boldsymbol{R}^n$ becomes the $\boldsymbol{R}$-linear operator ${}^{\boldsymbol{R}}(iE) = I : \boldsymbol{R}^{2n} \to \boldsymbol{R}^{2n}$ (Fig. 117). The operator I maps the real plane isomorphically onto the imaginary plane and the imaginary plane onto the real plane. The square of the operator I is the negative of the identity operator.

Problem 1. Let $(\boldsymbol{e}_1, \ldots, \boldsymbol{e}_n)$ be a basis of $\boldsymbol{R}^n$ and $(\boldsymbol{e}_1, \ldots, \boldsymbol{e}_n, i\boldsymbol{e}_1, \ldots, i\boldsymbol{e}_n)$ a basis of $\boldsymbol{R}^{2n} = {}^{\boldsymbol{RC}}\boldsymbol{R}^n$. Find the matrix of the operator I in this basis.

Answer. $(I) = \begin{pmatrix} 0 & -E \\ E & 0 \end{pmatrix}$.

We denote by $\sigma : \boldsymbol{R}^{2n} \to \boldsymbol{R}^{2n}$ (Fig. 118) the operator of complex conjugation: $\sigma(\boldsymbol{\xi} + i\boldsymbol{\eta}) = \boldsymbol{\xi} - i\boldsymbol{\eta}$. The action of σ is often denoted by a bar above the vector.

The operator σ coincides with the identity operator on the real plane and the negative of the identity operator on the imaginary plane. It is involutive: $\sigma^2 = E$.

Suppose $A : {}^{\boldsymbol{C}}\boldsymbol{R}^m \to {}^{\boldsymbol{C}}\boldsymbol{R}^n$ is a $\boldsymbol{C}$-linear operator. The *complex conjugate of the operator* A, denoted $\overline{A}$, is the operator $\overline{A} : {}^{\boldsymbol{C}}\boldsymbol{R}^m \to {}^{\boldsymbol{C}}\boldsymbol{R}^n$ defined by the relation

$$\overline{A}\boldsymbol{z} = \overline{A\overline{\boldsymbol{z}}} \text{ for every } \boldsymbol{z} \in {}^{\boldsymbol{C}}\boldsymbol{R}^m.$$

Problem 2. Prove that $\overline{A}$ is a $\boldsymbol{C}$-linear operator.

Problem 3. Prove that the matrix of the operator $\overline{A}$ *in a real basis* is the complex conjugate of the matrix of A in the same basis.

Problem 4. Prove that $\overline{A+B} = \overline{A} + \overline{B}$, $\overline{AB} = \overline{A}\,\overline{B}$, and $\overline{\lambda A} = \overline{\lambda}\,\overline{A}$.

Problem 5. Prove that the complex linear operator $A : {}^{\boldsymbol{C}}\boldsymbol{R}^m \to {}^{\boldsymbol{C}}\boldsymbol{R}^n$ is the complexification of a real operator if and only if $\overline{A} = A$.

4. The Exponential, Determinant, and Trace of a Complex Operator

The exponential, determinant, and trace of a complex operator are defined exactly as in the real case. They have the same properties as in the real case, the only difference being that the determinant, being a complex number, is not a volume.

Problem 1. Prove that the exponential has the following properties:

$${}^{\boldsymbol{R}}(e^A) = e^{{}^{\boldsymbol{R}}A}, \quad \overline{e^A} = e^{\overline{A}}, \quad {}^{\boldsymbol{C}}(e^A) = e^{{}^{\boldsymbol{C}}A}.$$

Problem 2. Prove that the determinant has the following properties:

$$\det {}^{\boldsymbol{R}}A = |\det A|^2, \quad \det \overline{A} = \overline{\det A}, \quad \det {}^{\boldsymbol{C}}A = \det A.$$

Problem 3. Prove that the trace has the following properties:

$$\operatorname{tr} {}^{\boldsymbol{R}}A = \operatorname{tr} A + \operatorname{tr} \overline{A}, \quad \operatorname{tr} \overline{A} = \overline{\operatorname{tr} A}, \quad \operatorname{tr} {}^{\boldsymbol{C}}A = \operatorname{tr} A.$$

Problem 4. Prove that the equality

$$\det e^A = e^{\operatorname{tr} A}$$

holds in the complex case also.

5. The Derivative of a Curve with Complex Values

Let $\varphi : I \to \boldsymbol{C}^n$ be a mapping from an interval I of the real t-axis into the complex vector space $\boldsymbol{C}^n$. We shall call φ a *curve*.

The *derivative* of the curve φ at the point $t_0 \in I$ is defined as usual: $\left.\frac{d\varphi}{dt}\right|_{t=t_0} = \lim_{h\to 0} \frac{\varphi(t_0+h)-\varphi(t_0)}{h}$. This is a vector of the space $\boldsymbol{C}^n$.

Example 1. Let $n = 1$ and $\varphi(t) = e^{it}$ (Fig. 119). Then $\left.\frac{d\varphi}{dt}\right|_{t=0} = i$.

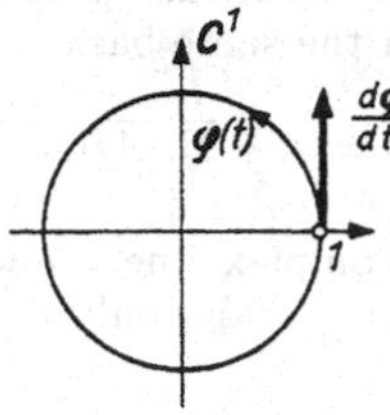

Fig. 119. The derivative of the mapping $t \mapsto e^{it}$ at the point 0 equals i

Let us consider the case $n = 1$ in more detail. Since multiplication is defined in $\boldsymbol{C}$, curves with values in $\boldsymbol{C}$ can be not only added but also multiplied:

$$(\varphi_1 + \varphi_2)(t) = \varphi_1(t) + \varphi_2(t), \quad (\varphi_1\varphi_2)(t) = \varphi_1(t)\varphi_2(t), \quad t \in I.$$

Problem 1. Prove that the derivative has the following properties:

$$\frac{d}{dt}(\varphi_1 + \varphi_2) = \frac{d\varphi_1}{dt} + \frac{d\varphi_2}{dt}, \quad \frac{d}{dt}(\varphi_1\varphi_2) = \frac{d\varphi_1}{dt}\varphi_2 + \varphi_1\frac{d\varphi_2}{dt}.$$

In particular the derivative of a polynomial with complex coefficients is given by the same formula as in the case of real coefficients.

If $n > 1$, it is impossible to multiply two curves with values in $\boldsymbol{C}^n$. However, since $\boldsymbol{C}^n$ is a $\boldsymbol{C}$-module, it is possible to multiply a curve $\varphi : I \to \boldsymbol{C}^n$ by a function $f : I \to \boldsymbol{C}$:

$$(f\varphi)(t) = f(t)\varphi(t).$$

Problem 2. Prove that the derivative has the following properties:

$$\frac{d({}^{\boldsymbol{R}}\varphi)}{dt} = {}^{\boldsymbol{R}}\frac{d\varphi}{dt}, \quad \frac{d}{dt}({}^{\boldsymbol{C}}\varphi) = {}^{\boldsymbol{C}}\frac{d\varphi}{dt}, \quad \frac{d\overline{\varphi}}{dt} = \overline{\frac{d\varphi}{dt}}.$$

$$\frac{d(\varphi_1 + \varphi_2)}{dt} = \frac{d\varphi_1}{dt} + \frac{d\varphi_2}{dt}, \quad \frac{d(f\varphi)}{dt} = \frac{df}{dt}\varphi + f\frac{d\varphi}{dt}.$$

Of course it is assumed here that the derivatives exist.

Theorem. *Let $A : \boldsymbol{C}^n \to \boldsymbol{C}^n$ be a $\boldsymbol{C}$-linear operator. Then for each $t \in \boldsymbol{R}$ the following $\boldsymbol{C}$-linear operator from $\boldsymbol{C}^n$ into $\boldsymbol{C}^n$ exists:*

$$\frac{d}{dt}e^{tA} = Ae^{tA}.$$

Proof. This can be proved exactly as in the real case, but one can also reduce it to the real case. For by realifying $\boldsymbol{C}^n$ we obtain

$$^{\boldsymbol{R}}\left(\frac{d}{dt}e^{tA}\right) = \frac{d}{dt}\,^{\boldsymbol{R}}(e^{tA}) = \frac{d}{dt}e^{t(^{\boldsymbol{R}}A)} = (^{\boldsymbol{R}}A)(e^{t^{\boldsymbol{R}}A}) = {}^{\boldsymbol{R}}(Ae^{tA}).$$

§ 19. The Linear Equation with a Complex Phase Space

The complex case, as frequently happens, is simpler than the real case. It is intrinsically important; moreover the study of of the complex case will help us to investigate the real case.

1. Definitions

Let $A : \boldsymbol{C}^n \to \boldsymbol{C}^n$ be a $\boldsymbol{C}$-linear operator. We define a *linear equation*[8] with the phase space $\boldsymbol{C}^n$ to be an equation

$$\dot{\boldsymbol{z}} = A\boldsymbol{z}, \quad \boldsymbol{z} \in \boldsymbol{C}^n. \tag{1}$$

A mapping $\varphi : I \to \boldsymbol{C}^n$ of the interval I of the real t-axis into $\boldsymbol{C}^n$ is a *solution* of Eq. (1) with the initial condition $\varphi(t_0) = \boldsymbol{z}_0$, $t_0 \in \boldsymbol{R}$, $\boldsymbol{z}_0 \in \boldsymbol{C}^n$ if

1) for every $\tau \in I$ $\left.\frac{d\varphi}{dt}\right|_{t=\tau} = A\varphi(\tau)$;

2) $t_0 \in I$ and $\varphi(t_0) = \boldsymbol{z}_0$.

In other words a mapping $\varphi : I \to \boldsymbol{C}^n$ is called a solution of Eq. (1) if it is a solution of the equation obtained by realifying the space $\boldsymbol{C}^n$ and the operator A. This equation has a $2n$-dimensional real phase space and is given by $\dot{\boldsymbol{z}} = {}^{\boldsymbol{R}}A\boldsymbol{z}$, $\boldsymbol{z} \in \boldsymbol{R}^{2n} = {}^{\boldsymbol{R}}\boldsymbol{C}^n$.

2. The Fundamental Theorem

The following theorems are proved exactly as in the real case (cf. § 15, Sects. 2 and 3):

[8] The full name is *a system of n first-order linear homogeneous ordinary differential equations with constant complex coefficients.*

Theorem. *The solution of Eq.* (1) *with initial condition* $\varphi(0) = z_0$ *is given by the formula* $\varphi(t) = e^{At}z_0$.

Theorem. *Every one-parameter group* $\{g^t\ (t \in \boldsymbol{R})\}$ *of* $\boldsymbol{C}$*-linear mappings of the space* $\boldsymbol{C}^n$ *has the form* $g^t = e^{At}$, *where* $A : \boldsymbol{C}^n \to \boldsymbol{C}^n$ *is a* $\boldsymbol{C}$*-linear operator.*

Our immediate goal is to study e^{At} and compute it explicitly.

3. The Diagonalizable Case

Let $A : \boldsymbol{C}^n \to \boldsymbol{C}^n$ be a $\boldsymbol{C}$-linear operator. Consider the characteristic equation

$$\det |A - \lambda E| = 0. \tag{2}$$

Theorem. *If the* n *roots* $\lambda_1, \dots, \lambda_n$ *of the characteristic equation are pairwise distinct, then* $\boldsymbol{C}^n$ *decomposes into the direct sum of one-dimensional subspaces that are invariant with respect to* A *and* e^{At}: $\boldsymbol{C}^n = \boldsymbol{C}^1_1 \dotplus \cdots \dotplus \boldsymbol{C}^1_n$. *Moreover in each one-dimensional invariant subspace* $\boldsymbol{C}^1_k$ *the operator* e^{At} *reduces to multiplication by the complex number* $e^{\lambda_k t}$.

Indeed the operator A has[9] n linearly independent one-dimensional eigenspaces: $\boldsymbol{C}^n = \boldsymbol{C}^1_1 \dotplus \cdots \dotplus \boldsymbol{C}^1_n$. On the line $\boldsymbol{C}^1_k$ the operator A acts as multiplication by λ_k, and so the operator e^{At} acts as multiplication by $e^{\lambda_k t}$.

Let us now examine the case $n = 1$ in more detail.

4. Example: A Linear Equation whose Phase Space is the Complex Line

Such an equation has the form

$$\frac{dz}{dt} = \lambda z, \quad z \in \boldsymbol{C}, \quad \lambda \in \boldsymbol{C}, \quad t \in \boldsymbol{R}. \tag{3}$$

We already know the solution of this equation: $\varphi(t) = e^{\lambda t}z_0$. Let us study the complex function $e^{\lambda t}$ of a real variable t:

$$e^{\lambda t} : \boldsymbol{R} \to \boldsymbol{C}.$$

If λ is real, then the function $e^{\lambda t}$ is real (Fig. 120).

In this case the phase flow of Eq. (3) consists of dilations by the factors $e^{\lambda t}$. If λ is a pure imaginary, $\lambda = i\omega$, then by Euler's formula

$$e^{\lambda t} = e^{i\omega t} = \cos\omega t + i \sin\omega t.$$

---

[9] This it the only place where the complex case differs from the real case. The reason the real case is more complicated is that the field $\boldsymbol{R}$ is not algebraically closed.

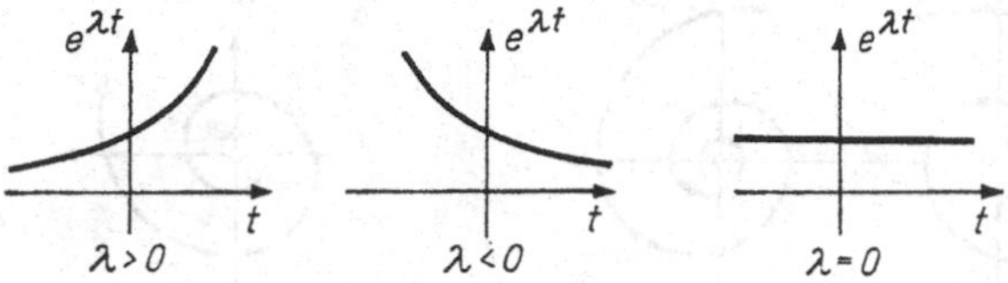

Fig. 120. The graphs of the functions $e^{\lambda t}$ for real λ

In this case the phase flow of Eq. (3) is a family of rotations $\{g^t\}$ through the angle ωt (Fig. 121). Finally, in the general case $\lambda = \alpha + i\omega$, and multiplication by $e^{\lambda t}$ is the product of a multiplication by $e^{\alpha t}$ and multiplication by $e^{i\omega t}$ (cf. § 15, Sect. 5):

$$e^{\lambda t} = e^{(\alpha + i\omega)t} = e^{\alpha t} \cdot e^{i\omega t}. \tag{4}$$

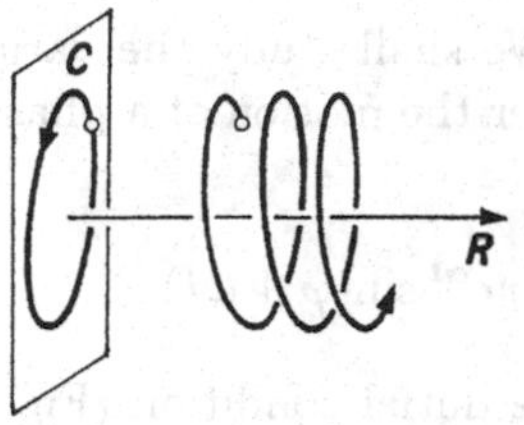

Fig. 121. Phase and integral curves of the equation $\dot{z} = \lambda z$ for purely imaginary λ

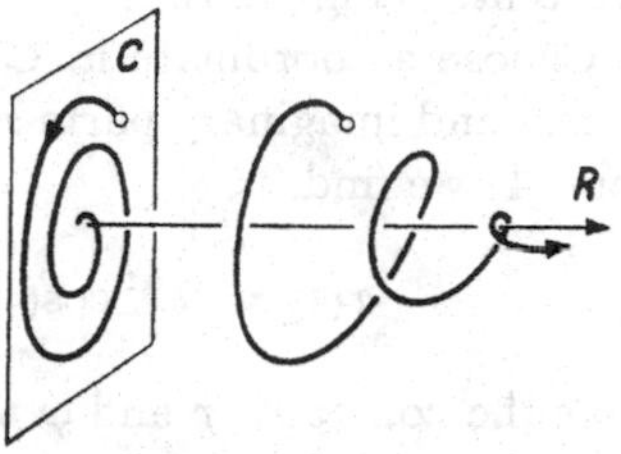

Fig. 122. Phase and integral curves of the equation $\dot{z} = \lambda z$ for $\lambda = \alpha + i\omega$, $\alpha < 0$, $\omega > 0$

Thus the transformation g^t of the phase flow of Eq. (3) is a dilation by a factor of $e^{\alpha t}$ and a simultaneous rotation through the angle ωt.

Let us now consider the phase curves. Suppose, for example, $\alpha < 0$ and $\omega > 0$ (Fig. 122). In such a case as t increases the phase point $e^{\lambda t} z_0$ will approach the origin, winding around it in a counterclockwise direction (i.e., from 1 to i).

In polar coordinates for a suitable choice of a ray from which angles are measured the phase curve is given by the equation

$$r = e^{k\varphi} \quad \text{or} \quad \varphi = k^{-1} \ln r, \quad \left(k = \frac{\alpha}{\omega}\right).$$

Such a curve is called a *logarithmic spiral*.

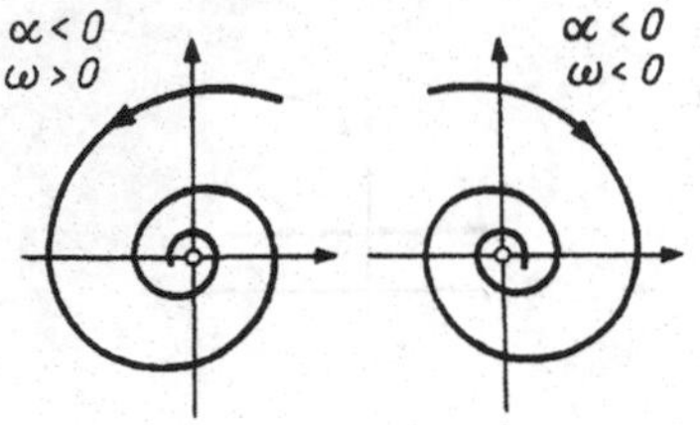

Fig. 123. Stable foci

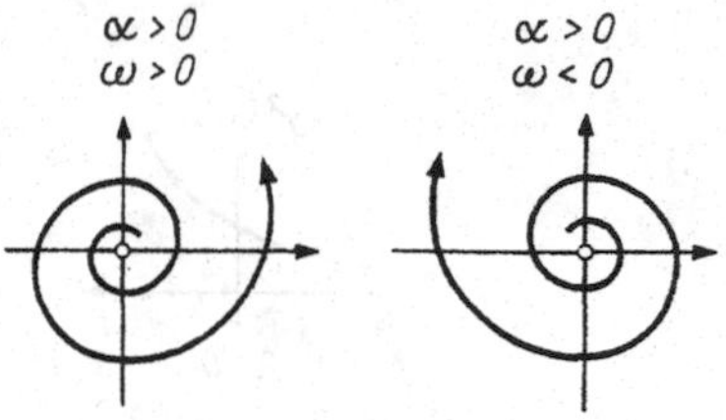

Fig. 124. Unstable foci

The curves are also logarithmic spirals with other combinations of signs of α and ω (Figs. 123 and 124).

In all cases (except $\lambda = 0$) the point $z = 0$ is the only fixed point of the phase flow (and the only singular point of the vector field corresponding to Eq. (3)).

This singular point is called a *focus* (we are assuming that $\alpha \neq 0$ and $\omega \neq 0$). If $\alpha < 0$, then $\varphi(t) \to \mathbf{o}$ as $t \to +\infty$, and the focus is said to be *stable*, while if $\alpha > 0$ it is *unstable*.

For $\alpha = 0$ and $\omega \neq 0$ the phase curves are circles and the singular point is their center (Fig. 125).

Choose a coordinate in $\boldsymbol{C}^1$: $z = x + iy$. We shall study the variation of the real and imaginary parts $x(t)$ and $y(t)$ under the motion of a phase point. From (4) we find

$$x(t) = re^{\alpha t}\cos(\varphi + \omega t), \quad y(t) = re^{\alpha t}\sin(\varphi + \omega t),$$

where the constants r and φ are defined by the initial condition (Fig. 126).

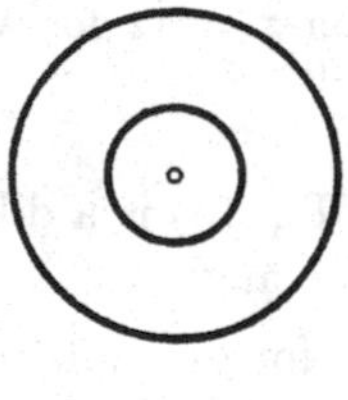

Fig. 125. A center

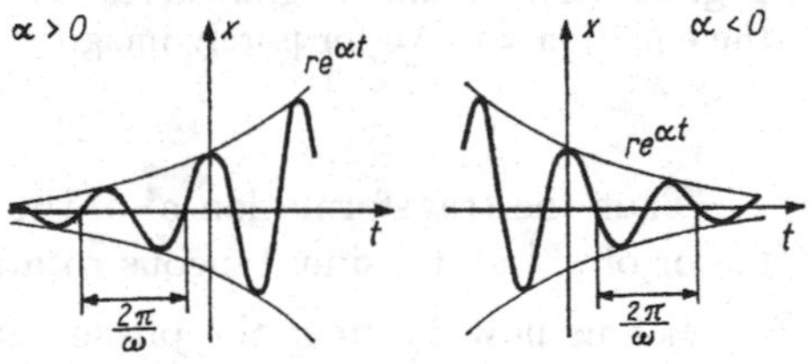

Fig. 126. The real part of $e^{\lambda t}$ as a function of time

Thus for $\alpha > 0$ the coordinates $x(t)$ and $y(t)$ undergo "harmonic oscillations with frequency ω and with exponentially increasing amplitude $r = e^{\alpha t}$," while with $\alpha < 0$ they undergo damped oscillations.

The variation of x or y with time can also be written as $Ae^{\alpha t}\cos\omega t + Be^{\alpha t}\sin\omega t$, where the constants A and B are determined by the initial conditions.

Remark 1. In studying Eq. (3), we have thereby simultaneously studied all the one-parameter groups of $\boldsymbol{C}$-linear transformations of the complex line.

Remark 2. At the same time we have studied the system of linear equations on the real plane

$$\dot{x} = \alpha x - \omega y, \quad \dot{y} = \omega x + \alpha y,$$

which Eq. (3) becomes after realification.

From the theorems of Sects. 2 and 3 and the computations of Sect. 4 there follows an explicit formula for the solutions of Eq. (1).

5. Corollary

Suppose the n roots $\lambda_1, \ldots, \lambda_n$ of the characteristic equation (2) *are pairwise distinct. Then every solution φ of Eq.* (1) *has the form*

$$\varphi(t) = \sum_{k=1}^{n} c_k e^{\lambda_k t} \boldsymbol{\xi}_k, \tag{5}$$

where $\boldsymbol{\xi}_k$ are constant vectors independent of the initial conditions and c_k are complex constants depending on the initial conditions. For any choice of these constants formula (5) *gives a solution of Eq.* (1).

If $z_1, \ldots, z_n$ is a linear coordinate system in $\boldsymbol{C}^n$, then the real (or imaginary) part of each coordinate $z_l = x_l + iy_l$ will vary with time like a linear combination of the functions $e^{\alpha_k t} \cos \omega_k t$, $e^{\alpha_k t} \sin \omega_k t$;

$$x_l = \sum_{k=1}^{n} r_{k,l} e^{\alpha_k t} \cos(\theta_{k,l} + \omega_k t) = \sum_{k=1}^{n} A_{k,l} e^{\alpha_k t} \cos \omega_k t + B_{k,l} e^{\alpha_k t} \sin \omega_k t, \tag{6}$$

where $\lambda_k = \alpha_k + i\omega_k$ and r, θ, A, and B are real constants depending on the initial conditions.

To prove this it suffices to expand the initial condition in a basis of eigenvectors: $\varphi(0) = c_k \boldsymbol{\xi}_1 + \cdots + c_n \boldsymbol{\xi}_n$.

§ 20. The Complexification of a Real Linear Equation

We shall now use the results of the study of the complex equation to study the real case.

1. The Complexified Equation

Let $A : \boldsymbol{R}^n \to \boldsymbol{R}^n$ be a linear operator defining a linear equation

$$\dot{\boldsymbol{x}} = A\boldsymbol{x}, \quad \boldsymbol{x} \in \boldsymbol{R}^n. \tag{1}$$

The complexification of Eq. (1) is the equation with complex phase space

$$\dot{\boldsymbol{z}} = {}^{C}\!A\boldsymbol{z}, \quad \boldsymbol{z} \in \boldsymbol{C}^n = {}^{C}\boldsymbol{R}^n. \tag{2}$$

Lemma 1. *The solutions of Eq.* (2) *with complex conjugate initial conditions are complex conjugates of each other.*

Proof. Let φ be the solution with initial condition $\varphi(t_0) = z_0$ (Fig. 127). Then $\overline{\varphi}(t_0) = \overline{z}_0$. We shall show that $\overline{\varphi}$ is a solution. The lemma will then be proved (in view of the uniqueness of the solution).

For any value of t we have

$$\frac{d\overline{\varphi}}{dt} = \overline{\frac{d\varphi}{dt}} = \overline{{}^{C}A\varphi} = \overline{{}^{C}A}\overline{\varphi} = {}^{C}A\overline{\varphi},$$

which was to be proved. □

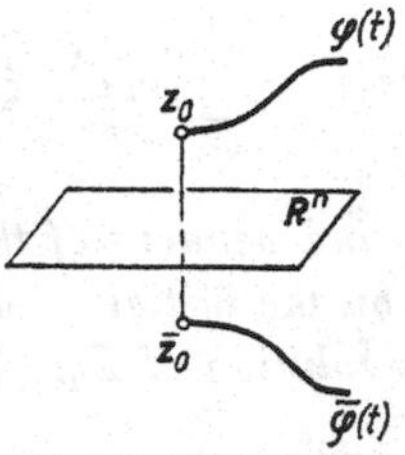

Fig. 127. Complex conjugate solutions

Remark. Instead of Eq. (2) we could have taken the more general equation

$$\dot{z} = F(z, t), \quad z \in {}^{C}R^n,$$

whose right-hand side assumes complex conjugate values at complex conjugate points: $F(\overline{z}, t) = \overline{F(z,t)}$.

For example, this condition is satisfied by any polynomial in the coordinates z_k of the vector z whose coefficients in a real basis are real-valued functions of t.

Corollary. *The solution of Eq.* (2) *with a real initial condition is real and satisfies Eq.* (1).

For if $\overline{\varphi} \neq \varphi$ (Fig. 128), the uniqueness theorem would be violated. □

In the following lemma it is essential that the equation be linear.

Lemma 2. *The function* $z = \varphi(t)$ *is a solution of the complexified equation* (2) *if and only if its real and imaginary parts satisfy the original equation* (1).

Indeed ${}^{C}A(x + iy) = Ax + iAy$, so that the realification of Eq. (2) decomposes into a direct product:

$$\begin{cases} \dot{x} = Ax, & x \in R^n, \\ \dot{y} = Ay, & y \in R^n. \end{cases}$$

One can see from Lemmas 1 and 2 how, knowing the complex solutions of Eq. (2), one can find the real solutions of Eq. (1) and conversely. In particular *formulas* (6) *of Sec.* 5 *of* § 19 *give the explicit form of the solution in the case when the characteristic equation has simple roots.*

2. The Invariant Subspaces of a Real Operator

Let $A : \boldsymbol{R}^n \to \boldsymbol{R}^n$ be a real linear operator. Let λ be one of the roots (in general complex) of the characteristic equation $\det|A-\lambda E| = 0$. The following lemma is obvious.

Lemma 3. *If* $\boldsymbol{\xi} \in \boldsymbol{C}^n = {}^{\boldsymbol{C}}\boldsymbol{R}^n$ *is an eigenvector of the operator* ${}^{\boldsymbol{C}}A$ *with eigenvalue* λ, *then* $\overline{\boldsymbol{\xi}}$ *is an eigenvector with eigenvalue* $\overline{\lambda}$. *The multiplicities of the eigenvalues* λ *and* $\overline{\lambda}$ *are the same.*

Indeed, since $\overline{{}^{C}A} = {}^{C}A$, the equation ${}^{C}A\boldsymbol{\xi} = \lambda\boldsymbol{\xi}$ is equivalent to ${}^{C}A\overline{\boldsymbol{\xi}} = \overline{\lambda}\,\overline{\boldsymbol{\xi}}$, and the characteristic equation has real coefficients.

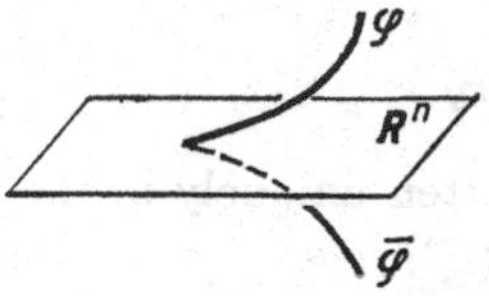

Fig. 128. The solution with a real initial condition cannot assume complex values

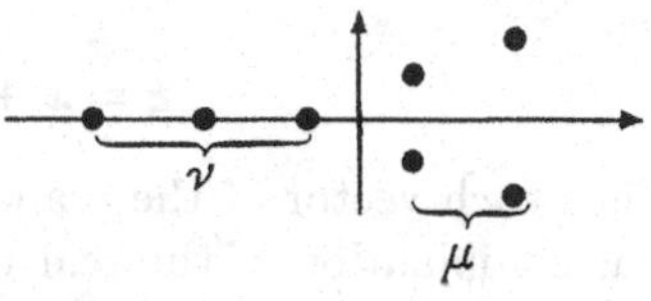

Fig. 129. The eigenvalues of a real operator

Let us now assume that the eigenvalues $\lambda_1, \ldots, \lambda_n \in \boldsymbol{C}$ *of the operator* $A : \boldsymbol{R}^n \to \boldsymbol{R}^n$ *are pairwise distinct* (Fig. 129). Among these eigenvalues there will be a certain number ν of real values and a certain number μ of complex conjugate pairs (and $\nu + 2\mu = n$, so that the parity of the number of real eigenvalues is the same as the parity of n). The following proposition is easily proved.

Theorem. *The space* $\boldsymbol{R}^n$ *decomposes into the direct sum of* ν *one-dimensional subspaces and* μ *two-dimensional subspaces, all invariant with respect to* A.

Indeed to a real eigenvalue there corresponds a real eigenvector and hence a one-dimensional subspace in $\boldsymbol{R}^n$.

Let λ and $\overline{\lambda}$ be a pair of complex conjugate eigenvalues. To the eigenvalue λ there corresponds an eigenvector $\boldsymbol{\xi} \in \boldsymbol{C}^n = {}^{C}\boldsymbol{R}^n$ of the complexified operator ${}^{C}A$.

By Lemma 3 the conjugate vector $\overline{\boldsymbol{\xi}}$ is also an eigenvector corresponding to the eigenvalue $\overline{\lambda}$.

The complex plane $\boldsymbol{C}^2$ spanned by the eigenvectors $\boldsymbol{\xi}$ and $\overline{\boldsymbol{\xi}}$ is invariant with respect to the operator ${}^{C}A$, The real subspace $\boldsymbol{R}^n \subset {}^{C}\boldsymbol{R}^n$ is also invariant. Therefore their intersection is also invariant with respect to ${}^{C}A$. We shall show that this intersection is a two-dimensional real plane $\boldsymbol{R}^2$ (Fig. 130).

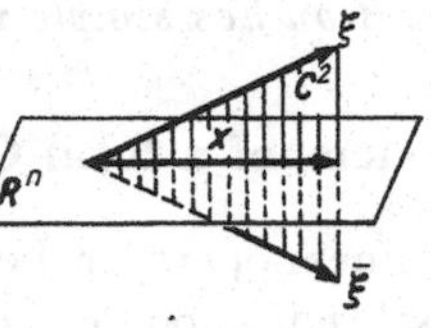

Fig. 130. The real part of a complex eigenvector belongs to an invariant real plane

Indeed, consider the real and imaginary parts of the eigenvector $\boldsymbol{\xi}$:

$$\boldsymbol{x} = \frac{1}{2}(\boldsymbol{\xi} + \overline{\boldsymbol{\xi}}) \in \boldsymbol{R}^n, \quad \boldsymbol{y} = \frac{1}{2i}(\boldsymbol{\xi} - \overline{\boldsymbol{\xi}}) \in \boldsymbol{R}^n.$$

Being $\boldsymbol{C}$-linear combinations of the vectors $\boldsymbol{\xi}$ and $\overline{\boldsymbol{\xi}}$, the vectors $\boldsymbol{x}$ and $\boldsymbol{y}$ belong to the intersection $\boldsymbol{C}^2 \cap \boldsymbol{R}^n$. The vectors $\boldsymbol{x}$ and $\boldsymbol{y}$ are $\boldsymbol{C}$-linearly independent, since the $\boldsymbol{C}$-linearly independent vectors $\boldsymbol{\xi}$ and $\overline{\boldsymbol{\xi}}$ can be expressed in terms of them:

$$\boldsymbol{\xi} = \boldsymbol{x} + i\boldsymbol{y}, \quad \overline{\boldsymbol{\xi}} = \boldsymbol{x} - i\boldsymbol{y}.$$

Thus each vector of the plane $\boldsymbol{C}^2$ can be written uniquely in the form of a linear combination of the real vectors $\boldsymbol{x}$ and $\boldsymbol{y}$:

$$\boldsymbol{\eta} = a\boldsymbol{x} + b\boldsymbol{y}, \quad a \in \boldsymbol{C}, \quad b \in \boldsymbol{C}.$$

Such a vector is real ($\boldsymbol{\eta} = \overline{\boldsymbol{\eta}}$) if and only if $\bar{a}\boldsymbol{x} + \bar{b}\boldsymbol{y} = a\boldsymbol{x} + b\boldsymbol{y}$, i.e., a and b are real. Thus *the intersection $\boldsymbol{C}^2 \cap \boldsymbol{R}^n$ is the two-dimensional real plane $\boldsymbol{R}^2$ spanned by the vectors $\boldsymbol{x}$ and $\boldsymbol{y}$, which are the real and imaginary parts of the eigenvector $\boldsymbol{\xi}$.*

The eigenvalues of the restriction of the operator A to the plane $\boldsymbol{R}^2$ are λ and $\overline{\lambda}$.

Indeed, complexification does not change the eigenvalues. After complexification of the restriction of A to $\boldsymbol{R}^2$ we obtain the restriction of ${}^{C}A$ to $\boldsymbol{C}^2$. But the plane $\boldsymbol{C}^2$ is spanned by the eigenvectors of the operator ${}^{C}A$ with eigenvalues λ and $\overline{\lambda}$. Thus the eigenvalues of $A|\boldsymbol{R}^2$ are λ and $\overline{\lambda}$.

It remains to show that the one- and two-dimensional invariant subspaces of the space $\boldsymbol{R}^n$ are $\boldsymbol{R}$-linearly independent. This follows immediately from the fact that the n eigenvectors of the operator ${}^{C}A$ are $\boldsymbol{C}$-linearly independent and can be expressed as linear combinations of our vectors $\boldsymbol{\xi}_k$, $(k = 1, \ldots, \nu)$ and $\boldsymbol{x}_k$ and $\boldsymbol{y}_k$ $(k = 1, \ldots, \mu)$.

The theorem is now proved. □

Thus *in the case when all the eigenvalues of the operator $A : \boldsymbol{R}^n \to \boldsymbol{R}^n$ are simple, the linear differential equation $\dot{\boldsymbol{x}} = A\boldsymbol{x}$, $\boldsymbol{x} \in \boldsymbol{R}^n$, decomposes into the direct product of equations with one- and two-dimensional phase spaces.*

We remark that a generic polynomial does not have multiple roots. Hence to study linear differential equations it is necessary first of all to consider linear equations on the line (which we have already done) and on the plane.

3. The Linear Equation on the Plane

Theorem. *Let* $A : \boldsymbol{R}^2 \to \boldsymbol{R}^2$ *be a linear operator with nonreal eigenvalues* λ *and* $\overline{\lambda}$.

Then A *is the realification of the operator* $\Lambda : \boldsymbol{C}^1 \to \boldsymbol{C}^1$ *of multiplication by the complex number* λ.

More precisely, the plane $\boldsymbol{R}^2$ can be endowed with the structure of the complex line $\boldsymbol{C}^1$, so that $\boldsymbol{R}^2 = {}^{\boldsymbol{R}}\boldsymbol{C}^1$ and $A = {}^{\boldsymbol{R}}\Lambda$.

Proof. The proof is a rather mysterious computation[10]. Let $\boldsymbol{x} + i\boldsymbol{y} \in {}^{\boldsymbol{C}}\boldsymbol{R}^2$ be a complex eigenvector of the operator ${}^{\boldsymbol{C}}A$ with eigenvalue $\lambda = \alpha + i\omega$. The vectors $\boldsymbol{x}$ and $\boldsymbol{y}$ form a basis in $\boldsymbol{R}^2$. We have, on the one hand,

$$ {}^{\boldsymbol{C}}A(\boldsymbol{x} + i\boldsymbol{y}) = (\alpha + i\omega)(\boldsymbol{x} + i\boldsymbol{y}) = \alpha\boldsymbol{x} - \omega\boldsymbol{y} + i(\omega\boldsymbol{x} + \alpha\boldsymbol{y}) $$

and on the other hand ${}^{\boldsymbol{C}}A(\boldsymbol{x} + i\boldsymbol{y}) = A\boldsymbol{x} + iA\boldsymbol{y}$, whence $A\boldsymbol{x} = \alpha\boldsymbol{x} - \omega\boldsymbol{y}$ and $A\boldsymbol{y} = \omega\boldsymbol{x} + \alpha\boldsymbol{y}$, i.e., in the basis $\boldsymbol{x}$, $\boldsymbol{y}$ the operator $A : \boldsymbol{R}^2 \to \boldsymbol{R}^2$ has the same matrix

$$ \begin{pmatrix} \alpha & \omega \\ -\omega & \alpha \end{pmatrix}, $$

as the operator ${}^{\boldsymbol{R}}\Lambda$ of multiplication by $\lambda = \alpha + i\omega$ in the basis $1, -i$. Thus the desired complex structure on $\boldsymbol{R}^2$ results from taking $\boldsymbol{x}$ as 1 and $-\boldsymbol{y}$ as i.

Corollary 1. *Let* $A : \boldsymbol{R}^2 \to \boldsymbol{R}^2$ *be a linear transformation of the Euclidean plane with nonreal eigenvalues* λ *and* $\overline{\lambda}$. *Then the transformation* A *is similar to a dilation by a factor of* $|\lambda|$ *with a rotation through the angle* $\arg \lambda$.

Corollary 2. *The phase flow of a linear equation* (1) *on the Euclidean plane* $\boldsymbol{R}^2$ *with nonreal eigenvalues* $\lambda, \overline{\lambda} = \alpha \pm i\omega$ *is similar to a family of dilations by factors of* $e^{\alpha t}$ *with simultaneous rotations through angles* ωt.

In particular the singular point 0 is a focus, and the phase curves are the affine images of logarithmic spirals approaching the origin as $t \to +\infty$ in the case when the real part α of the eigenvalues $\lambda, \overline{\lambda}$ is negative and receding from it in the case when $\alpha > 0$ (Fig. 131).

---

[10] The computation can be replaced by the following reasoning. Let $\lambda = \alpha + i\omega$. Define an operator $I : \boldsymbol{R}^2 \to \boldsymbol{R}^2$ by the condition $A = \alpha E + \omega I$. Such an operator I exists since $\omega \neq 0$ by hypothesis. Then $I^2 = -E$, since the operator A satisfies its characteristic equation. Taking I as multiplication by i, we obtain the required complex structure on $\boldsymbol{R}^2$.

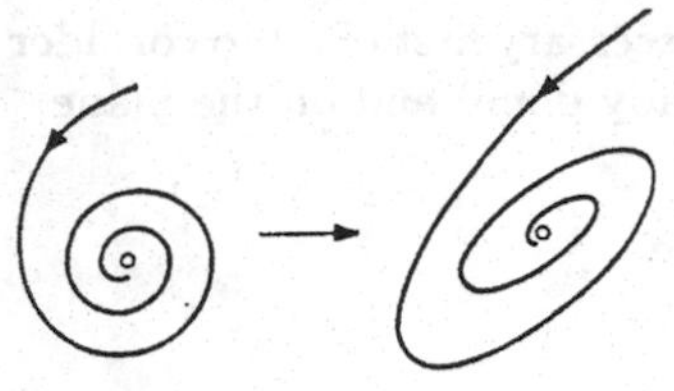

Fig. 131. The image of a logarithmic spiral under a linear transformation

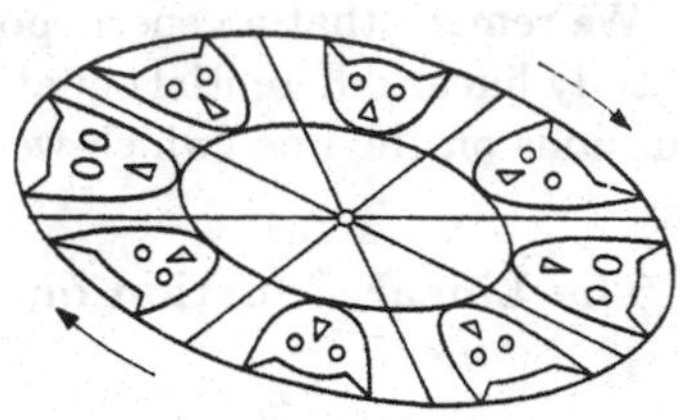

Fig. 132. An elliptic rotation

In the case $\alpha = 0$ (Fig. 132) the phase curves form a family of concentric ellipses, and the singular point is their *center*. In this case the phase flow transformations are called *elliptic rotations.*

4. The Classification of Singular Points in the Plane

Now let

$$\dot{\boldsymbol{x}} = A\boldsymbol{x}, \quad \boldsymbol{x} \in \boldsymbol{R}^2, \quad A : \boldsymbol{R}^2 \to \boldsymbol{R}^2,$$

be an arbitrary linear transformation in the plane. Let the roots λ_1 and λ_2 of the characteristic equation be distinct. If they are real and $\lambda_1 < \lambda_2$, the equation decomposes into two one-dimensional equations, and we obtain one of the cases already studied in Chapt. 1 (Figs. 133, 134, 135).

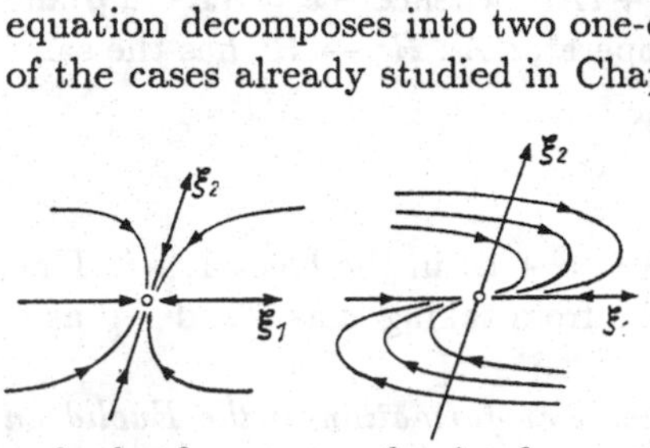

Fig. 133. Stable nodes

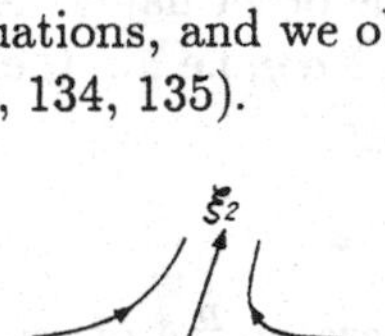

Fig. 134. A saddle point

Here we are omitting the borderline cases when one of λ_1 and λ_2 is zero. These cases are much less interesting, since they are rarely encountered and are lost under arbitrarily small perturbations. The study of these cases presents no difficulty.

If the roots are complex, $\lambda_{1,2} = \alpha \pm i\omega$, then depending on the sign of α one of the cases depicted in Figs. 136, 137, and 138 may occur.

The case of a center is exceptional, but it occurs, for example, in conservative systems (cf. § 12). The cases of multiple roots are also exceptional. It is left to the reader to verify that the case depicted in Fig. 133 ($\lambda_1 = \lambda_2 < 0$, the so-called degenerate node) corresponds to a Jordan block.

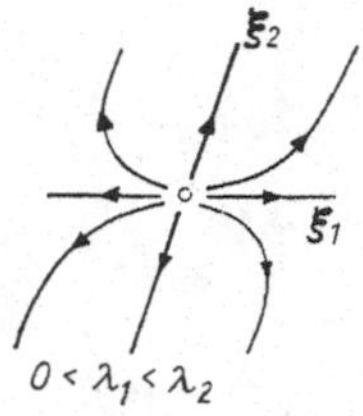

Fig. 135. An unstable node

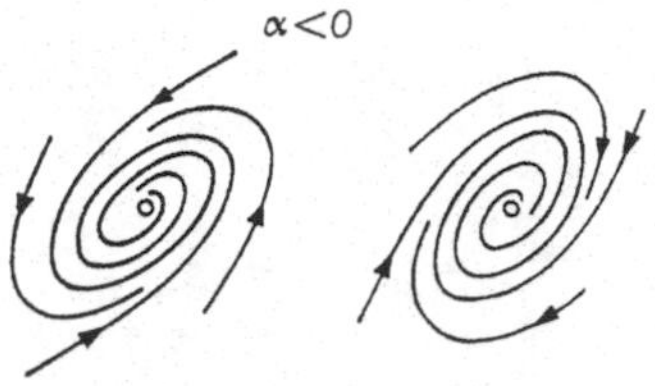

Fig. 136. Stable foci

5. Example: The Pendulum with Friction

Let us apply what has just been said to the equation of small oscillations of a pendulum with friction, $\ddot{x} = -x - k\dot{x}$ (k is the coefficient of friction). We form the equivalent system

$$\dot{x}_1 = x_2, \quad \dot{x}_2 = -x_1 - kx_2.$$

Let us study the characteristic equation. The matrix of the system

$$\begin{pmatrix} 0 & 1 \\ -1 & -k \end{pmatrix}$$

has determinant 1 and trace $-k$. The roots of the characteristic equation $\lambda^2 + k\lambda + 1 = 0$ are complex for $|k| < 2$, i.e., when the friction is not too large[11].

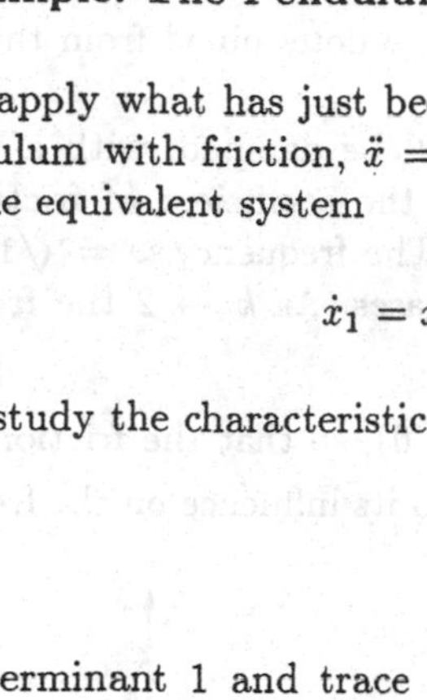

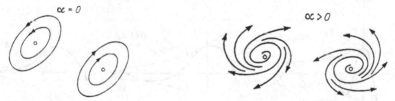

Fig. 137. Centers

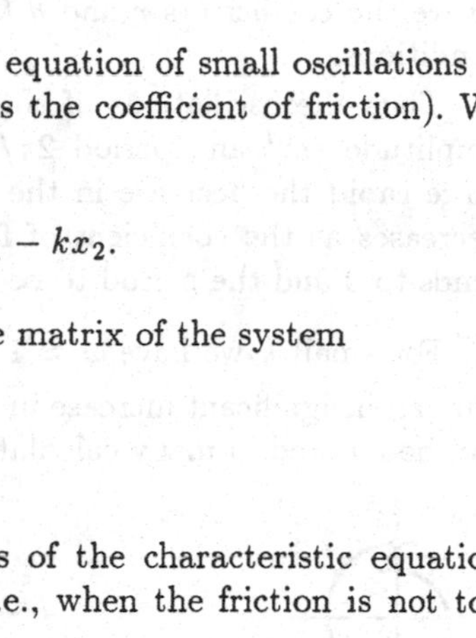

Fig. 138. Unstable foci

The real part of each of the complex roots $\lambda_{1,2} = \alpha \pm i\omega$ is $-k/2$. In other words *for a positive, not too large, coefficient of friction ($0 < k < 2$) the lower equilibrium position of the pendulum ($x_1 = x_2 = 0$) is a stable focus.*

As $k \to 0$ the focus changes into a center: the smaller the coefficient of friction, the more slowly the phase point approaches the equilibrium position as $t \to +\infty$ (Fig. 139). The explicit formulas for the variation of $x_1 = x$ with time are obtained from Corollary 2 of Sect. 3 and the formulas of Sect. 4 of § 19:

$$x(t) = r^{\alpha t}\cos(\theta - \omega t) = Ae^{\alpha t}\cos\omega t + Be^{\alpha t}\sin\omega t,$$

[11] The case of real roots was studied in § 17, Sect. 2.

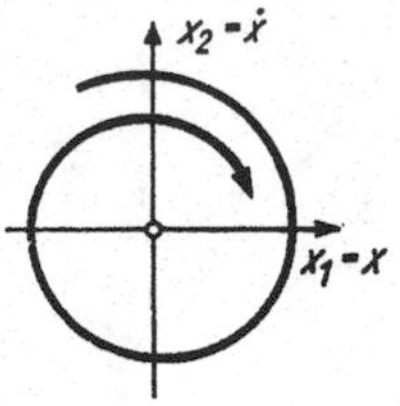

Fig. 139. The phase plane of a pendulum with small friction

where the coefficients r and θ (or A and B) are determined from the initial conditions.

Thus the oscillations of the pendulum will be damped, with a variable amplitude $re^{\alpha t}$ and period $2\pi/\omega$. The larger the coefficient of friction, the more rapid the decrease in the amplitude[12]. The frequency $\omega = \sqrt{1 - k^2/4}$ decreases as the coefficient of friction k increases. As $k \to 2$ the frequency tends to 0 and the period to ∞ (Fig. 140).

For small k we have $\omega \approx 1 - \dfrac{k^2}{8}$ (as $k \to 0$), so that the friction causes only an insignificant increase in the period, and its influence on the frequency can be ignored in many calculations.

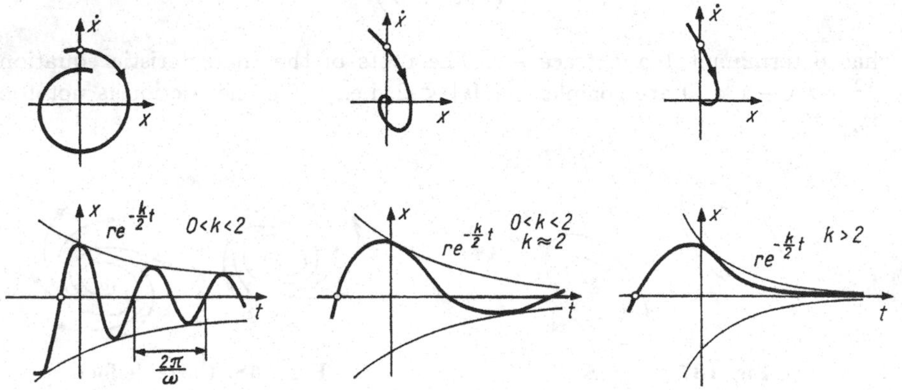

Fig. 140. The transition from damped oscillations to a nonoscillatory motion of the pendulum: the phase curves and the graphs of the solutions for three values of the coefficient of friction

Problem 1. Sketch the phase curves of the unlinearized pendulum with friction, $\ddot{x} = -\sin x - k\dot{x}$ (Fig. 141).

Hint. Calculate the derivative of the total energy along a phase curve.

[12] And yet for any value $k < 2$ the pendulum will perform an infinite number of swings. But if $k > 2$, the pendulum makes at most one change of direction.

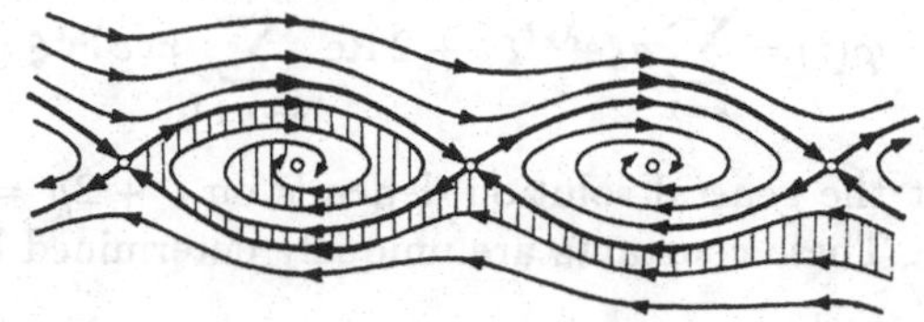

Fig. 141. After several revolutions the pendulum begins to swing around the lower equilibrium position

6. The General Solution of a Linear Equation in the Case when the Characteristic Equation Has Only Simple Roots

We already know that every solution φ of the complexified equation is a linear combination of exponentials (cf. § 19, Sect. 5):

$$\varphi(t) = \sum_{k=1}^{n} c_k e^{\lambda_k t} \xi_k,$$

where ξ_k is an eigenvector corresponding to the eigenvalue λ_k. *We shall choose real eigenvectors for real eigenvalues, and complex conjugate pairs of eigenvectors corresponding to complex conjugate pairs of eigenvalues.*

We already know that the solutions of a real equation are solutions of its complexification with real initial conditions. A necessary and sufficient condition for the vector $\varphi(0)$ to be real is that

$$\sum_{k=1}^{n} c_k \xi_k = \sum_{k=1}^{n} \bar{c}_k \bar{\xi}_k.$$

For this to happen *the coefficients of complex conjugate pairs of vectors must be complex conjugates and those of real vectors must be real.*

We remark that the n complex constants c_k (for a fixed choice of eigenvectors) are determined uniquely by the solution of the complex equation. We have thus proved the following theorem.

Theorem. *Every solution of a real equation can be uniquely written (for a fixed choice of eigenvectors) in the form*

$$\varphi(t) = \sum_{k=1}^{\nu} a_k e^{\lambda_k t} \xi_k + \sum_{k=\nu+1}^{\nu+\mu} c_k e^{\lambda_k t} \xi_k + \bar{c}_k e^{\bar{\lambda}_k t} \bar{\xi}_k, \tag{1}$$

where a_k are real constants and c_k are complex constants.

Formula (1) is called the *general solution* of the equation. It can be rewritten in the form

$$\varphi(t) = \sum_{k=1}^{\nu} a_k e^{\lambda_k t}\xi_k + 2\,\mathrm{Re}\sum_{k=\nu+1}^{\nu+\mu} c_k e^{\lambda_k t}\xi_k.$$

We remark that the general solution depends on $\nu + 2\mu = n$ real constants a_k, $\mathrm{Re}\, c_k$, $\mathrm{Im}\, c_k$. These constants are uniquely determined by the initial conditions.

Corollary 1. *Let $\varphi = (\varphi_1, \ldots, \varphi_n)$ be a solution of a system of n first-order real linear differential equations with the matrix A. Suppose all the roots of the characteristic equation of the matrix A are simple. Then each of the functions φ_m is a linear combination of the functions $e^{\lambda_k t}$, $e^{\alpha_k t}\cos\omega_k t$, and $e^{\alpha_k t}\sin\omega_k t$, where λ_k are the real roots and $\alpha_k \pm i\omega_k$ the complex roots of the characteristic equation.*

Proof. We expand the general solution of (1) in a coordinate basis: $\varphi = \varphi_1 \mathbf{e}_1 + \cdots + \varphi_n \mathbf{e}_n$. Taking into account the relation $e^{(\alpha_k \pm i\omega_k)t} = e^{\alpha_k t}(\cos\omega_k t \pm i\sin\omega_k t)$, we obtain the required result. □

In the practical solution of linear systems, having found the eigenvalues, one may seek the solution in the form of a linear combination of the functions $e^{\lambda_k t}$, $e^{\alpha_k t}\cos\omega_k t$, and $e^{\alpha_k t}\sin\omega_k t$ by the method of undetermined coefficients.

Corollary 2. *Let A be a real square matrix having no multiple eigenvalues. Then each of the elements of the matrix e^{At} is a linear combination of the functions $e^{\lambda_k t}$, $e^{\alpha_k t}\cos\omega_k t$, $e^{\alpha_k t}\sin\omega_k t$, where λ_k are the real roots and $\alpha_k \pm i\omega_k$ the complex roots of the characteristic equation of A.*

Proof. Each column of the matrix e^{At} is composed of the coordinates of the image of a basis vector under the action of a phase flow of the system of differential equations with matrix A. □

Remark. Everything that has been said carries over immediately to equations and systems of equations of order higher than 1, since the latter reduce to systems of first order (cf. § 8).

Problem 1. Find all real solutions of the equations $x^{(iv)} + 4x = 0$, $x^{(iv)} = x$, $\dddot{x} + x = 0$.

§ 21. The Classification of Singular Points of Linear Systems

We have seen above that in the general case (when the characteristic equation has no multiple roots) a real linear system decomposes into the direct product of one- and two-dimensional systems. Since we have already studied the one- and two-dimensional systems, we can now study multi-dimensional systems.

1. Example: Singular Points in Three-dimensional Space

The characteristic equation is a real cubic. A real cubic equation may have three real roots or one real and two complex roots. Many different cases are possible, depending on the location of the roots λ_1, λ_2, λ_3 in the plane of the complex variable λ.

We call attention to the order and sign of the real parts. There are 10 "robust" cases (Fig. 142) and a number of "degenerate" cases (cf., for example, Fig. 143), when the real part of one of the roots is zero or equal to the real part of a root that is not its complex conjugate (we are not considering the case of multiple roots at present). The study of the behavior of the phase curves in each of these cases presents no difficulty.

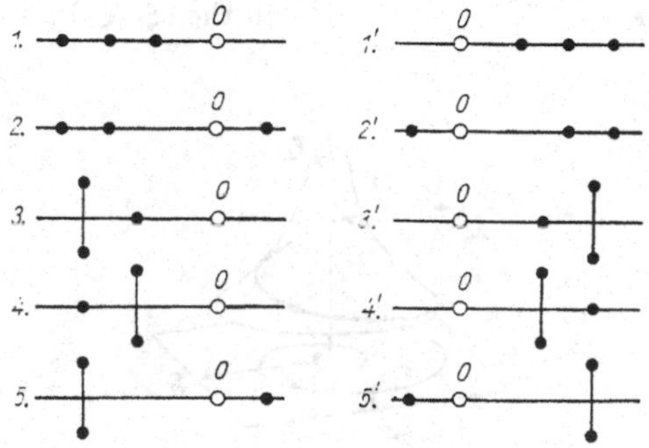

Fig. 142. The eigenvalues of a real operator $A : \boldsymbol{R}^3 \to \boldsymbol{R}^3$. Robust cases

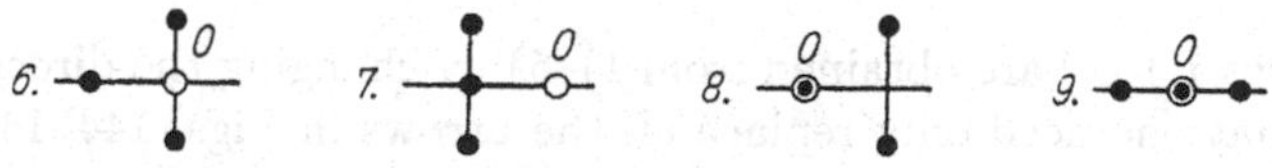

Fig. 143. Some degenerate cases

Taking account of the fact that $e^{\lambda t}$ tends to zero as $t \to +\infty$ (when $\operatorname{Re}\lambda < 0$), and the smaller $\operatorname{Re}\lambda$ the faster the vanishing, we obtain the phase curves depicted in Figs. 144–148:

$$\varphi(t) = \operatorname{Re}(c_1 e^{\lambda_1 t}\boldsymbol{\xi}_1 + c_2 e^{\lambda_2 t}\boldsymbol{\xi}_2 + c_3 e^{\lambda_3 t}\boldsymbol{\xi}_3).$$

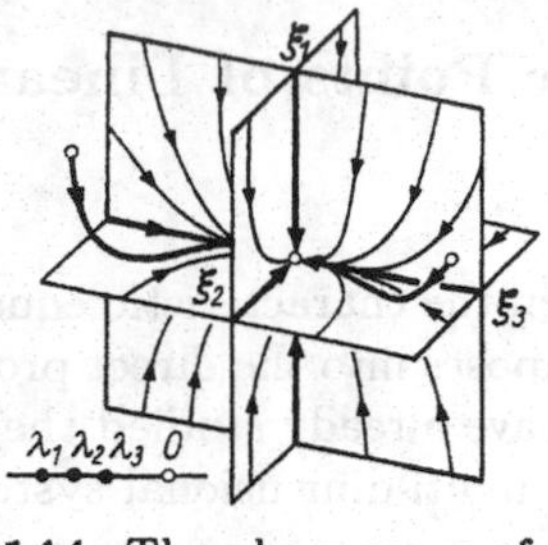

Fig. 144. The phase space of a linear equation in the case $\lambda_1 < \lambda_2 < 0$. The phase flow consists of contractions along three directions

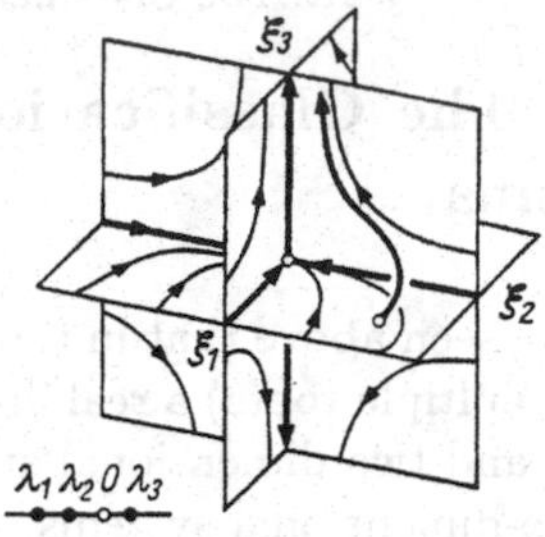

Fig. 145. The case $\lambda_1 < \lambda_2 < 0 < \lambda_3$. Contraction along two directions and dilation along a third

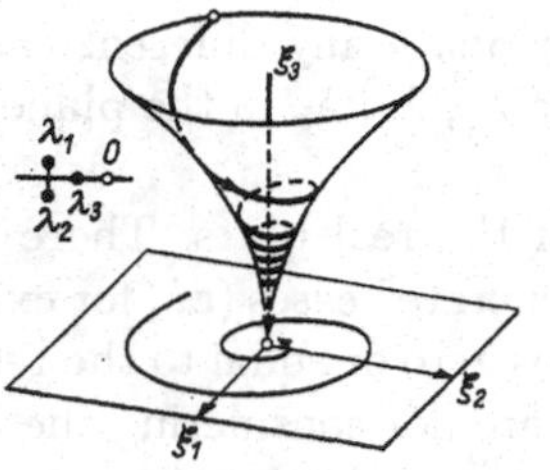

Fig. 146. The case $\operatorname{Re}\lambda_{1,2} < \lambda_3 < 0$. Contraction along the direction of $\boldsymbol{\xi}_3$, rotation with a more rapid contraction in the $(\boldsymbol{\xi}_1, \boldsymbol{\xi}_2)$-plane

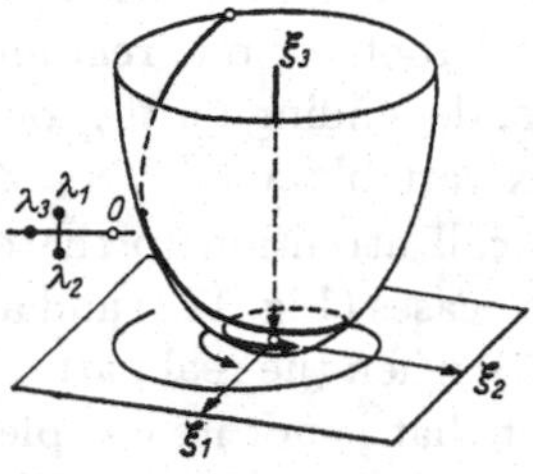

Fig. 147. The case $\lambda_3 < \operatorname{Re}\lambda_{1,2} < 0$. Contraction along the direction of $\boldsymbol{\xi}_3$, rotation with a less rapid contraction in the $(\boldsymbol{\xi}_1, \boldsymbol{\xi}_2)$-plane

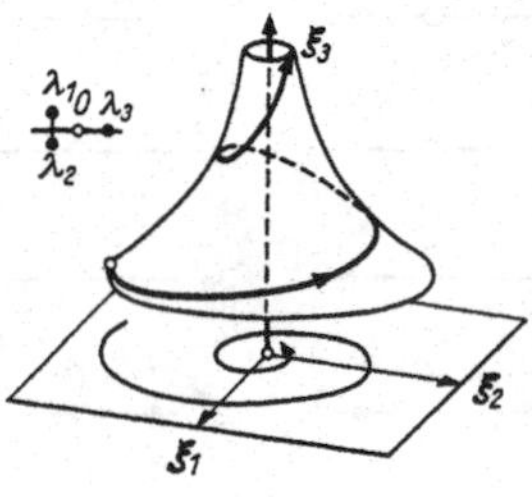

Fig. 148. The case $\operatorname{Re}\lambda_{1,2} < 0 < \lambda_3$. Dilation along the direction of $\boldsymbol{\xi}_3$, rotation with contraction in the $(\boldsymbol{\xi}_1, \boldsymbol{\xi}_2)$-plane

The cases 1′)–5′) are obtained from 1)–5) by changing the direction of the t-axis, so that one need only replace all the arrows in Figs. 144–148 by their opposites.

Problem 1. Sketch the phase curves in cases 6), 7), 8), and 9) of Fig. 143.

2. Linear, Differentiable, and Topological Equivalence

Every classification is based on some equivalence relation. There exist at least three reasonable equivalence relations for linear systems: they correspond to the algebraic, differential and topological approaches.

Let $\{f^t\}, \{g^t\} : \boldsymbol{R}^n \to \boldsymbol{R}^n$ be phase flows.

Definition. The flows $\{f^t\}$ and $\{g^t\}$ are *equivalent*[13] if there exists a one-to-one onto mapping $h : \boldsymbol{R}^n \to \boldsymbol{R}^n$ taking the flow $\{f^t\}$ to the flow $\{g^t\}$, so that $h \circ f^t = g^t \circ h$ for any $t \in \boldsymbol{R}$ (Fig. 149). We can say that the flow $\{f^t\}$ becomes $\{g^t\}$ under the change of coordinates h.

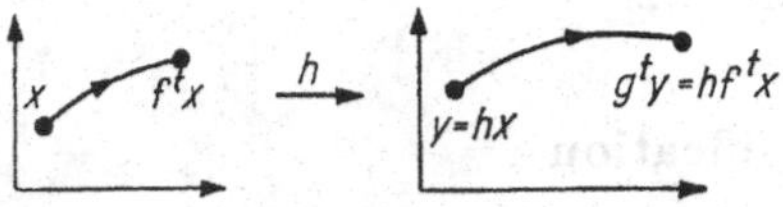

Fig. 149. Equivalent flows

In this context, flows are called:

1) *linearly equivalent* if there exists such a mapping $h : \boldsymbol{R}^n \to \boldsymbol{R}^n$ that is a *linear isomorphism*, $h \in \mathrm{GL}(\boldsymbol{R}^n)$;

2) *differentiably equivalent* if there exists such a mapping $h : \boldsymbol{R}^n \to \boldsymbol{R}^n$ that is a *diffeomorphism*;

3) *topologically equivalent* if there exists such a mapping $h : \boldsymbol{R}^n \to \boldsymbol{R}^n$ that is a *homeomorphism*, i.e., a continuous mapping that is one-to-one, onto, and has a continuous inverse.

Problem 1. Prove that linear equivalence implies differentiable equivalence, and that differentiable equivalence implies topological equivalence.

We remark that the mapping h maps the phase curves of the flow $\{f^t\}$ into the phase curves of the flow $\{g^t\}$.

Problem 2. Does every linear automorphism $h \in \mathrm{GL}(\boldsymbol{R}^n)$ that maps the phase curves of the flow $\{f^t\}$ into the phase curves of the flow $\{g^t\}$ realize a linear equivalence of flows?

Answer. No.

[13] The equivalence relation introduced here is also called *conjugacy* and *similarity*.

Hint. Consider $n = 1$, $f^t x = e^t x$, $g^t x = e^{2t} x$.

Problem 3. Prove that the relations of linear, differentiable, and topological equivalence are genuine equivalence relations, i.e.

$$f \sim f, \quad (f \sim g) \Rightarrow (g \sim f), \quad (f \sim g, g \sim k) \Rightarrow (f \sim k).$$

In particular everything that has been said applies to the phase flows of linear systems. For brevity we shall speak about equivalence of the systems themselves.

Thus we have partitioned all linear systems into equivalence classes by three different methods (linear, differentiable, and topological). Let us study these classes in more detail.

3. The Linear Classification

Theorem. *Let $A, B : \boldsymbol{R}^n \to \boldsymbol{R}^n$ be linear operators all of whose eigenvalues are simple. Then the systems*

$$\dot{\boldsymbol{x}} = A\boldsymbol{x}, \quad \boldsymbol{x} \in \boldsymbol{R}^n, \text{ and } \dot{\boldsymbol{y}} = B\boldsymbol{y}, \quad \boldsymbol{y} \in \boldsymbol{R}^n,$$

are linearly equivalent if and only if the eigenvalues of the operators A and B coincide.

Proof. A necessary and sufficient condition for equivalence of the linear systems is that $B = hAh^{-1}$ for some $h \in \mathrm{GL}(\boldsymbol{R}^n)$ (Fig. 150) (for $\dot{\boldsymbol{y}} = h\dot{\boldsymbol{x}} = hA\boldsymbol{x} = hAh^{-1}\boldsymbol{y}$). The eigenvalues of the operators A and hAh^{-1} coincide. (It is not essential here that the eigenvalues be simple.)

Conversely, suppose the eigenvalues of A are simple and coincide with those of B. Then A and B decompose into the direct products of the same number of (linearly equivalent) one- and two-dimensional systems, according to § 20; therefore they are linearly equivalent. □

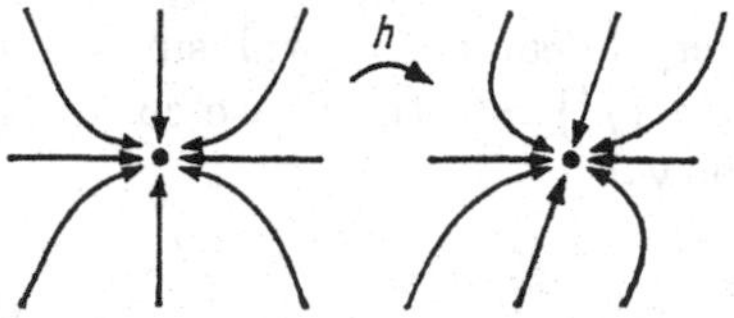

Fig. 150. Linearly equivalent systems

Problem 1. Show that the systems $\dot{x}_1 = x_1$, $\dot{x}_2 = x_2$, and $\dot{x}_1 = x_1 + x_2$, $\dot{x}_2 = x_2$ are linearly inequivalent, even though they have the same eigenvalues.

4. The Differentiable Classification

The following fact is obvious.

Theorem. *Two linear systems*

$$\dot{\boldsymbol{x}} = A\boldsymbol{x}, \quad \dot{\boldsymbol{x}} = B\boldsymbol{x}, \quad \boldsymbol{x} \in \boldsymbol{R}^n,$$

are differentiably equivalent if and only if they are linearly equivalent[14].

Proof. Let $h : \boldsymbol{R}^n \to \boldsymbol{R}^n$ be a diffeomorphism that maps the phase flow of the system A to the phase flow of the system B. The point $\boldsymbol{x} = \mathbf{o}$ is fixed for the phase flow of the system A. Therefore h maps $\mathbf{o}$ to one of the fixed points $\boldsymbol{c}$ of the flow of the system B, so that $B\boldsymbol{c} = \mathbf{o}$. The diffeomorphism $d : \boldsymbol{R}^n \to \boldsymbol{R}^n$ of translation by $\boldsymbol{c}$ $(d\boldsymbol{x} = \boldsymbol{x} - \boldsymbol{c})$ maps the phase flow of B into itself: $((\boldsymbol{x}-\boldsymbol{c})^{\cdot} = \dot{\boldsymbol{x}} = B\boldsymbol{x} = B(\boldsymbol{x}-\boldsymbol{c}))$. The diffeomorphism $h_1 = d \circ h : \boldsymbol{R}^n \to \boldsymbol{R}^n$ maps the flow of A to the flow of B and leaves $\mathbf{o}$ fixed: $h_1(\mathbf{o}) = \mathbf{o}$.

Let us denote by $H : \boldsymbol{R}^n \to \boldsymbol{R}^n$ the derivative of the diffeomorphism h_1 at $\mathbf{o}$. The diffeomorphisms $h_1 \circ e^{At} = e^{Bt} \circ h_1$ coincide for any t. Therefore for any t their derivatives also coincide at $\boldsymbol{x} = \mathbf{o}$:

$$He^{At} = e^{Bt}H,$$

which was to be proved. □

§ 22. The Topological Classification of Singular Points

Consider the two linear systems:

$$\dot{\boldsymbol{x}} = A\boldsymbol{x}, \quad \dot{\boldsymbol{x}} = B\boldsymbol{x}, \quad \boldsymbol{x} \in \boldsymbol{R}^n,$$

and suppose that the real parts of all their eigenvalues are nonzero. We shall denote by m_- the number of eigenvalues with negative real part and by m_+ the number of eigenvalues with positive real part, so that $m_- + m_+ = n$.

1. Theorem

A necessary and sufficient condition for topological equivalence of two linear systems having no eigenvalues with real part zero is that the number of eigenvalues with negative (or positive) real part be the same for both systems:

$$m_-(A) = m_-(B), \quad m_+(A) = m_+(B).$$

This theorem asserts, for example, that stable nodes and foci (Fig. 151) are topologically equivalent to each other ($m_- = 2$) but not equivalent to a saddle point ($m_- = m_+ = 1$).

[14] It should not be thought that every diffeomorphism that establishes the equivalence is linear. For example, consider $A = B = 0$.

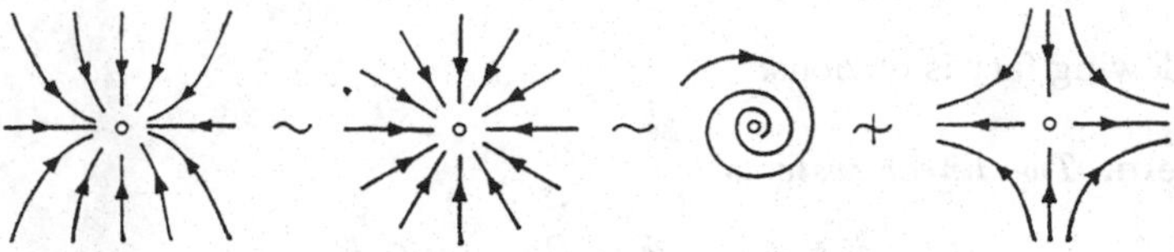

Fig. 151. Topologically equivalent and inequivalent systems

Like the signature of a nondegenerate quadratic form, the number m_- is the only topological invariant of the system.

Remark. An analogous proposition holds *locally* (in a neighborhood of a fixed point) for *nonlinear* systems whose linear parts have no purely imaginary eigenvalues. In particular in a neighborhood of a fixed point such a system is topologically equivalent to its linear part (Fig. 152). We cannot take the time to prove this proposition, which is quite important for the study of nonlinear systems.

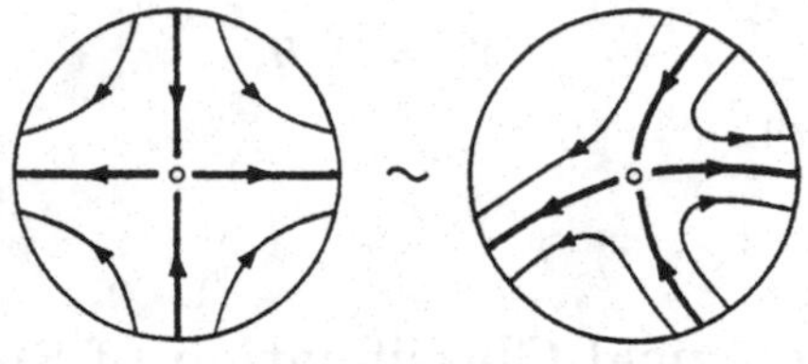

Fig. 152. Topological equivalence of a system and its linearization

2. Reduction to the Case $m_- = 0$

The topological equivalence of linear systems with the same m_- and m_+ follows from the following three lemmas:

Lemma 1. *The direct products of topologically equivalent systems are topologically equivalent.*

That is, if the systems given by the operators $A_1, B_1 : \boldsymbol{R}^{m_1} \to \boldsymbol{R}^{m_1}$ and $A_2, B_2 : \boldsymbol{R}^{m_2} \to \boldsymbol{R}^{m_2}$ map into each other by homeomorphisms $h_1 : \boldsymbol{R}^{m_1} \to \boldsymbol{R}^{m_1}$ and $h_2 : \boldsymbol{R}^{m_2} \to \boldsymbol{R}^{m_2}$, there exists a homeomorphism $h : \boldsymbol{R}^{m_1} + \boldsymbol{R}^{m_2} \to \boldsymbol{R}^{m_1} + \boldsymbol{R}^{m_2}$ mapping a phase flow of the product-system $\dot{\boldsymbol{x}}_1 = A_1\boldsymbol{x}_1$, $\dot{\boldsymbol{x}}_2 = A_2\boldsymbol{x}_2$ to the phase flow of the product-system $\dot{\boldsymbol{x}}_1 = B_1\boldsymbol{x}_1$, $\dot{\boldsymbol{x}}_2 = B_2\boldsymbol{x}_2$.

The proof is obvious: one may set $\boldsymbol{h}(\boldsymbol{x}_1, \boldsymbol{x}_2) = (h_1(\boldsymbol{x}_1), h_2(\boldsymbol{x}_2))$.

The following lemma is known from linear algebra.

Lemma 2. *If the operator $A : \boldsymbol{R}^n \to \boldsymbol{R}^n$ has no purely imaginary eigenvalues, then the space $\boldsymbol{R}^n$ decomposes into the direct sum of two A-invariant*

subspaces, $\boldsymbol{R}^n = \boldsymbol{R}^{m_-} \dotplus \boldsymbol{R}^{m_+}$, *so that all the eigenvalues of the restriction of* A *to* $\boldsymbol{R}^{m_-}$ *have negative real part and those of the restriction of* A *to* $\boldsymbol{R}^{m_+}$ *have positive real part.* (Fig. 153).

This follows, for example from the Jordan normal form.

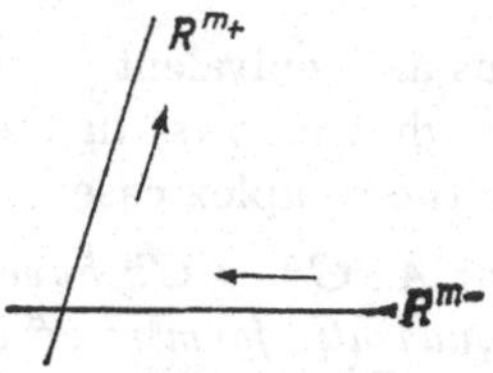

Fig. 153. Invariant subspaces of an operator having no purely imaginary eigenvalues

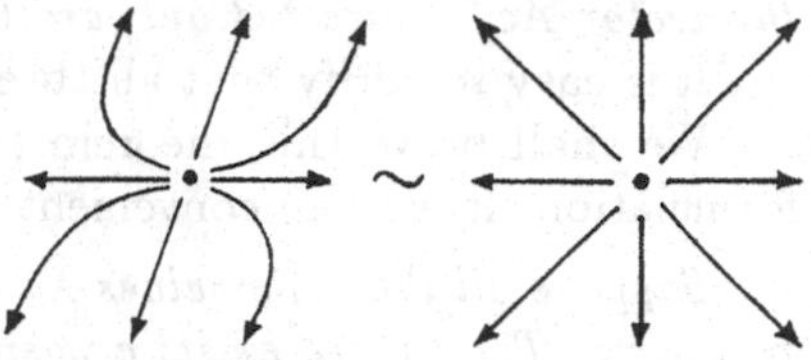

Fig. 154. All unstable nodes are topologically equivalent

Lemmas 1 and 2 reduce the proof of topological equivalence to the following special case:

Lemma 3. *Let* $A : \boldsymbol{R}^n \to \boldsymbol{R}^n$ *be a linear operator, all of whose eigenvalues have positive real part* (Fig. 154). *Then the system*

$$\dot{\boldsymbol{x}} = A\boldsymbol{x}, \quad \boldsymbol{x} \in \boldsymbol{R}^n,$$

is topologically equivalent to the standard system (Fig. 154):

$$\dot{\boldsymbol{x}} = \boldsymbol{x}, \quad \boldsymbol{x} \in \boldsymbol{R}^n.$$

This lemma is almost obvious in the one-dimensional case and in the case of a focus in the plane, hence also – by Lemma 1 – in any system having no multiple roots.

Let us now carry out the proof of Lemma 3 in the general case.

3. The Lyapunov Function

The proof of Lemma 3 is based on the construction of a special quadratic function – the so-called *Lyapunov function.*

Theorem. *Let* $A : \boldsymbol{R}^n \to \boldsymbol{R}^n$ *be a linear operator all of whose eigenvalues have positive real part. Then there exists a Euclidean structure on* $\boldsymbol{R}^n$ *such that the vector* $A\boldsymbol{x}$ *forms an acute angle with the radius-vector* $\boldsymbol{x}$ *at each point* $\boldsymbol{x} \neq \mathbf{o}$.

In other words:

There exists a positive-definite quadratic form r^2 *on* $\boldsymbol{R}^n$ *whose derivative in the direction of the vector field* $A\boldsymbol{x}$ *is positive:*

$$L_{Ax} r^2 > 0 \text{ for } x \neq \mathrm{o}. \tag{1}$$

Or again:

There exists an ellipsoid in $\boldsymbol{R}^n$ with center at o *such that at each point $\boldsymbol{x}$ the vector $A\boldsymbol{x}$ is directed outward* (Fig. 155).

It is easy to verify that all three formulations are equivalent.

We shall prove this theorem (and use it in what follows) in the second formulation. It is more convenient to prove it in the complex case:

Suppose all the eigenvalues λ_k of the operator $A : \boldsymbol{C}^n \to \boldsymbol{C}^n$ have positive real parts. Then there exists a positive-definite quadratic form $r^2 : {}^{\boldsymbol{R}}\boldsymbol{C}^n \to \boldsymbol{R}$ whose derivative in the direction of the vector field ${}^{\boldsymbol{R}}A\boldsymbol{z}$ is a positive-definite quadratic form:

$$L_{{}^{\boldsymbol{R}}A\boldsymbol{z}} r^2 > 0 \text{ for } \boldsymbol{z} \neq \mathrm{o}. \tag{2}$$

Applying inequality (2) in the case when the operator A is the complexification of a real operator and $\boldsymbol{z}$ belongs to the real subspace (Fig. 156), we obtain the real theorem (1). □

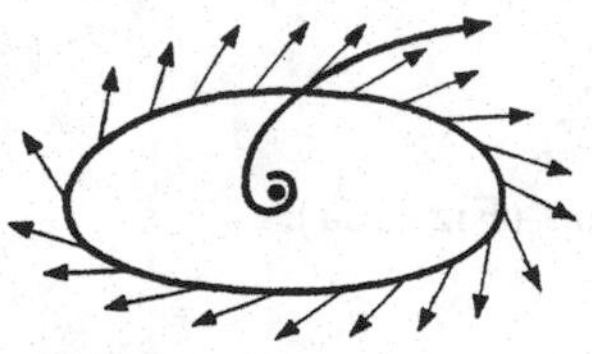

Fig. 155. The level surface of a Lyapunov function

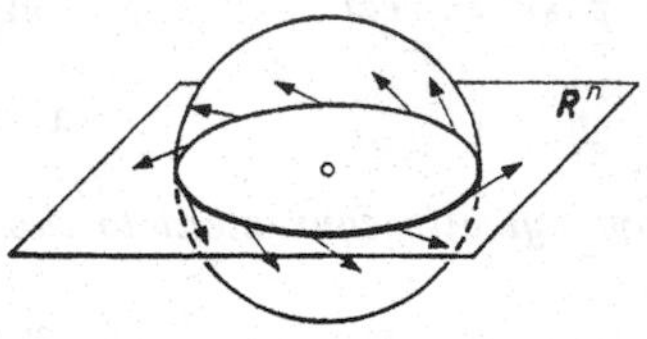

Fig. 156. The level surface of a Lyapunov function in $\boldsymbol{C}^n$

4. Construction of the Lyapunov Function

As the Lyapunov function r^2 we shall take the sum of the squares of the absolute values of the coordinates in a suitable complex basis: $r^2 = (\boldsymbol{z}, \overline{\boldsymbol{z}}) = \sum_{k=1}^{n} z_k \overline{z}_k$. For a fixed basis we can identify the vector $\boldsymbol{z}$ with the set of numbers $z_1, \ldots, z_n$ and the operator $A : \boldsymbol{C}^n \to \boldsymbol{C}^n$ with the matrix (a_{kl}). Computation shows that *the derivative is a quadratic form*

$$L_{{}^{\boldsymbol{R}}A\boldsymbol{z}}(\boldsymbol{z}, \overline{\boldsymbol{z}}) = (A\boldsymbol{z}, \overline{\boldsymbol{z}}) + (\boldsymbol{z}, \overline{A\boldsymbol{z}}) = 2\mathrm{Re}\,(A\boldsymbol{z}, \overline{\boldsymbol{z}}). \tag{3}$$

If the basis is proper, then the form obtained is positive-definite (Fig. 157). Indeed, in this case

$$2\,\mathrm{Re}\,(A\boldsymbol{z}, \overline{\boldsymbol{z}}) = 2\sum_{k=1}^{n} \mathrm{Re}\,\lambda_k |z_k|^2. \tag{4}$$

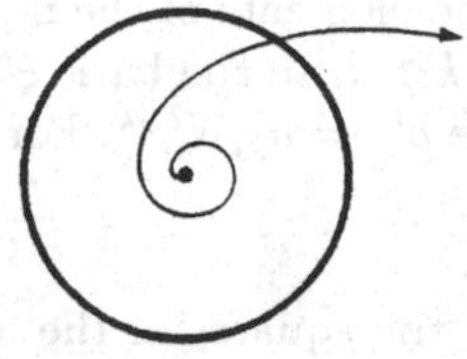

Fig. 157. The positive-definiteness of the form (4) in the case $n = 1$

By hypothesis all the real parts of the eigenvalues λ_k are positive. Therefore the form (4) is positive-definite.

If the operator A has no proper basis, it has an almost proper basis that can be used with equal success to construct a Lyapunov function.

More precisely, we have the following lemma.

Lemma 4. *Let $A : \mathbf{C}^n \to \mathbf{C}^n$ be a $\mathbf{C}$-linear operator and $\varepsilon > 0$. Then a basis $\xi_1, \ldots, \xi_n$ of $\mathbf{C}^n$ can be chosen in which the matrix of A is upper triangular and all the elements above the diagonal are less than ε in absolute value:*

$$(A) = \begin{pmatrix} \lambda_1 & & < \varepsilon \\ & \ddots & \\ 0 & & \lambda_n \end{pmatrix}.$$

Proof. The existence of a basis in which the matrix is upper-triangular follows, for example, from the Jordan normal form.

Such a basis is easily constructed by induction on n using only the existence of an eigenvector for every linear operator $A : \mathbf{C}^n \to \mathbf{C}^n$. Let ξ_1 be this vector (Fig. 158). Consider the quotient space $\mathbf{C}^n/\mathbf{C}\xi_1 \cong \mathbf{C}^{n-1}$. The operator A defines an operator $\tilde{A} : \mathbf{C}^{n-1} \to \mathbf{C}^{n-1}$ on the quotient space. Let $\eta_2, \ldots \eta_n$ be a basis of $\mathbf{C}^{n-1}$ in which the matrix of the operator $\tilde{A}$ is upper triangular. We denote by $\xi_2, \ldots, \xi_n$ any representatives of the classes $\eta_2, \ldots, \eta_n$ in $\mathbf{C}^n$. Then the basis $\xi_1, \ldots, \xi_n$ is the one required.

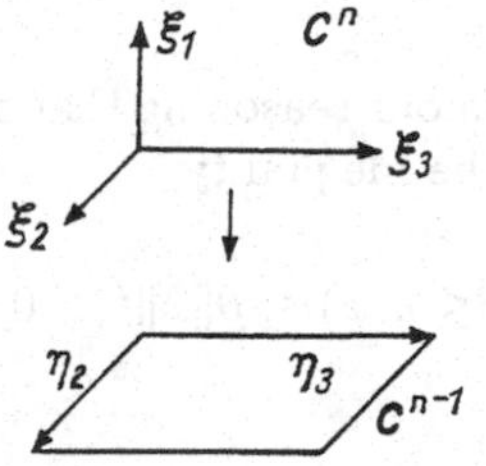

Fig. 158. Construction of a basis in which the matrix of an operator is triangular

Suppose the matrix of the operator A is upper triangular in the basis $\xi_1, \ldots, \xi_n$. We shall show that *the elements above the diagonal can be made arbitrarily small by replacing the vectors of the basis with vectors proportional to them.* Indeed let a_{kl} be the elements of the matrix of the operator A in the basis ξ_k, so that $a_{kl} = 0$ for $k > l$. In the basis $\xi'_k = N^k \xi_k$ the elements of the matrix of the operator A are $a'_{kl} = a_{kl} N^{l-k}$. For sufficiently small N we shall have $|a'_{kl}| < \varepsilon$ for all $l > k$.

Lemma 4 is now proved. □

We shall use the sum of the squares of the absolute values of the coordinates in the "ε-almost proper" basis as the Lyapunov function (for sufficiently small ε).

5. An Estimate of the Derivative

Consider the set of all quadratic forms on $\boldsymbol{R}^m$. This set has a natural vector space structure as $\boldsymbol{R}^{\frac{m(m+1)}{2}}$.

The following lemma is obvious.

Lemma 5. *The set of positive-definite quadratic forms on $\boldsymbol{R}^m$ is open in $\boldsymbol{R}^{\frac{m(m+1)}{2}}$.*

That is, if the form $a = \sum_{k,l=1}^{m} a_{kl} x_k x_l$ is positive-definite, there exists $\varepsilon > 0$ such that every form $a + b$ with $|b_{kl}| < \varepsilon$ (for all k, l with $1 \leq k, l \leq m$) is also positive-definite.

Proof. The form a is positive at all points of the unit sphere $\sum_{k=1}^{m} x_k^2 = 1$. The sphere is compact and the form is continuous. Therefore the lower bound is attained, and so $a(x) \geq \alpha > 0$ everywhere on the sphere.

If $|b_{kl}| < \varepsilon$, then $|b(x)| \leq \sum |b_{kl}| \leq m^2 \varepsilon$ on the sphere.

Therefore for $\varepsilon < \alpha/m^2$ the form $a + b$ is positive on the sphere and hence positive-definite. The lemma is now proved. □

Remark. It also follows from our reasoning that any positive-definite quadratic form satisfies everywhere the inequality

$$\alpha \|x\|^2 \leq a(x) \leq \beta \|x\|^2, \quad 0 < \alpha < \beta. \tag{5}$$

Problem 1. Prove that the set of nondegenerate quadratic forms of a given signature is open.

Example 1. The space of quadratic forms of two variables $ax^2 + 2bxy + cy^2$ is a three-dimensional space with coordinates a, b, and c (Fig. 159). The cone $b^2 = ac$ divides this space into three open parts corresponding to the signatures.

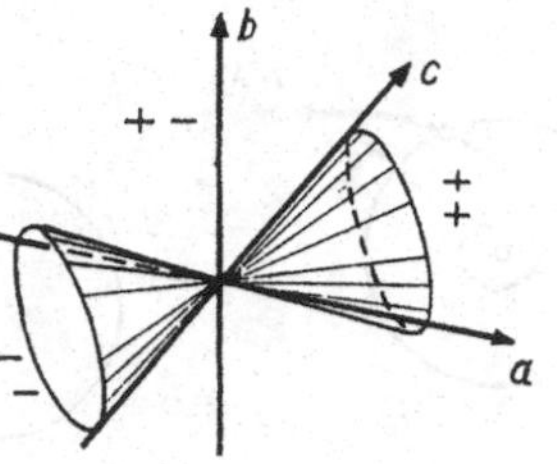

Fig. 159. The space of quadratic forms

We shall use Lemma 5 to prove the following: for sufficiently small ε the derivative of the sum of the squares of the absolute values of the coordinates in the ε-almost proper basis chosen in Lemma 4 in the direction of the vector field ${}^{R}Az$ is positive-definite.

According to formula (3) this derivative is the quadratic form of the real and imaginary parts of the coordinates $z_k = x_k + iy_k$.

Let us display the terms in formula (3) with diagonal and superdiagonal elements of the matrix (A):

$$L_{{}^{R}Az} r^2 = P + Q, \text{ where } P = 2\,\mathrm{Re} \sum_{k=l} a_{kl} z_k \overline{z}_l, \quad Q = 2\,\mathrm{Re} \sum_{k<l} a_{kl} z_k \overline{z}_l.$$

We remark that the diagonal terms of the triangular matrix (A) are the eigenvalues λ_k of the operator A. Therefore *the quadratic form* $P = \sum_{k=1}^{n} 2\,\mathrm{Re}\,\lambda_k (x_k^2 + y_k^2)$ *of the variables* x_k *and* y_k *is positive-definite and independent of the choice of the basis*[15].

By Lemma 5 we conclude that for sufficiently small ε the form $P+Q$ (near P) is also positive-definite. For the coefficients of the variables x_k and y_k in the form Q become arbitrarily small for sufficiently small ε (since $|a_{kl}| < \varepsilon$ for $k < l$).

Inequality (2), and with it (1) also, is now proved.

Remark. Since $L_{Ax} r^2$ is a positive-definite quadratic form, an inequality of the form (5) holds:

$$\alpha r^2 \leq L_{Ax} r^2 \leq \beta r^2, \qquad (5')$$

where $\beta > \alpha > 0$ are constants.

Thus the theorem on the Lyapunov function stated in Sect. 3 is now proved. □

The following series of problems leads to another proof of this theorem.

Problem 2. Prove that differentiation in the direction of the vector field Ax in $\boldsymbol{R}^n$ defines a linear operator $L_A : \boldsymbol{R}^{n(n+1)/2} \to \boldsymbol{R}^{n(n+1)/2}$ from the space of quadratic forms on $\boldsymbol{R}^n$ into itself.

---

[15] It should be noted that the mapping ${}^{R}\mathbb{C}^n \to \boldsymbol{R}$ defined by the form P *does depend* on the choice of the basis.

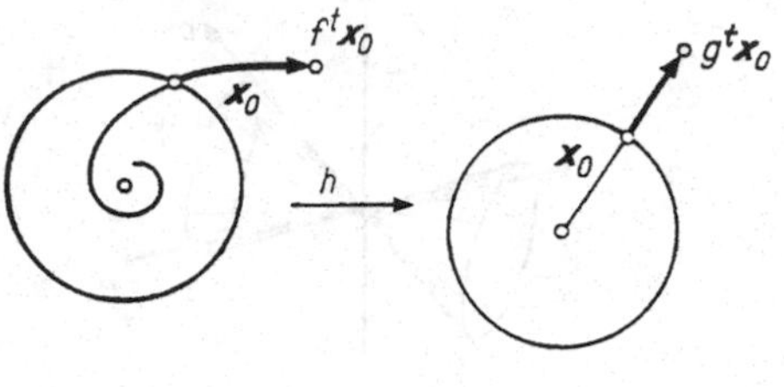

Fig. 160. Construction of the homeomorphism h

Problem 3. Knowing the eigenvalues λ_i of the operator A, find the eigenvalues of the operator L_A.

Answer. $\lambda_i + \lambda_j,\ 1 \le i, j \le n$.

Hint. Let A have a proper basis. Then the eigenvectors of L_A are the quadratic forms equal to the pairwise products of the linear forms that are the eigenvectors of the operator dual to A.

Problem 4. Prove that the operator L_A is an isomorphism if no two of the eigenvalues of A are negatives of each other. In particular, if the real parts of all the eigenvalues of A have the same sign, then every quadratic form on $\boldsymbol{R}^n$ is the derivative of some quadratic form in the direction of the vector field $A\boldsymbol{x}$.

Problem 5. Prove that if the real parts of all the eigenvalues of the operator A are positive, then a form whose derivative in the direction of the field $A\boldsymbol{x}$ is positive-definite is itself positive-definite, and consequently satisfies all the hypotheses of the theorem just proved.

Hint. Represent the form as the integral of its derivative along the phase curves.

6. Construction of the Homeomorphism h

We now proceed to the proof of Lemma 3. A homeomorphism $h : \boldsymbol{R}^n \to \boldsymbol{R}^n$ mapping the phase flow $\{f^t\}$ of the equation $\dot{\boldsymbol{x}} = A\boldsymbol{x}$ ($\operatorname{Re} \lambda_k > 0$) to the phase flow $\{g^t\}$ of the equation $\dot{\boldsymbol{x}} = \boldsymbol{x}$ will be constructed as follows. Consider the sphere[16]

$$S = \{\boldsymbol{x} \in \boldsymbol{R}^n : r^2(\boldsymbol{x}) = 1\},$$

where r^2 is the Lyapunov function of (1).

The homeomorphism h will leave fixed the points of this sphere. Let $\boldsymbol{x}_0$ be a point of the sphere (Fig. 160). The mapping h will map the point $f^t\boldsymbol{x}_0$ of the phase trajectory of the equation $\dot{\boldsymbol{x}} = A\boldsymbol{x}$ to the point $g^t\boldsymbol{x}_0$ of the phase trajectory of the equation $\dot{\boldsymbol{x}} = \boldsymbol{x}$:

$$h(f^t\boldsymbol{x}_0) = g^t\boldsymbol{x}_0 \quad \forall t \in \boldsymbol{R}, \quad \boldsymbol{x}_0 \in S, \quad h(0) = 0. \tag{6}$$

[16] Actually an ellipsoid.

We must verify the following:

1) the formula (6) determines the value of h unambiguously at every point $\boldsymbol{x} \in \boldsymbol{R}^n$;

2) the mapping $h^t : \boldsymbol{R}^n \to \boldsymbol{R}^n$ is one-to-one, onto, and continuous and has a continuous inverse;

3) $h \circ f^t = g^t \circ h$.

The proofs of all of these assertions are obvious.

7. Proof of Lemma 3

Lemma 6. *Let $\varphi : \boldsymbol{R}^n \to \boldsymbol{R}^n$ be a nonzero solution of the equation $\dot{\boldsymbol{x}} = A\boldsymbol{x}$. We form a real-valued function of a real variable t:*

$$\rho(t) = \ln r^2(\varphi(t)).$$

Then the mapping $\rho : \boldsymbol{R} \to \boldsymbol{R}$ is a diffeomorphism, and

$$\alpha \leq d\rho/dt \leq \beta.$$

Proof. By the uniqueness theorem $r^2(\varphi(t)) \neq 0$ for all $t \in \boldsymbol{R}$. According to (5′) we find for $d\rho/dt = L_{A\boldsymbol{x}}r^2/r^2$ the estimate $\alpha \leq d\rho/dt \leq \beta$, which was to be proved. □

It follows from Lemma 6 that:

1) *Each point $\boldsymbol{x} \neq \mathbf{o}$ can be represented in the form $\boldsymbol{x} = f^t\boldsymbol{x}_0$, where $\boldsymbol{x}_0 \in S$, $t \in \boldsymbol{R}$, and $\{f^t\}$ is the phase flow for the equation $\boldsymbol{x} = A\boldsymbol{x}$.*

Indeed, consider the solution φ with initial condition $\varphi(0) = \boldsymbol{x}$. By Lemma 6 for some τ we shall have $r^2(\varphi(\tau)) = 1$. The point $\boldsymbol{x}_0 = \varphi(\tau)$ belongs to S. Setting $t = -\tau$, we obtain $\boldsymbol{x} = f^t\boldsymbol{x}_0$.

2) *Such a representation is unique.*

Indeed, the phase curve emanating from $\boldsymbol{x}$ (Fig. 160), is unique and intersects the sphere in a single point $\boldsymbol{x}_0$ (by Lemma 6); the uniqueness of t also follows from the fact that ρ is monotonic (Lemma 6).

Thus we have constructed a one-to-one correspondence between the direct product of the line and the sphere and the Euclidean space with one point removed

$$F : \boldsymbol{R} \times S^{n-1} \to \boldsymbol{R}^n \setminus \mathbf{o}, \quad F(t, \boldsymbol{x}_0) = f^t\boldsymbol{x}_0.$$

It follows from the theorem on dependence of the solution on the initial conditions that both the mapping F and the inverse mapping are continuous (even diffeomorphisms).

We now remark that for the standard equation $\dot{\boldsymbol{x}} = \boldsymbol{x}$ we have $d\rho/dt = 2$. Therefore the mapping $G : \boldsymbol{R} \times S^{n-1} \to \boldsymbol{R}^n \setminus \mathbf{o}$ given by $G(t, \boldsymbol{x}_0) = g^t\boldsymbol{x}_0$ is also a one-to-one correspondence that is continuous in both directions. By definition (6) the mapping h coincides with the mapping $G \circ F^{-1}$: we have proved that $h : \boldsymbol{R}^n \to \boldsymbol{R}^n$ is a one-to-one correspondence.

The continuity of h and h^{-1} everywhere except at the point $\mathbf{o}$ follows from the continuity of F, F^{-1}, and G, G^{-1} (actually h is a diffeomorphism everywhere except at the point $\mathbf{o}$, cf. Fig. 161).

The continuity of h and h^{-1} at the point $\mathbf{o}$ follows from Lemma 6. This lemma even makes it possible to obtain an explicit estimate of $r^2(h(\boldsymbol{x}))$ in terms of $r^2(\boldsymbol{x})$ for $\|\boldsymbol{x}\| \leq 1$:

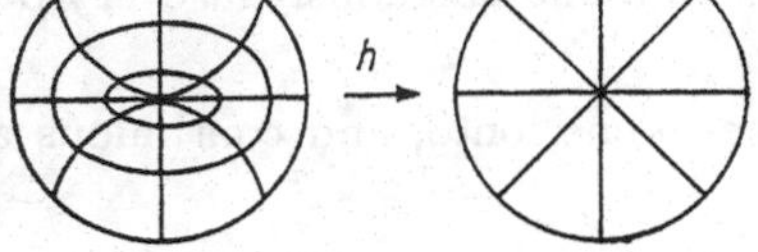

Fig. 161. The homeomorphism h is a diffeomorphism everywhere except at **o**

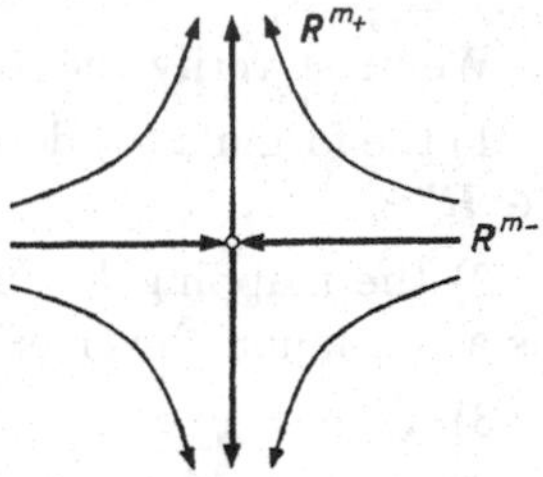

Fig. 162. The standard saddle point

$$(r^2(\boldsymbol{x}))^{2/\alpha} \le r^2(h(\boldsymbol{x})) \le (r^2(\boldsymbol{x}))^{\frac{2}{\beta}}.$$

Indeed, suppose $\boldsymbol{x} = F(t, \boldsymbol{x}_0)$ for $t \le 0$. Then $\beta t \le \ln r^2(\boldsymbol{x}) \le \alpha t$ and $\ln r^2(h(\boldsymbol{x})) = 2t$. Finally for $\boldsymbol{x} \ne \mathbf{o}$ we have $\boldsymbol{x} = f^s \boldsymbol{x}_0$, so that

$$(h \circ f^t)(\boldsymbol{x}) = h(f^t(f^s(\boldsymbol{x}_0))) = h(f^{t+s}(\boldsymbol{x}_0)) = g^{t+s}(\boldsymbol{x}_0) =$$
$$= g^t(g^s(\boldsymbol{x}_0)) = g^t(h(\boldsymbol{x})) = (g^t \circ h)(\boldsymbol{x}).$$

For $\boldsymbol{x} = \mathbf{o}$ we also have $(h \circ f^t)(\boldsymbol{x}) = (g^t \circ h)(\boldsymbol{x})$. Thus assertions 1), 2), and 3) of Sect. 6 are proved. The proof of Lemma 3 is now complete.

8. Proof of the Topological Classification Theorem

It follows from Lemmas 1, 2, and 3 that every linear system $\dot{\boldsymbol{x}} = A\boldsymbol{x}$ for which the operator $A : \boldsymbol{R}^n \to \boldsymbol{R}^n$ has no eigenvalues with real part zero is topologically equivalent to the standard multidimensional saddle point (Fig. 162):

$$\dot{\boldsymbol{x}}_1 = -\boldsymbol{x}_1, \quad \dot{\boldsymbol{x}}_2 = \boldsymbol{x}_2, \quad \boldsymbol{x}_1 \in \boldsymbol{R}^{m_-}, \quad \boldsymbol{x}_2 \in \boldsymbol{R}^{m_+}.$$

Consequently two such systems having the same numbers m_- and m_+ are topologically equivalent.

We remark that the subspaces $\boldsymbol{R}^{m_-}$ and $\boldsymbol{R}^{m_+}$ are invariant with respect to the phase flow $\{g^t\}$. As t increases each point of $\boldsymbol{R}^{m_-}$ approaches **o**.

Problem 1. Prove that $g^t\boldsymbol{x} \to \mathbf{o}$ as $t \to +\infty$ if and only if $\boldsymbol{x} \in \boldsymbol{R}^{m_-}$.

For this reason $\boldsymbol{R}^{m_-}$ is called the *incoming* (or *contracting*) *manifold* of the saddle. In exactly the same way $\boldsymbol{R}^{m_+}$ is called the *outgoing* (or *dilating*) *manifold*. The outgoing manifold is defined by the condition $g^t\boldsymbol{x} \to \mathbf{o}$ as $t \to -\infty$.

Let us now prove the second part of the topological classification theorem: *topologically equivalent systems have the same number of eigenvalues with negative real part.*

This number is the dimension m_- of the incoming manifold. Thus it suffices to prove that *the dimensions of the incoming manifolds of equivalent saddles are the same.*

We remark that every homeomorphism h that maps the phase flow of one saddle point to the phase flow of a second must map the incoming manifold of the one to the incoming manifold of the other (since a homeomorphism preserves the relation of tending to $\mathbf{o}$ as $t \to +\infty$). Therefore the homeomorphism h also provides a homeomorphic mapping of the incoming manifold of one saddle to the incoming manifold of the other.

The agreement between the dimensions of the manifolds is now a consequence of the following topological proposition:

The dimension of the space $\boldsymbol{R}^n$ *is a topological invariant.* In other words *there exists a homeomorphism* $h : \boldsymbol{R}^m \to \boldsymbol{R}^n$ *only between spaces of the same dimension.*

Although this proposition seems obvious[17], its proof is not easy and will not be given here.

Problem 2. Prove that the four saddle points with three-dimensional phase space and with $(m_-, m_+) = (3,0), (2,1), (1,2), (0,3)$ are not topologically equivalent (without using the unproved topological proposition just stated).

Hint. A one-dimensional invariant manifold consists of three phase curves, while an invariant manifold of more than one dimension consists of infinitely many (Fig. 163).

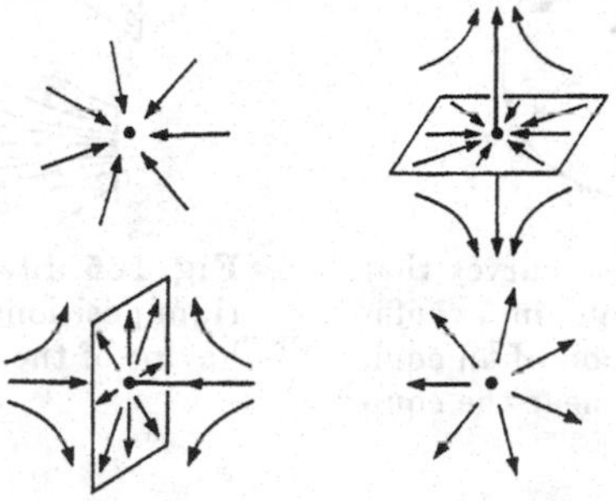

Fig. 163. The invariant manifolds of three-dimensional saddles

Thus the topological classification of linear systems whose eigenvalues have nonzero real parts in $\boldsymbol{R}^1$, $\boldsymbol{R}^2$, and $\boldsymbol{R}^3$ has now been carried out completely, while in $\boldsymbol{R}^n$ with $n > 3$ we are obliged to refer to an unproved proposition on the topological invariance of dimension.

Problem 3. Carry out the topological classification of the linear operators $A : \boldsymbol{R}^n \to \boldsymbol{R}^n$ having no eigenvalues of absolute value 1.

[17] There exist, however, one-to-one correspondences $\boldsymbol{R}^m \to \boldsymbol{R}^n$ and also continuous mappings of $\boldsymbol{R}^m$ onto $\boldsymbol{R}^n$ with $m < n$ (for example, continuous mappings of $\boldsymbol{R}^1$ onto $\boldsymbol{R}^2$).

§ 23. Stability of Equilibrium Positions

The problem of the stability of an equilibrium position of a nonlinear system is solved just as for the linearized system, provided the latter has no eigenvalues on the imaginary axis.

1. Lyapunov Stability

Consider the equation

$$\dot{\boldsymbol{x}} = \boldsymbol{v}(\boldsymbol{x}), \quad \boldsymbol{x} \in U \subset \boldsymbol{R}^n, \tag{1}$$

where $\boldsymbol{v}$ is a vector field that is differentiable r times ($r > 2$) in a domain U. Assume that Eq. (1) has an equilibrium position (Fig. 164). Choose the coordinates x_i so that the equilibrium position is the origin: $\boldsymbol{v}(\mathbf{o}) = \mathbf{o}$.

The solution with the initial condition $\varphi(t_0) = \mathbf{o}$ is $\varphi = \mathbf{o}$. We are interested in the behavior of the solutions with nearby initial conditions.

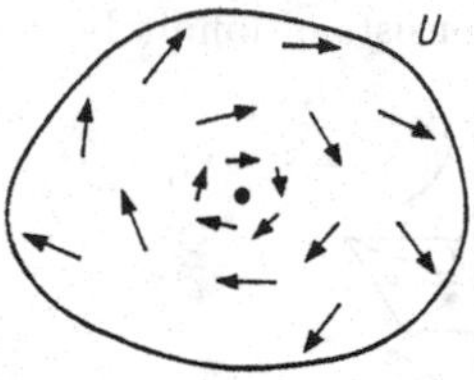

Fig. 164. Will the phase curves that emanate from the points in a sufficiently small neighborhood of an equilibrium position remain near the equilibrium position?

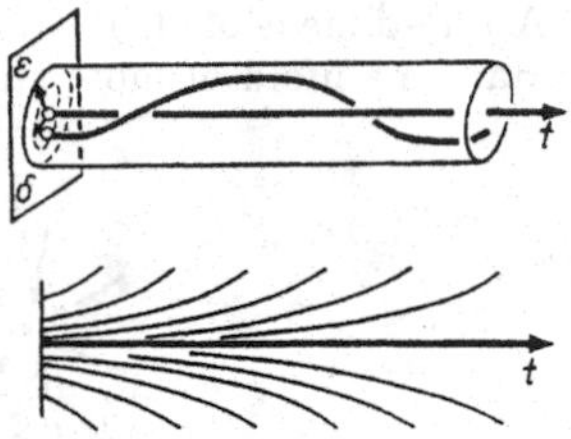

Fig. 165. Stable and unstable equilibrium positions: the difference in the behavior of the integral curves

Definition. The equilibrium position $\boldsymbol{x} = \mathbf{o}$ of Eq. (1) is called *stable* (or *Lyapunov stable*) if for every $\varepsilon > 0$ there exists $\delta > 0$ (depending only on ε and not on the value of the t that will be introduced below) such that for every $\boldsymbol{x}_0$ for which[18] $\|\boldsymbol{x}_0\| < \delta$ the solution φ of Eq. (1) with initial condition $\varphi(0) = \boldsymbol{x}_0$ can be extended to the entire half-line $t > 0$ and satisfies the inequality $\|\varphi(t)\| < \varepsilon$ for all $t > 0$ (Fig. 165).

In other words *Lyapunov stability of an equilibrium position means that the solutions whose initial values tend to the equilibrium position in question converge uniformly on the interval* $[0, +\infty)$ *(to a constant solution).* The convergence of the values of the solutions for fixed t is guaranteed by the theorem on continuous dependence of the solution on the initial condition; what is important is the uniform convergence, i.e., the independence of δ from t.

[18] If $\boldsymbol{x} = (x_1, \ldots, x_n)$, then $\|\boldsymbol{x}\|^2 = x_1^2 + \cdots + x_n^2$.

Problem 1. Study the stability of the equilibrium positions for the following equations:

1) $\dot{x} = 0$;
2) $\dot{x} = x$;
3) $\begin{cases} \dot{x}_1 = x_2, \\ \dot{x}_2 = -x_1; \end{cases}$
4) $\begin{cases} \dot{x}_1 = x_1; \\ \dot{x}_2 = -x_2; \end{cases}$
5) $\begin{cases} \dot{x}_1 = x_2, \\ \dot{x}_2 = -\sin x_1. \end{cases}$

Problem 2. Prove that the definition given above is unambiguous, i.e., that the stability of an equilibrium position is independent of the system of coordinates occurring in the definition.

Problem 3. Suppose it is known that for any $N > 0$ and $\varepsilon > 0$ there exists a solution φ of Eq. (1) such that for some $t > 0$ the inequality $\|\varphi(t)\| > N\|\varphi(0)\|$ holds, and $\|\varphi(0)\| < \varepsilon$. Does it follow from this that the equilibrium position $\boldsymbol{x} = \mathbf{o}$ is unstable?

Hint. Consider the soft pendulum equation $\ddot{x} = -x^3$.

2. Asymptotic Stability

Definition. The equilibrium position $\boldsymbol{x} = \mathbf{o}$ of Eq. (1) is *asymptotically stable* if it is (Lyapunov) stable and

$$\lim_{t \to +\infty} \varphi(t) = \mathbf{o}$$

for every solution φ with initial condition $\varphi(0)$ lying in a sufficiently small neighborhood of zero (Fig. 166).

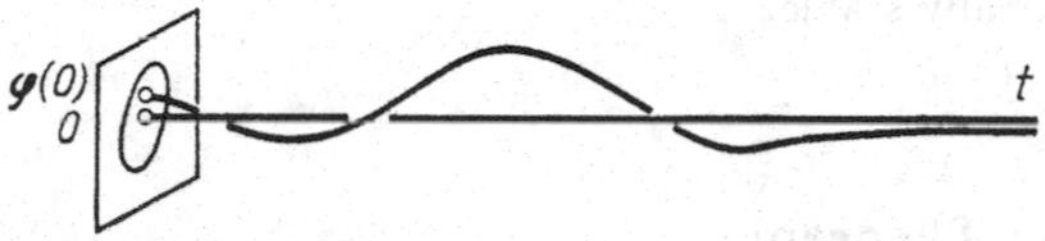

Fig. 166. An asymptotically stable equilibrium position: the integral curves

Problem 1. Solve Problems 1), 2), and 3) of Sect. 1 with stability replaced by asymptotic stability throughout.

Problem 2. Does Lyapunov stability of an equilibrium position follow if every solution tends to this equilibrium position as $t \to +\infty$?

Answer. No.

3. A Theorem on Stability in First Approximation

Along with (1) we shall consider the linearized equation (Fig. 167)

$$\dot{\boldsymbol{x}} = A\boldsymbol{x}, \quad A : \boldsymbol{R}^n \to \boldsymbol{R}^n. \tag{2}$$

Then $\boldsymbol{v}(\boldsymbol{x}) = \boldsymbol{v}_1 + \boldsymbol{v}_2$, where $\boldsymbol{v}_1(\boldsymbol{x}) = A\boldsymbol{x}$ and $\boldsymbol{v}_2(\boldsymbol{x}) = O(\|\boldsymbol{x}\|^2)$.

Theorem. *Suppose all the eigenvalues λ of the operator A lie in the left half-plane:* $\operatorname{Re}\lambda < 0$ (Fig. 168). *Then the equilibrium position $\boldsymbol{x} = \mathbf{o}$ of Eq.* (1) *is asymptotically stable.*

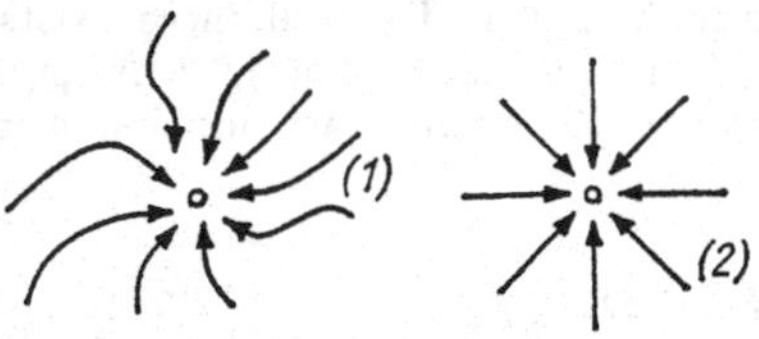

Fig. 167. The phase curves of Eqs. (1) and (2)

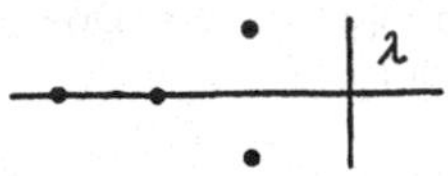

Fig. 168. The eigenvalues of the operator A

Problem 1. Give an example of a Lyapunov unstable equilibrium position of Eq. (1) for which $\operatorname{Re}\lambda \leq 0$ for all λ.

Remark. It can be shown that if the real part of at least one eigenvalue λ is *positive*, then the equilibrium position is unstable. In the case of zero real parts stability depends on the terms of the Taylor series of degree higher than the first.

Problem 2. Is the origin a (Lyapunov and asymptotically) stable equilibrium position for the system $\dot{x}_1 = x_2$, $\dot{x}_2 = -x_1^n$?

Answer. If n is even it is (Lyapunov) unstable; if n is odd, it is Lyapunov stable, but not asymptotically stable.

4. Proof of the Theorem

According to § 22, Sect. 3, there exists a Lyapunov function: a positive-definite quadratic form r^2 whose derivative in the direction of the linear field $\boldsymbol{v}_1$ is negative-definite:

$$L_{\boldsymbol{v}_1} r^2 \leq -2\gamma r^2,$$

where γ is a positive constant (Fig. 169).

Lemma. *In a sufficiently small neighborhood of the point $\boldsymbol{x} = \mathbf{o}$ the derivative of the Lyapunov function in the direction of a nonlinear field $\boldsymbol{v}$ satisfies the inequality*

$$L_{\boldsymbol{v}} r^2 \leq -\gamma r^2. \tag{3}$$

Indeed, $L_{\boldsymbol{v}} r^2 = L_{\boldsymbol{v}_1} r^2 + L_{\boldsymbol{v}_2} r^2$. We shall show that for small r the second term is much smaller than the first:

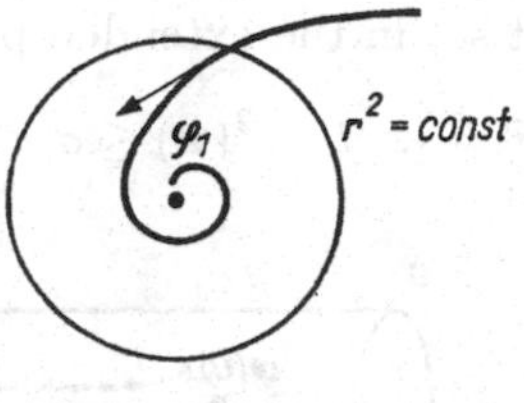

Fig. 169. The level surface of a Lyapunov function

$$L_{v_2} r^2 = O(r^3). \tag{4}$$

In fact, for any field u and any function f

$$L_u f = \sum_{i=1}^{n} \frac{\partial f}{\partial x_i} u_i.$$

In our case ($u = v_2$, $f = r^2$) we have $u_i = O(r^2)$ and $\dfrac{\partial f}{\partial x_i} = O(r)$ (why?), from which relation (4) follows.

Thus there exist $C > 0$ and $\sigma > 0$ such that for all x with $\|x\| < \sigma$ the inequality $|L_{v_2} r^2|_x \leq C|r^2(x)|^{3/2}$ holds. The right-hand side is at most γr^2 for sufficiently small $\|x\|$, so that in some neighborhood of the point $x = o$ we have

$$L_v r^2 \leq -2\gamma r^2 + \gamma r^2 = -\gamma r^2.$$

The lemma is now proved.

Let φ be a non-zero solution of Eq. (1) with initial condition in a sufficiently small neighborhood of the point $x = o$. We define a function ρ of time by the relation

$$\rho(t) = \ln r^2(\varphi(t)), \quad t \geq 0.$$

By the uniqueness theorem we have $r^2(\varphi(t)) \neq 0$, so that the function ρ is defined and differentiable. According to inequality (3)

$$\dot{\rho} = \frac{1}{r^2 \circ \varphi} \frac{d}{dt} r^2 \circ \varphi = \frac{L_v r^2}{r^2} \leq -\gamma.$$

It follows from this that $r^2(\varphi(t))$ decreases monotonically to 0 as $t \to +\infty$:

$$\rho(t) \leq \rho(0) - \gamma t, \quad r^2(\varphi(t)) \leq r^2(\varphi(0)) e^{-\gamma t} \to 0. \tag{5}$$

which was to be proved.

Problem 1. Point out the gap in the preceding proof.

Solution. We have not proved that the solution φ can be extended indefinitely.

Consider a $\sigma > 0$ such that inequality (3) holds for $\|\boldsymbol{x}\| < \sigma$.

Consider a compact set in the extended phase space (Fig. 170):

$$F = \{\boldsymbol{x}, t : r^2(\boldsymbol{x}) \le \sigma, \quad |t| \le T\}.$$

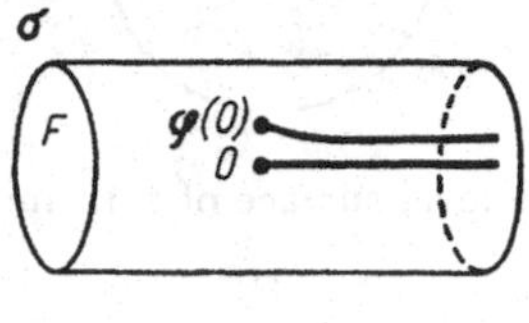

Fig. 170. Indefinite forward extensibility of the solution

Consider the solution φ with the initial condition $\varphi(0)$, where $r^2(\varphi(0)) < \sigma$. By the extension theorem φ can be extended forward to the boundary of the cylinder F. But while the point $(t, \varphi(t))$ belongs to F, the derivative of the function $r^2(\varphi(t))$ is negative. Therefore the solution cannot exit through the lateral surface of the cylinder F (where $r^2 = \sigma^2$) and hence it can be extended to the top $t = T$.

Since T is arbitrary (and independent of σ) the solution φ can be extended forward indefinitely with $r^2(\varphi(t)) < \sigma^2$ and so inequality (3) holds for all $t \ge 0$.

Remark 1. We have proved more than the asymptotic stability of the equilibrium position. It can be seen from inequality (5) that the convergence $\varphi(t) \to \mathbf{o}$ is uniform (with respect to initial conditions $\boldsymbol{x}_0$ sufficiently close to $\mathbf{o}$).

Moreover inequality (5) indicates the rate of convergence (exponential).

In essence this theorem asserts that the uniform exponential convergence of the solutions of the linear equation (2) to zero is not violated under a nonlinear perturbation $v_2(\boldsymbol{x}) = O(\|\boldsymbol{x}\|^2)$ of the right-hand side of the equation. A similar assertion holds for various perturbations of a more general nature. For example, one could consider a nonautonomous perturbation $\boldsymbol{v}_2(\boldsymbol{x}, t)$ for which $\|\boldsymbol{v}_2(\boldsymbol{x}, t)\| \le \varphi(\|\boldsymbol{x}\|)$, where $\varphi(\|\boldsymbol{x}\|) = o(\|\boldsymbol{x}\|)$ as $\boldsymbol{x} \to \mathbf{o}$.

Problem 2. Prove that under the hypotheses of the theorem Eqs. (1) and (2) are topologically equivalent in neighborhoods of the equilibrium position.

Remark 2. In connection with the theorem just proved we arrive at the following algebraic problem (the so-called *Routh-Hurwitz problem*):

It is required to determine *whether all the roots of a given polynomial lie in the left half-plane.*

This question can be solved in a finite number of arithmetic operations on the coefficients of the polynomial. The corresponding algorithms are described in courses of algebra (*Hurwitz' criterion, Sturm's method*) and functions of a complex variable (*the argument principle, the methods of Vyshegradskii, Nyquist, and Mikhailov*). See, for example, the book by A. G. Kurosh, *A*

Course of Higher Algebra [Russian], Nauka, Moscow, 1968, Chapter 9, or the book of M. A. Lavrent'ev and B. V. Shabat, *Methods of the Theory of Functions of a Complex Variable* [Russian], Fizmatgiz, Moscow, 1958), Chapter V; also see the book of M. M. Postnikov, *Stable Polynomials* [Russian], Nauka, Moscow, 1981. We shall return to the Routh-Hurwitz problem in Sect. 5 of § 36.

§ 24. The Case of Purely Imaginary Eigenvalues

Linear equations having no purely imaginary eigenvalues have been studied in detail in §§ 21 and 22. Their phase curves behave rather simply (the saddle point, § 22, Sect. 8).

Linear equations with purely imaginary eigenvalues provide us with examples of more complicated behavior in the phase curves.

Such equations occur, for example, in the theory of oscillations of conservative systems (cf. § 25, Sect. 6).

1. The Topological Classification

Suppose all the eigenvalues $\lambda_1, \ldots, \lambda_n$ of the linear equation

$$\dot{\boldsymbol{x}} = A\boldsymbol{x}, \quad \boldsymbol{x} \in \boldsymbol{R}^n, \quad A : \boldsymbol{R}^n \to \boldsymbol{R}^n, \tag{1}$$

are purely imaginary.

In which cases are two equations of the form (1) topologically equivalent?

Problem 1. Prove that in the case of the plane ($n = 2$, $\lambda_{1,2} = \pm i\omega \neq 0$) a necessary and sufficient condition for topological equivalence of two equations of the form (1) is algebraic equivalence, i.e., having the same eigenvalues.

A similar result has now been proved for $n > 2$ also.

2. An Example

Consider the following equation in $\boldsymbol{R}^4$:

$$\begin{cases} \dot{x}_1 = \omega_1 x_2, & \lambda_{1,2} = \pm i\omega_1, \\ \dot{x}_2 = -\omega_1 x_1, & \\ \dot{x}_3 = \omega_2 x_4, & \lambda_{3,4} = \pm i\omega_2. \\ \dot{x}_4 = -\omega_2 x_3, & \end{cases} \tag{2}$$

The space $\boldsymbol{R}^4$ decomposes into the direct sum of two invariant planes (Fig. 171):

$$\boldsymbol{R}^4 = \boldsymbol{R}_{1,2} + \boldsymbol{R}_{3,4}.$$

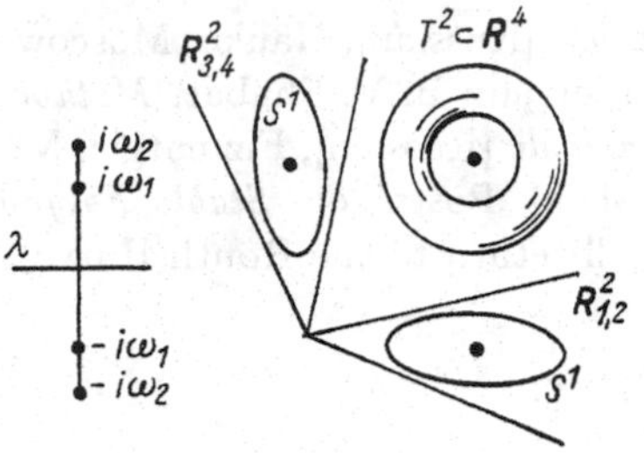

Fig. 171. The phase space of the system (2)

The system (2) decomposes into two independent systems:

$$\begin{cases} \dot{x}_1 = \omega_1 x_2, \\ \dot{x}_2 = -\omega_1 x_2, \end{cases} \quad (x_1, x_2) \in \boldsymbol{R}_{1,2}$$
$$\begin{cases} \dot{x}_3 = \omega_2 x_4, \\ \dot{x}_4 = -\omega_2 x_3. \end{cases} \quad (x_3, x_4) \in \boldsymbol{R}_{3,4} \tag{3}$$

In each of the planes the phase curves are circles

$$S^1 = \{\boldsymbol{x} \in \boldsymbol{R}_{1,2} : x_1^2 + x_2^2 = C > 0\}$$

or a point ($C = 0$), and the phase flow consists of rotations (by the angles $\omega_1 t$ and $\omega_2 t$ respectively).

Each phase curve of Eq. (2) belongs to the direct product of the phase curves in the planes $\boldsymbol{R}_{1,2}$ and $\boldsymbol{R}_{3,4}$. Suppose these two curves are both circles.

The direct product of two circles

$$T^2 = S^1 \times S^1 = \{\boldsymbol{x} \in \boldsymbol{R}^4 : x_1^2 + x_2^2 = C,\, x_3^2 + x_4^2 = C\}$$

is called a *two-dimensional torus*.

To get a better picture of the torus T^2 one can proceed as follows. Consider the surface of an anchor ring in $\boldsymbol{R}^3$ (Fig. 172) obtained by rotating a circle about an axis lying in the plane of the circle but not intersecting it. A point of such a surface is defined by two angular coordinates $\varphi_1, \varphi_2 \bmod 2\pi$. The coordinates φ_1 and φ_2 define a diffeomorphism of the surface of the anchor ring and the direct product T^2 of two circles.

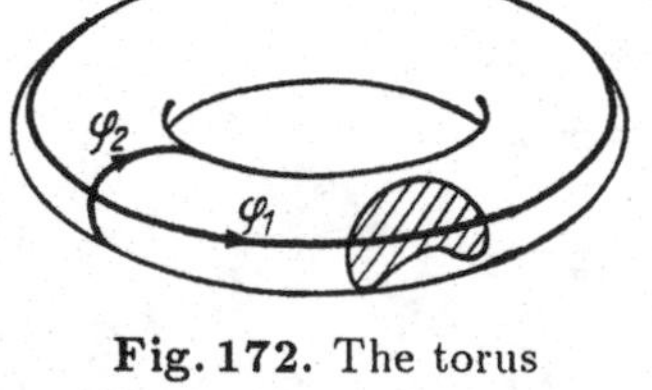

Fig. 172. The torus

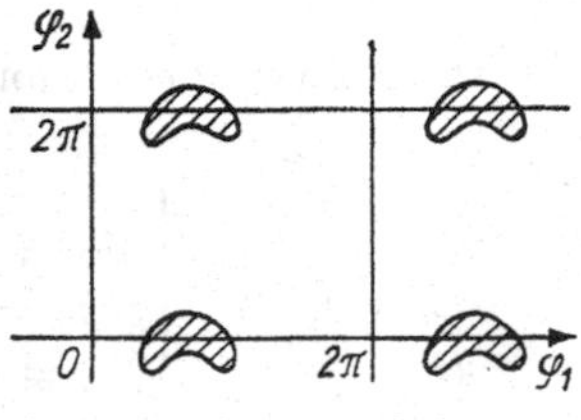

Fig. 173. A map of the torus

The coordinates φ_1 and φ_2 can be called *longitude* and *latitude*. A map of the torus T^2 (cf. Fig. 173) can be represented on the square $0 \leq \varphi_1 \leq 2\pi$,

$0 \leq \varphi_2 \leq 2\pi$ of the (φ_1, φ_2)-plane by "gluing together" the points $(\varphi_1, 0)$ and $(\varphi_1, 2\pi)$ and the points $(0, \varphi_2)$ and $(2\pi, \varphi_2)$. One can also regard the entire (φ_1, φ_2)-plane as a map, but then each point of the torus will have an infinite number of images on the map (like the two images of Chukotka on maps of the hemispheres).

The phase flow of Eq. (2) leaves the torus $T^2 \subset \boldsymbol{R}^4$ fixed. The phase curves of Eq. (2) lie on the surface T^2. If φ_1 is the polar angle of the plane $\boldsymbol{R}_{1,2}$ measured from the unit vector of the x_2-axis in the direction of the unit vector of the x_1-axis, then according to (3) we have $\dot{\varphi}_1 = \omega_1$. Similarly, measuring φ_2 from x_4 to x_3, we obtain $\dot{\varphi}_2 = \omega_2$. Thus:

The phase trajectories of the flow (2) *on the surface* T^2 *satisfy the differential equation*

$$\dot{\varphi}_1 = \omega_1, \quad \dot{\varphi}_2 = \omega_2. \tag{4}$$

The latitude and longitude of the phase point vary uniformly, and on a map of the torus the motion is represented by a straight line, while on the surface of the anchor ring a "solenoid" is obtained (Fig. 174).

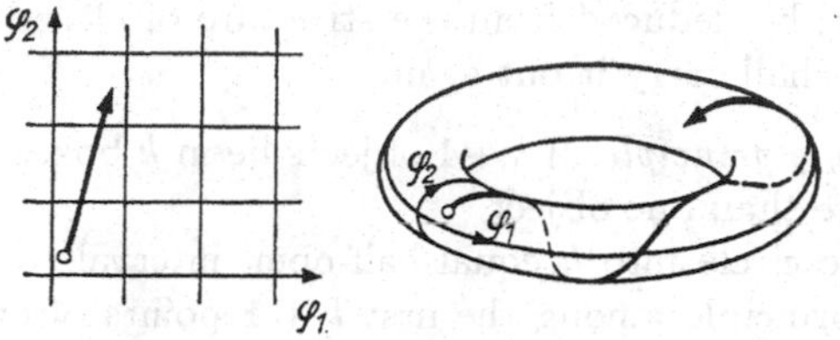

Fig. 174. A solenoid on the torus

3. The Phase Curves of Eq. (4) on the Torus

The numbers ω_1 and ω_2 are called *rationally independent* if the relation $k_1\omega_1 + k_2\omega_2 = 0$ with integers k_1 and k_2 implies $k_1 = k_2 = 0$. For example $\sqrt{2}$ and $\sqrt{8}$ are rationally dependent, while $\sqrt{6}$ and $\sqrt{8}$ are rationally independent.

Theorem. *If* ω_1 *and* ω_2 *are rationally dependent, then every phase curve of Eq.* (4) *on the torus is closed. If* ω_1 *and* ω_2 *are rationally independent, then every phase curve of Eq.* (4) *is everywhere dense*[19] *on the torus* T^2 (Fig. 175).

In other words if on each square of an infinite chessboard there is an identical hare (and identically situated) and a hunter shoots in a direction of irrational slope with respect to the lines of the chessboard, he will hit at least one hare. (It is clear that if the slope is rational, sufficiently small hares could be placed so that the hunter will miss all of them.)

[19] A set A is *everywhere dense* in a space B if there is a point of the set A in every arbitrarily small neighborhood of any point of the space B.

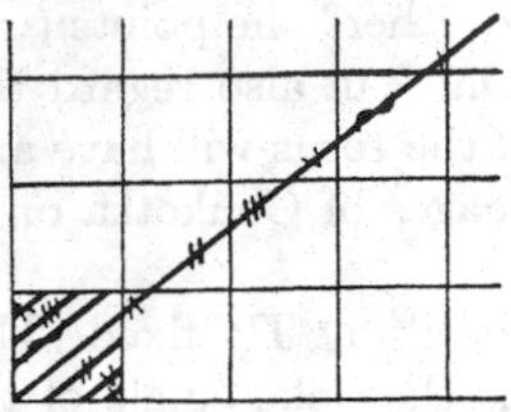

Fig. 175. An everywhere-dense curve on the torus

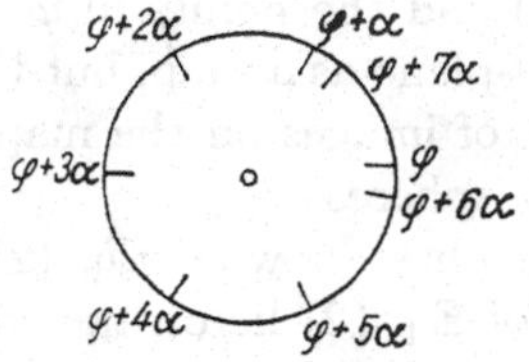

Fig. 176. The images of a point of the circle under repeated rotations through the angle α

Lemma. *Consider a rotation of the circle S^1 by an angle α that is incommensurable with 2π (Fig. 176). Then the images of any point on the circle under the repeated rotations*

$$\varphi,\ \varphi+\alpha,\ \varphi+2\alpha,\ \varphi+3\alpha, \ldots (\mathrm{mod}\, 2\pi)$$

form a set that is everywhere dense on the circle.

The proof can be deduced from the structure of closed subgroups of the line (cf. § 9). We shall carry it out again.

The pigeon-hole principle. If $k+1$ objects lie in k boxes, then at least one box contains more than one object.

We divide the circle into k equal half-open intervals of length $2\pi/k$. By the pigeon-hole principle among the first $k+1$ points of our sequence there are two lying on the same half-open interval. Let these points be $\varphi + p\alpha$ and $\varphi + q\alpha$ with $p > q$. Consider $s = p - q$. The angle of rotation $s\alpha$ differs from a multiple of 2π by less than $2\pi/k$. In the sequence of points φ, $\varphi + s\alpha$, $\varphi + 2s\alpha$, $\varphi + 3s\alpha$, ... (mod 2π) (Fig. 177) each pair of adjacent points are at the same distance, less than $2\pi/k$, from each other. Let $\varepsilon > 0$ be given. Choosing k sufficiently large, we can get $2\pi/k < \varepsilon$. In any ε-neighborhood of any point of S^1 there are points of the sequence $\varphi + Ns\alpha$ (mod 2π).

The lemma is now proved. □

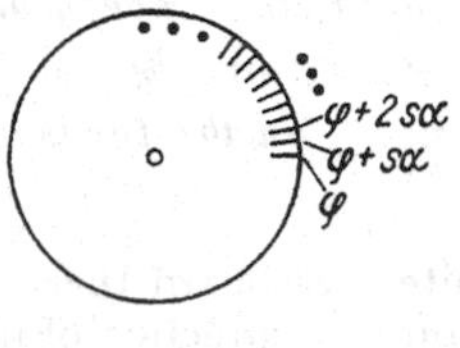

Fig. 177. The points $\varphi + Ns\alpha$

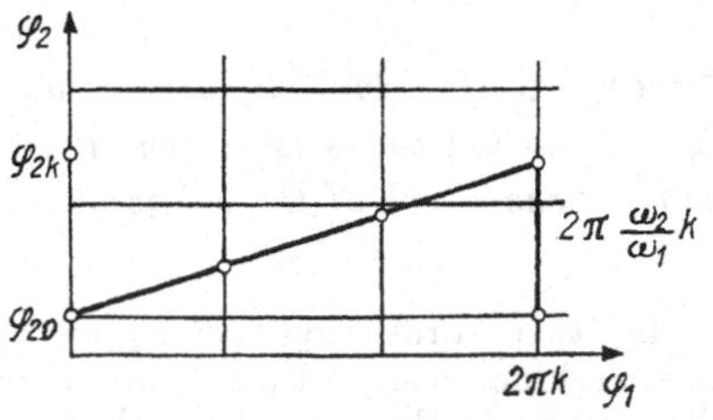

Fig. 178. Reduction of the theorem to the lemma

Remark. We have not used the incommensurability of α with 2π. But it is obvious that the lemma is not true for α commensurable with 2π.

Problem 1. Find and repair the gap in the proof of the lemma.

Proof of the theorem. A solution of Eq. (4) has the form

$$\varphi_1(t) = \varphi_1(0) + \omega_1 t, \quad \varphi_2(t) = \varphi_2(0) + \omega_2 t. \tag{5}$$

Let ω_1 and ω_2 be rationally dependent: $k_1\omega_1 + k_2\omega_2 = 0$, $k_1^2 + k_2^2 \neq 0$. The equations in the unknown T

$$\omega_1 T = 2\pi k_2, \quad \omega_2 T = -2\pi k_1$$

are consistent. A solution T of these equations is a period of the closed phase curve (5).

Let ω_1 and ω_2 be rationally independent. Then ω_1/ω_2 is an irrational number. Consider the successive points of intersection of the phase curve (5) with the meridian $\varphi_1 = 0 \pmod{2\pi}$ (Fig. 178). The latitudes of these points are

$$\varphi_{2,k} = \varphi_{2,0} + 2\pi\frac{\omega_2}{\omega_1}k \pmod{2\pi}.$$

By the lemma the set of points of intersection is everywhere dense on the meridian. We remark that the lines drawn from the points of a set that is everywhere dense on a line lying in the plane in a direction not coinciding with the direction of this line form an everywhere dense subset of the plane. Therefore the image

$$\tilde{\varphi}_1(t) = \varphi_1(t) - 2\pi\left[\frac{\varphi_1(t)}{2\pi}\right], \quad \tilde{\varphi}_2(t) = \varphi_2(t) - 2\pi\left[\frac{\varphi_2(t)}{2\pi}\right]$$

of the phase curve (5) on the square $0 \le \tilde{\varphi}_1 < 2\pi$, $0 \le \tilde{\varphi}_2 < 2\pi$ is everywhere dense. Hence a phase curve of Eq. (4) (and hence of Eq. (2)) is everywhere dense on the torus.

4. Corollaries

A number of simple corollaries of the theorem just proved go beyond the scope of the theory of ordinary differential equations.

Problem 1. Consider the sequence of first digits of powers of 2:

$$1, 2, 4, 8, 1, 3, 6, 1, 2, 5, 1, 2, 4, 8, \ldots$$

Does the digit 7 occur in this sequence? In general, can the number 2^n begin with any combination of digits?

Problem 2. Prove that $\sup\limits_{0<t<\infty} \cos t + \sin\sqrt{2}t = 2$.

5. The Multidimensional Case

Suppose the eigenvalues of Eq. (1) in $\boldsymbol{R}^{2m}$ are simple and of the form

$$\lambda = \pm i\omega_1, \pm i\omega_2, \dots, \pm i\omega_m.$$

Reasoning as in the example of Sect. 2, we shall show that the phase curves lie on the m-dimensional torus

$$T^m = S^1 \times \cdots \times S^1 = \{(\varphi_1, \dots, \varphi_m) \bmod 2\pi\} \cong \boldsymbol{R}^m/\boldsymbol{Z}^m$$

and satisfy the equations $\dot{\varphi}_1 = \omega_1$, $\dot{\varphi}_2 = \omega_2$, ..., $\dot{\varphi}_m = \omega_m$. The numbers $\omega_1, \dots, \omega_m$ are *rationally independent* if for integers k

$$(k_1\omega_1 + \cdots + k_m\omega_m = 0) \Rightarrow (k_1 = \cdots = k_m = 0).$$

Problem *1. Prove that if the frequencies $\omega_1, \dots, \omega_m$ are rationally independent, then each phase curve of Eq. (1) lying on the torus T^m is everywhere dense in the torus.

Corollary. *Suppose a horse makes jumps of size* $(\sqrt{2}, \sqrt{3})$ *across a field* (Fig. 179), *where corn is sown in a square-nest pattern. Then the horse will necessarily knock down at least one plant.*

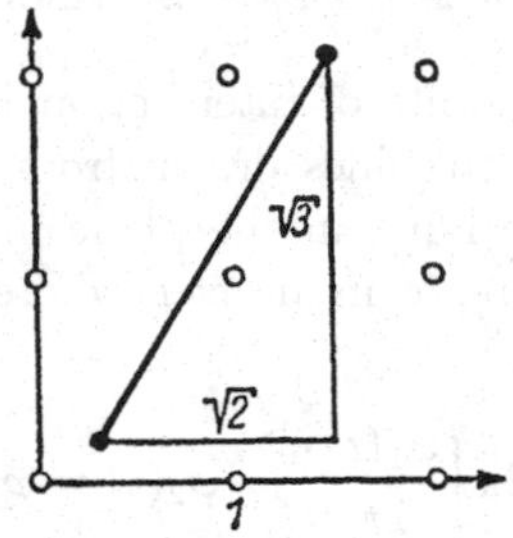

Fig. 179. The phase curve of the system $\dot{\varphi}_1 = 1$, $\dot{\varphi}_2 = \sqrt{2}$, $\dot{\varphi}_3 = \sqrt{3}$ is everywhere dense on the three-dimensional torus

6. The Uniform Distribution

The everywhere-dense curves considered above possess the remarkable property of being uniformly distributed over the surface of the tori. We shall state the corresponding theorem in the simplest case. Consider the sequence of points $\varphi_1, \varphi_2, \dots$ on the circle $S^1 = \{\varphi \bmod 2\pi\}$. The sequence is *uniformly distributed* if for any arc $\Delta \subset S^1$ the number $N(\Delta, k)$ of points of a long segment of the sequence $(\varphi_1, \dots, \varphi_k)$ in Δ is asymptotically proportional to the length of Δ:

$$\lim_{k\to\infty} \frac{N(\Delta, k)}{k} = \frac{|\Delta|}{2\pi}.$$

Problem *1. Prove that the sequence φ, $\varphi + \alpha$, $\varphi + 2\alpha, \dots$, where α is an angle incommensurable with 2π, is uniformly distributed on S^1.

Corollary. *The numbers* 2^n *begin with* 7 *more often than with* 8. *If* $N_7(k)$ *and* $N_8(k)$ *are the numbers of elements among* $(1, 2, 4, \dots, 2^k)$ *beginning with* 7 *and* 8 *respectively, then the following limit exists:*

$$\lim_{k\to\infty} \frac{N_7(k)}{N_8(k)}.$$

Problem 2. Find this limit and verify that it is larger than 1.

Remark. The initial segment of the sequence (cf. Sect. 4) seems to indicate that there are fewer 7's. This is connected with the fact that the irrational number $\log 2 = 0.3010\ldots$ is very close to the rational number $\frac{3}{10}$[20].

§ 25. The Case of Multiple Eigenvalues

Solving a linear equation with constant coefficients reduces to computing the matrix e^{At}. If the eigenvalues of the matrix A are pairwise distinct, then the explicit form of the matrix e^{At} is indicated in § 19, Sect. 5 and in § 20, Sect. 6. To find the explicit form of the matrix e^{At} in the case of multiple eigenvalues, we shall use the Jordan normal form.

1. The Computation of e^{At}, where A is a Jordan Block

One method of computing e^{At}, where A is a Jordan block:

$$\begin{pmatrix} \lambda & 1 & & \\ & \lambda & \ddots & \\ & & \ddots & 1 \\ & & & \lambda \end{pmatrix} : \boldsymbol{R}^n \to \boldsymbol{R}^n,$$

was shown in § 14: A is the matrix of the operator of differentiation in the basis

$$e_k = t^k e^{\lambda t}/k!, \quad 0 \le k < n,$$

of the space of quasi-polynomials of degree less than n with exponent λ. By Taylor's formula e^{As} is the matrix of the translation operator $f(\,\cdot\,) \mapsto f(\cdot + s)$ in the same basis.

Another method is based on the following lemma:

Lemma. *Let A and B be linear operators from $\boldsymbol{R}^n$ to $\boldsymbol{R}^n$. If they commute, then $e^{A+B} = e^A e^B$.*

Proof. We compare the formal series

[20] The first digits of the powers of 3 and those of the populations of the countries of the world are distributed according to the same law.

$$e^A e^B = \left(E + A + \frac{A^2}{2} + \cdots\right)\left(E + B + \frac{B^2}{2} + \cdots\right) =$$
$$= E + (A+B) + \frac{1}{2}(A^2 + 2AB + B^2) + \cdots,$$
$$e^{A+B} = E + (A+B) + \frac{1}{2}(A+B)^2 + \cdots =$$
$$= E + (A+B) + \frac{1}{2}(A^2 + 2AB + B^2) + \cdots$$

If $AB = BA$, then the series coincide (since $e^{x+y} = e^x e^y$ for numbers). Since the series converge absolutely, $e^{A+B} = e^A e^B$, which was to be proved.

□

We represent A in the form $A = \lambda E + \Delta$, where Δ is a nilpotent Jordan block:

$$\Delta = \begin{pmatrix} 0 & 1 & & \\ & 0 & \ddots & \\ & & \ddots & 1 \\ & & & 0 \end{pmatrix}.$$

Since λE commutes with any operator, we have $e^{At} = e^{t(\lambda E + \Delta)} = e^{\lambda t} e^{\Delta t}$. We compute the matrix

$$e^{\Delta t} = E + \Delta t + \frac{\Delta^2 t^2}{2} + \cdots + \frac{\Delta^{n-1} t^{n-1}}{(n-1)!} \quad (\Delta^n = 0).$$

We remark that the action of Δ on the basis $e_1, \ldots, e_n$ is that of a shift: $0 \leftarrow\!\shortmid e_1 \leftarrow\!\shortmid e_2 \leftarrow\!\shortmid \cdots \leftarrow\!\shortmid e_n$. Therefore Δ^k acts as translation by k positions and has the matrix

$$\begin{pmatrix} 0 & \ldots & 1 & & \\ & & & \ddots & \\ & & & & 1 \\ & & & & \vdots \\ & & & & 0 \end{pmatrix}$$

Thus we have proved the following result.

Theorem.

$$e^{\Delta t} = \begin{pmatrix} 1 & t & t^2/2 & \ldots & t^{n-1}/(n-1)! \\ & 1 & t & \ldots & \vdots \\ & & 1 & \ldots & t^2/2 \\ & & & \ddots & t \\ & & & \ddots & 1 \end{pmatrix},$$

$$e^{At} = \begin{pmatrix} e^{\lambda t} & te^{\lambda t} & \ldots & t^{n-1}e^{\lambda t}/(n-1)! \\ & e^{\lambda t} & \ddots & \vdots \\ & & \ddots & te^{\lambda t} \\ & & & e^{\lambda t} \end{pmatrix}.$$

Our computations carry through without any changes in the complex case ($\lambda \in \boldsymbol{C}, A : \boldsymbol{C}^n \to \boldsymbol{C}^n$).

2. Applications

The following result is an immediate consequence of formula (1).

Corollary 1. *Let $A : \boldsymbol{C}^n \to \boldsymbol{C}^n$ be a linear operator with eigenvalues $\lambda_1, \dots, \lambda_k$ of respective multiplicities $\nu_1, \dots, \nu_k$, and let $t \in \boldsymbol{R}$. Then each element of the matrix of e^{At} (in any fixed basis) is the sum of quasi-polynomials in t with exponents λ_l of degree less than ν_l respectively ($l = 1, \dots, k$).*

Proof. Consider the matrix of the operator e^{At} in a basis in which the matrix A has Jordan form. Our assertion then follows from (1). The elements of the matrix of the operator e^{At} in any other basis are linear combinations (with constant coefficients) of the elements of the matrix of the operator e^{At} in the given basis. □

Corollary 2. *Let φ be a solution of the differential equation $\dot{\boldsymbol{x}} = A\boldsymbol{x}$, $\boldsymbol{x} \in \boldsymbol{C}^n$, and $A : \boldsymbol{C}^n \to \boldsymbol{C}^n$. Then each component of the vector φ (in any fixed basis) is a sum of quasi-polynomials in t with exponents λ_l of degrees less than ν_l respectively: $\varphi_j(t) = \sum_{l=1}^{k} e^{\lambda_l t} p_{j,l}(t)$, where $p_{j,l}$ is a polynomial of degree less than ν_l.*

Indeed $\varphi(t) = e^{At}\varphi(0)$.

Corollary 3. *Let $A : \boldsymbol{R}^n \to \boldsymbol{R}^n$ be a linear operator with real eigenvalues λ_l ($1 \le l \le k$) of respective multiplicities ν_l and complex eigenvalues $\alpha_l \pm i\omega_l$, ($1 \le l \le m$) of multiplicities μ_l. Then each element of the matrix of e^{At} and each component of the solution of the equation $\dot{\boldsymbol{x}} = A\boldsymbol{x}$, $\boldsymbol{x} \in \boldsymbol{R}^n$, is a sum of complex quasi-polynomials with exponents λ_l and $\alpha_l \pm i\omega_l$ of degrees less than ν_l and μ_l respectively.*

Such a sum can also be written in the less convenient form:

$$\varphi_j(t) = \sum_{l=1}^{k} e^{\lambda_l t} p_{j,l} + \sum_{l=1}^{m} e^{\alpha_l t}[q_{j,l}(t)\cos\omega_l t + r_{j,l}(t)\sin\omega_l t],$$

where p, q, and r are polynomials of degrees less than ν_l, μ_l, and μ_l respectively, with *real* coefficients.

Indeed, if $z = x + iy$ and $\lambda = \alpha + i\omega$, then

$$\operatorname{Re} z e^{\lambda t} = \operatorname{Re} e^{\alpha t}(x + iy)(\cos\omega t + i\sin\omega t) = e^{\alpha t}(x\cos\omega t - y\sin\omega t).$$

Incidentally it can be seen from these formulas that if the real parts of all eigenvalues are negative, then all solutions tend to zero as $t \to +\infty$ (as must be the case according to § 22 and § 23).

3. Applications to Systems of Equations of Order Higher than the First

By writing the system as a system of first-order equations we reduce the problem to the one considered above, and we can solve it by bringing the matrix to Jordan form. In practice it is often convenient to proceed in another way. First of all, the eigenvalues of the equivalent first-order system can be found without writing out its matrix.

Indeed, to an eigenvalue λ there corresponds an eigenvector, hence a solution $\varphi(t) = e^{\lambda t}\varphi(0)$ of the equivalent first-order system. But then the original system also has a solution of the form $\psi(t) = e^{\lambda t}\psi(0)$. In the original system we make the substitution $\psi = e^{\lambda t}\boldsymbol{\xi}$. The system admits such a solution (nonzero) if and only if λ satisfies an algebraic equation from which we can find the eigenvalues λ_l.

The solutions themselves can then be sought in the form of sums of quasi-polynomials with exponents λ_l and undetermined coefficients.

Example 1. $x^{(\text{iv})} = x$.

We make the substitution $x = e^{\lambda t}\xi$. We find $\lambda^4 e^{\lambda t}\xi = e^{\lambda t}\xi$, $\lambda^4 = 1$, $\lambda_{1,2,3,4} = 1, -1, i, -i$.

Every solution of our equation has the form

$$x = C_1 e^t + C_2 e^{-t} + C_3 \cos t + C_4 \sin t.$$

Example 2. $\ddot{x}_1 = x_2$, $\ddot{x}_2 = x_1$.

We substitute $\boldsymbol{x} = e^{\lambda t}\boldsymbol{\xi}$. We find $\lambda^2\xi_1 = \xi_2$, $\lambda^2\xi_2 = \xi_1$. This system of linear equations in ξ_1 and ξ_2 has a nontrivial solution if and only if $\lambda^4 = 1$. Every solution of our system has the form

$$x_1 = C_1 e^t + C_2 e^{-t} + C_3 \cos t + C_4 \sin t, \quad x_2 = D_1 e^t + D_2 e^{-t} + D_3 \cos t + D_4 \sin t.$$

Substituting into the system gives $D_1 = C_1$, $D_2 = C_2$, $D_3 = -C_3$, $D_4 = -C_4$.

Example 3. $x^{(iv)} - 2\ddot{x} + x = 0$.

We substitute $x = e^{\lambda t}\xi$. We find

$$\lambda^4 - 2\lambda^2 + 1 = 0, \quad \lambda^2 = 1, \quad \lambda_{1,2,3,4} = 1, 1, -1, -1.$$

Every solution of the original equation has the form

$$(C_1 t + C_2)e^t + (C_3 t + C_4)e^{-t}.$$

Problem 1. Find the Jordan normal form of the 4×4 matrix corresponding to this equation.

4. The Case of a Single *n*th-order Equation

We remark that the multiplicities of the eigenvalues in general do not determine the dimensions of the Jordan blocks. The situation is simpler in the case of a linear operator A corresponding to a single nth-order differential equation:

$$x^{(n)} = a_1 x^{(n-1)} + \cdots + a_n x, \quad a_k \in C. \tag{2}$$

The following result is a consequence of Corollary 2 of Sect. 2.

Corollary 4. *Every solution of Eq.* (2) *has the form*

$$\varphi(t) = \sum_{l=1}^{k} e^{\lambda_l t} p_l(t), \tag{3}$$

where $\lambda_1, \ldots, \lambda_k$ *are the roots of the characteristic equation*

$$\lambda^n = a_1 \lambda^{n-1} + \cdots + a_n, \tag{4}$$

and p_l *is a polynomial of degree less than* ν_l *(where* ν_l *is the multiplicity of the root* λ_l*).*

Indeed Eq. (2) has a solution of the form $e^{\lambda t}(\boldsymbol{\xi})$ if and only if λ is a root of Eq. (4). Corollary (4) is now proved.

We now turn to the equivalent first-order system of equations:

$$\dot{x} = Ax, \quad A = \begin{pmatrix} 0 & 1 & & & \\ & 0 & 1 & & \\ & & & \ddots & \\ & & & \ddots & 1 \\ a_n & & & \ldots & a_1 \end{pmatrix}. \tag{5}$$

We obtain the following result.

Corollary 5. *If the operator* $A : C^n \to C^n$ *has a matrix of the form* (5), *then to each of its eigenvalues* λ *there corresponds exactly one Jordan block, whose dimension equals the multiplicity of* λ.

Indeed, according to formula (3), to each eigenvalue λ there corresponds one characteristic direction. In fact, let $\boldsymbol{\xi}$ be an eigenvector of the operator A. Then among the solutions of the form (3) is the first component $e^{\lambda t}\xi_0$ of the vector $e^{\lambda t}\boldsymbol{\xi}$. But then the remaining components are the derivatives: $\xi_k = \lambda^k \xi_0$. Therefore the number λ determines the direction of the vector $\boldsymbol{\xi}$ uniquely.

Since each Jordan block has its own characteristic direction, Corollary 5 is now proved. □

Problem 1. Is every linear combination of the quasi-polynomials (3) a solution of Eq. (2)?

5. On Recursive Sequences

Our study of exponentials with a continuous exponent e^{tA} is easily extended to an exponential with a discrete exponent A^n. In particular we can now study a recursive sequence defined by the relation

$$x_n = a_1 x_{n-1} + \cdots + a_k x_{n-k} \tag{6}$$

(for example, the sequence 0, 1, 2, 5, 12, 29, ... defined by the relation $x_n = 2x_{n-1} + x_{n-2}$ and the initial condition $x_0 = 0$, $x_1 = 1$).

Corollary 6. *As a function of n the nth term of a recursive sequence is a sum of quasi-polynomials in n:*

$$x_n = \sum_{l=1}^{m} \lambda_l^n p_l(n),$$

where λ_l are the eigenvalues of the matrix A corresponding to the sequence and p_l is a polynomial of degree less than ν_l (where ν_l is the multiplicity of λ_l).

We recall that the matrix A is the matrix of the operator $A : \boldsymbol{R}^k \to \boldsymbol{R}^k$ that takes a segment of length k of our sequence $\boldsymbol{\xi}_{n-1} = (x_{n-k}, \ldots, x_{n-1})$ to the succeeding segment of length k, $\boldsymbol{\xi}_n = (x_{n-k+1}, \ldots, x_n)$:

$$A\boldsymbol{\xi}_{n-1} = \begin{pmatrix} 0 & 1 & & & \\ & 0 & 1 & & \\ & & \ddots & \ddots & \\ & & & 0 & 1 \\ a_k & & \ldots & a_2 & a_1 \end{pmatrix} \begin{pmatrix} x_{n-k} \\ \vdots \\ x_{n-1} \end{pmatrix} = \begin{pmatrix} x_{n-k+1} \\ \vdots \\ x_n \end{pmatrix} = \boldsymbol{\xi}_n.$$

It is important to note that the operator A is independent of n. Therefore x_n is one of the components of the vector $A^n\boldsymbol{\xi}$, where $\boldsymbol{\xi}$ is a constant vector. The matrix A has the form (5). Using Corollary (5) and bringing A into Jordan form, we obtain Corollary 6. □

In the computations there is no need to write out the matrix or bring it to normal form. An eigenvector of the operator A corresponds to a solution of Eq. (6) of the form $x = \lambda^n$. Substituting into Eq. (6), we find for λ the equation

$$\lambda^k = a_1\lambda^{k-1} + \cdots + a_k.$$

It is easy to verify that this is the characteristic equation of the operator A.

Example 1. For the sequence 0, 1, 2, 5, 12, 29, ... ($x_n = 2x_{n-1} + x_{n-2}$) we find $\lambda^2 = 2\lambda + 1$, $\lambda_{1,2} = 1 \pm \sqrt{2}$. Therefore the relation $x_n = 2x_{n-1} + x_{n-2}$ is satisfied by the sequences $x_n = (1+\sqrt{2})^n$, $x_n = (1-\sqrt{2})^n$, and also any linear combinations of them (and only such linear combinations)

$$x_n = c_1(1+\sqrt{2})^n + c_2(1-\sqrt{2})^n.$$

Among these combinations it is easy to choose one for which $x_0 = 0$, $x_1 = 1$: $c_1 + c_2 = 0$, $\sqrt{2}(c_1 - c_2) = 1$, namely $x_n = [(1+\sqrt{2})^n - (1-\sqrt{2})^n]/(2\sqrt{2})$.

Remark. As $n \to +\infty$ the first term increases exponentially and the second decreases exponentially. Therefore for large n

$$x_n \approx (1+\sqrt{2})^n/(2\sqrt{2})$$

and in particular $x_{n+1}/x_n \approx 1+\sqrt{2}$. From this we find very good approximations for $\sqrt{2}$: $\sqrt{2} \approx (x_{n+1} - x_n)/x_n$. Substituting $x_n = 1, 2, 5, 12, 29, \ldots$, we find

$$\sqrt{2} \approx (2-1)/1 = 1; \quad \sqrt{2} \approx (5-2)/2 = 1.5;$$
$$\sqrt{2} \approx (12-5)/5 = 1.4; \sqrt{2} \approx (29-12)/12 = 17/12 \approx 1.417\ldots$$

These are the same approximations used to compute $\sqrt{2}$ in antiquity; they can also be obtained by expanding $\sqrt{2}$ in a continued fraction. Furthermore $(x_{n+1} - x_n)/x_n$ is the best rational approximation to $\sqrt{2}$ by a fraction whose denominator does not exceed x_n.

6. Small Oscillations

We have studied above the case when there is only one eigenvector corresponding to each root of the characteristic equation, whatever its multiplicity: the case of a single nth-order equation. There exists a case that is in a certain sense the opposite of this, when to each root there corresponds a number of linearly independent eigenvectors equal to the multiplicity of the root. This is the case for small oscillations of a conservative mechanical system.

In the *Euclidean* space $\boldsymbol{R}^n$ we consider a quadratic form U given by a symmetric operator A:

$$U(\boldsymbol{x}) = \frac{1}{2}(A\boldsymbol{x}, \boldsymbol{x}), \quad \boldsymbol{x} \in \boldsymbol{R}^n, \quad A: \boldsymbol{R}^n \to \boldsymbol{R}^n, \quad A' = A.$$

Consider the differential equation[21]

$$\ddot{\boldsymbol{x}} = -\operatorname{\mathbf{grad}} U \tag{7}$$

(here U is the potential energy).

In studying Eq. (7) it is useful to imagine a ball rolling along the graph of the potential energy (compare with § 12).

Equation (7) can be written in the form $\ddot{\boldsymbol{x}} = -A\boldsymbol{x}$ or in coordinate notation as a system of n second-order linear equations. By the general rule we seek a solution $\varphi = e^{\lambda t}\boldsymbol{\xi}$, and we find

[21] The vector field $\operatorname{\mathbf{grad}} U$ is defined by the condition "$dU(\boldsymbol{\xi}) = (\operatorname{\mathbf{grad}} U, \boldsymbol{\xi})$ for every vector $\boldsymbol{\xi} \in T\boldsymbol{R}^n_{\boldsymbol{x}}$." Here the parentheses denote the Euclidean inner product. In Cartesian (orthonormal) coordinates the vector field $\operatorname{\mathbf{grad}} U$ is given by the components $\left(\frac{\partial U}{\partial x_1}, \ldots, \frac{\partial U}{\partial x_n}\right)$.

$$\lambda^2 e^{\lambda t}\boldsymbol{\xi} = -A e^{\lambda t}\boldsymbol{\xi}, \quad (A+\lambda^2 E)\boldsymbol{\xi} = \mathbf{o}, \quad \det|A+\lambda^2 E| = 0.$$

From this we find n real (why?) values of λ^2 and $2n$ values of λ.

If they are all distinct, then every solution of Eq. (7) is a linear combination of exponentials. If there are multiple roots, the question of the Jordan blocks arises.

Theorem. *If the quadratic form U is nondegenerate, then to each eigenvalue λ there corresponds a number of linearly independent eigenvectors equal to its multiplicity, so that every solution of Eq. (7) can be written in the form of a sum of exponentials*[22]*: $\varphi(t) = \sum_{k=1}^{2n} e^{\lambda_k t}\xi_k$, $\xi_k \in C^n$.*

Proof. By an orthogonal transformation we can bring the form U to *principal axes*: there exists an orthonormal basis $\boldsymbol{e}_1, \ldots, \boldsymbol{e}_n$ in which U can be written in the form

$$U(\boldsymbol{x}) = \frac{1}{2}\sum_{k=1}^{n} a_k x_k^2, \quad \boldsymbol{x} = x_1\boldsymbol{e}_1 + \cdots + x_n\boldsymbol{e}_n.$$

The nondegeneracy of the form U says that none of the numbers a_k is zero. In the coordinates chosen Eq. (7) assumes the form

$$\ddot{x}_1 = -a_1 x_1, \quad \ddot{x}_2 = -a_2 x_2, \ldots, \quad \ddot{x}_n = -a_n x_n$$

whether or not there are multiple roots[23]. Our system has split into the direct product of n "pendulum equations." Each of these ($\ddot{x} = -ax$) can be solved immediately.

If $a > 0$, then $a = \omega^2$ and

$$x = C_1 \cos\omega t + C_2 \sin\omega t.$$

If $a < 0$, then $a = -\alpha^2$ and

$$x = C_1 \cosh\alpha t + C_2 \sinh\alpha t = D_1 e^{\alpha} t + D_2 e^{-\alpha t}.$$

These formulas contain, in particular, the assertion of the theorem. □

If the form U is positive-definite, then all the a_k are positive and the point $\boldsymbol{x}$ performs n independent oscillations along the n mutually perpendicular directions $\boldsymbol{e}_1, \ldots, \boldsymbol{e}_n$ (Fig. 180). These oscillations are called the *principal* or *characteristic* oscillations, and the numbers ω_k are called the *natural frequencies*. They satisfy the equation $\det|A - \omega^2 E| = 0$.

[22] It is of interest to note that Lagrange, who was the first to study the equation of small oscillations (7), at first made a mistake. He thought that in the case of multiple roots some "secular" terms of the form $te^{\lambda t}$ would be required (in the real case $t\sin\omega t$), as in Sects. 2, 4, and 5 above.

[23] We remark that we are making essential use of the orthonormality of the basis $\boldsymbol{e}_k$: if the basis were not orthonormal, the components of the vector $\mathbf{grad}\,\frac{1}{2}\sum a_k x_k^2$ *would not be equal to* $a_k x_k$.

The trajectory of a point $\boldsymbol{x} = \varphi(t)$ in $\boldsymbol{R}^n$ (where φ is a solution of Eq. (7)) lie in the parallelepiped $|x_k| \le X_k$, where X_k is the amplitude of the kth characteristic oscillation. In particular for $n = 2$ the point lies in a rectangle.

If the frequencies ω_1 and ω_2 are commensurable, then the trajectory is a closed curve. In this case it is called a *Lissajous curve* (Fig. 181).

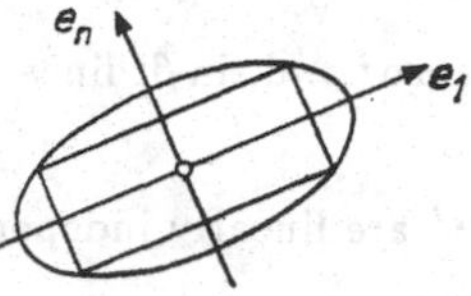

Fig. 180. The directions of the characteristic oscillations and the level lines of the potential energy

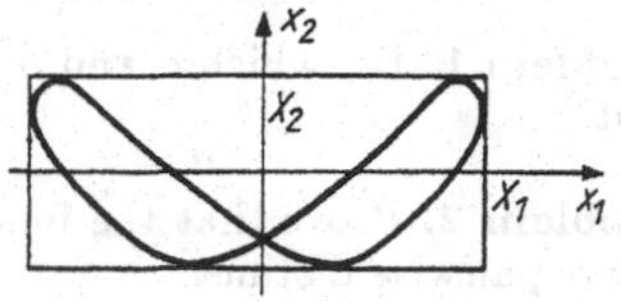

Fig. 181. One of the Lissajous curves with $\omega_2 = 2\omega_1$

If, on the other hand, ω_1 and ω_2 are incommensurable, then the trajectory fills up an everywhere dense subset of the rectangle. This follows from the theorem of § 24.

Problem 1. Sketch the Lissajous curves for $\omega_1 = 1, \omega_2 = 3$, and for $\omega_1 = 2, \omega_2 = 3$.

Problem 2. Prove that among the Lissajous curves with $\omega_2 = n\omega_1$ is the graph of a polynomial of degree n. This polynomial is called the *Chebyshev polynomial.*

$$T_n(x) = \cos(n \arccos x).$$

Problem 3. What does the trajectory of $\boldsymbol{x} = \varphi(t)$ look like in the case $U = x_1^2 - x_2^2$?

Problem 4. For which U is the equilibrium position $x = \dot{x} = 0$ of Eq. (7) stable a) in the sense of Lyapunov? b) asymptotically?

§ 26 Quasi-polynomials

In solving linear equations with constant coefficients we constantly encountered quasi-polynomials. We shall now explain the reason for this phenomenon and give some new applications of it.

1. A Linear Function Space

Consider the set F of all infinitely differentiable complex-valued functions on the real axis $\boldsymbol{R}$.

The set F has a natural complex vector space structure: if f_1 and f_2 are functions of F, the function $c_1 f_1 + c_2 f_2$ (where c_1 and c_2 are constants in $\boldsymbol{C}$) also belongs to F.

Definition. The functions $f_1, \ldots, f_n \in F$ are *linearly independent* if they are linearly independent as vectors of the vector space F, i.e., if

$$(c_1 f_1 + \cdots + c_n f_n \equiv 0) \Rightarrow (c_1 = \cdots = c_n = 0),$$

where $c_1, \ldots, c_n \in \boldsymbol{C}$.

Problem 1. For which α and β are the functions $\sin \alpha t$ and $\sin \beta t$ linearly independent?

Problem 2. Prove that the functions $e^{\lambda_1 t}, \ldots, e^{\lambda_k t}$ are linearly independent if the λ_k are pairwise distinct.

Hint. This follows from the existence of an nth-order linear equation with solutions $e^{\lambda_1 t}, \ldots, e^{\lambda_n t}$ (cf. Sect. 2 below).

Among the elements of the space F are the quasi-polynomials with exponent λ: $f(t) = e^{\lambda t} \sum_{k=0}^{\nu-1} c_k t^k$ and, more generally, finite sums of quasi-polynomials with different exponents

$$f(t) = \sum_{l=1}^{k} e^{\lambda_l t} \sum_{m=0}^{\nu_l - 1} c_{l,m} t^m, \quad \lambda_i \neq \lambda_j. \tag{1}$$

Problem 3. Prove that each function of the form (1) can be written uniquely in the form of a sum (1). In other words:

If the sum (1) *equals* 0, *then each coefficient* $c_{l,m}$ *equals* 0.

Hint. One possible solution can be found in Sect. 2 (the corollary below).

2. The Vector Space of Solutions of a Linear Equation

Theorem. *The set X of all solutions of the linear equation*

$$x^{(n)} + a_1 x^{(n-1)} + \cdots + a_n x = 0 \tag{2}$$

constitutes a subspace of F of finite dimension n.

Proof. Consider the operator $D : F \to F$ that maps each function to its derivative. The operator D is linear:

$$D(c_1 f_1 + c_2 f_2) = c_1 D f_1 + c_2 D f_2.$$

Consider a polynomial in the operator D:

$$A = a(D) = D^n + a_1 D^{n-1} + \cdots + a_n E.$$

The operator A is linear and $A : F \to F$. The solutions[24] of Eq. (2) form the kernel of the operator A. Thus $X = \operatorname{Ker} A$.

But the kernel $\operatorname{Ker} A$ of a linear operator is a vector space. Therefore X is a vector space. We shall show that X is isomorphic to C^n.

Let $\varphi \in X$. We shall assign to the function φ a set of n numbers: the set of values at the point $t = 0$ of the function φ and its derivatives $\varphi_0 = (\varphi(0), (D\varphi)(0), \ldots, (D^{n-1}\varphi)(0))$. We then obtain a mapping

$$B : X \to C^n, \quad B(\varphi) = \varphi_0,$$

This mapping is linear. The image of the mapping B is the entire space C^n. For by the existence theorem, there exists a solution $\varphi \in X$ with any given initial conditions φ_0.

The kernel of the mapping B is zero. For by the uniqueness theorem, the initial conditions $\varphi_0 = o$ determine the solution ($\varphi \equiv 0$) uniquely. Thus B is an isomorphism.

The theorem is now proved. □

Corollary. *Let $\lambda_1, \ldots, \lambda_k$ be the roots of the characteristic equation $a(\lambda) = 0$ of the differential equation* (2) *and $\nu_1, \ldots, \nu_k$ their respective multiplicities. Then each solution of Eq.* (2) *can be written uniquely in the form* (1) *and each sum of quasi-polynomials of the form* (1) *satisfies Eq.* (2).

Proof. Formula (1) defines a mapping $\Phi : C^n \to F$ that assigns to the set of n coefficients $c_{l,m}$ the function f. This mapping is linear. Its image contains the space X of solutions of Eq. (2). For according to § 25 each solution of Eq. (2) can be written in the form (1). By the theorem the dimension of the space X is n.

A linear mapping of the space C^n onto a space X of the same dimension is an isomorphism. Therefore Φ realizes an isomorphism of C^n and X. This is the assertion of the corollary. □

3. Translation-invariance

Theorem. *The space X of solutions of the differential equation* (2) *is invariant under the translations mapping the function $\varphi(t)$ into $\varphi(t + s)$.*

Indeed a translation of a solution will be a solution, as is the case for any autonomous equation (compare § 10).

Examples of translation-invariant subspaces of the space F are:

Example 1. The one-dimensional space $\{ce^{\lambda t}\}$.

[24] We know in advance that all solutions of Eq. (2) are infinitely differentiable, i.e., belong to F (cf. § 25, Sect. 4).

Example 2. The space of quasi-polynomials $\{e^{\lambda t} p_{<n}(t)\}$ of dimension n.

Example 3. The plane $\{c_1 \cos \omega t + c_2 \sin \omega t\}$.

Example 4. The space $\{p_{<n}(t) \cos \omega t + q_{<n}(t) \sin \omega t\}$ of dimension $2n$.

It can be shown that every finite-dimensional translation-invariant subspace of the space F is the space of solutions of some differential equation (2).

In other words, such a subspace always decomposes into the direct sum of spaces of quasi-polynomials. This explains the significance of the quasi-polynomials for the theory of linear equations with constant coefficients.

If an equation is invariant with respect to some group of transformations, then spaces of functions invariant under that group will play an important role in solving the equation. In this way various special functions arise in mathematics. For example, the spherical functions are connected with the group of rotations of the sphere; these are finite-dimensional spaces of functions on the sphere that are invariant under rotations.

Problem *1. Find all finite-dimensional subspaces of the space of smooth functions on the circle that are invariant with respect to rotations of the circle.

4. Historical Remark

The theory of linear differential equations with constant coefficients was founded by Euler and Lagrange before the Jordan normal form of a matrix was constructed.

They reasoned as follows. Let λ_1 and λ_2 be two roots of the characteristic equation. They correspond to the solutions $e^{\lambda_1 t}$ and $e^{\lambda_2 t}$ that span the two-dimensional plane $\{c_1 e^{\lambda_1 t} + c_2 e^{\lambda_2 t}\}$ (Fig. 182) in the space F. Now suppose the equation varies so that λ_2 approaches λ_1. Then $e^{\lambda_2 t}$ approaches $e^{\lambda_1 t}$, and for $\lambda_2 = \lambda_1$ the plane degenerates into a line.

The question arises: Does there exist a limiting position of the plane as $\lambda_2 \to \lambda_1$?

Instead of $e^{\lambda_1 t}$ and $e^{\lambda_2 t}$ the basis can be taken as $e^{\lambda_1 t}$ and $e^{\lambda_2 t} - e^{\lambda_1 t}$ (when $\lambda_2 \neq \lambda_1$). But $e^{\lambda_2 t} - e^{\lambda_1 t} \approx (\lambda_2 - \lambda_1) t e^{\lambda_1 t}$. Hence the basis $(e^{\lambda_1 t}, (e^{\lambda_1 t} - e^{\lambda_2 t}/(\lambda_2 - \lambda_1))$ of our plane becomes the basis $(e^{\lambda_1 t}, te^{\lambda_1 t})$ of the limiting plane as $\lambda_2 \to \lambda_1$. Hence it is natural to expect that the solution of the limiting equation (with multiple root $\lambda_2 = \lambda_1$ will lie in the limiting plane $\{c_1 e^{\lambda_1 t} + c_2 t e^{\lambda_2 t}\}$. Once the formula is written out, it can be verified by substituting into the equation.

The appearance of the solutions $t^k e^{\lambda t}$ $(k < \nu)$ in the case of a root of multiplicity ν is explained in the same way.

The reasoning just given can be made completely rigorous (for example, by citing the theorem on differentiable dependence of the solutions on a parameter).

5. Inhomogeneous Equations

Let $A : L_1 \to L_2$ be a linear transformation. Any pre-image $x \in L_1$ of the element $f \in L_2$ is called a *solution* of the inhomogeneous equation $Ax = f$ with right-hand side f (Fig. 183).

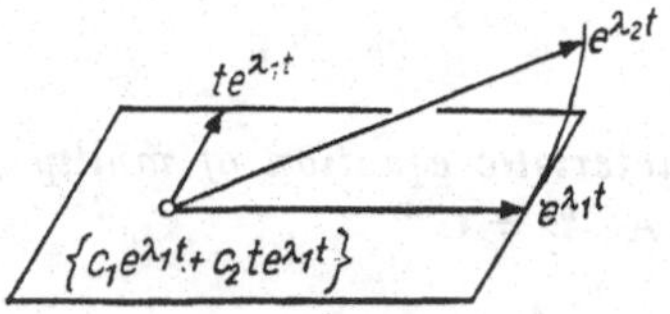

Fig. 182. The limiting position of the plane spanned by two exponentials

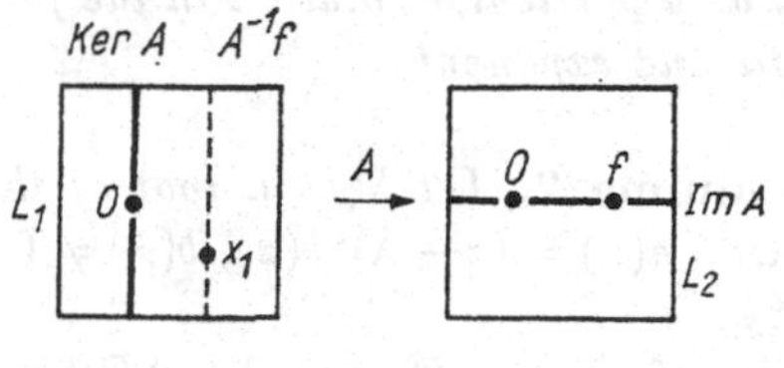

Fig. 183. The kernel and image of an operator A

Every solution of an inhomogeneous equation is the sum of a particular solution x_1 and the general solution of the homogeneous equation $Ax = 0$:

$$A^{-1}f = x_1 + \operatorname{Ker} A.$$

The inhomogeneous equation is solvable if f belongs to the vector space $\operatorname{Im} A = A(L_1) \subseteq L_2$.[25]

Consider, in particular, the differential equation

$$x^{(n)} + a_1 x^{(n-1)} + \cdots + a_n x = f(t) \tag{3}$$

(an nth-order inhomogeneous equation with constant coefficients).

Theorem. *Suppose the right-hand side $f(t)$ of Eq. (3) is a sum of quasi-polynomials. Then every solution of Eq. (3) is a sum of quasi-polynomials.*

Consider the space C^m of all quasi-polynomials

$$C^m = \{e^{\lambda t} p_{<m}(t)\}$$

of degree less than m with exponent λ. The linear operator D (mapping every function into its derivative) maps C^m into itself. Therefore the operator $A = a(D) = D^n + a_1 D^{n-1} + \cdots + a_n E : C^m \to C^m$ is also a linear transformation from C^m into itself. We can now write Eq. (3) in the form $Ax = f$. To study its solvability we must find the image $\operatorname{Im} A = A(C^m)$ of the mapping A.

Lemma 1. *Suppose λ is not a root of the characteristic equation, i.e., $a(\lambda) \neq 0$. Then $A : C^m \to C^m$ is an isomorphism.*

Proof. The matrix of the operator $D : C^m \to C^m$ in a suitable basis is a Jordan block with λ on the diagonal. In this same basis the operator A has a

[25] Here the notation Im denotes the image of the linear transformation A, not the imaginary part of a complex number. *Trans.*

triangular matrix with $a(\lambda)$ on the diagonal. Hence $\det A = (a(\lambda))^m \neq 0$, and A is an isomorphism. □

Corollary. *If λ is not a root of the characteristic equation, then Eq.* (3) *with right-hand side in the form of a quasi-polynomial of degree m and exponent λ has a particular solution in the form of a quasi-polynomial of degree less than m and exponent λ.*

Lemma 2. *Let λ be a root of the characteristic equation of multiplicity ν, i.e., $a(z) = (z-\lambda)^\nu b(z)$, $b(\lambda) \neq 0$. Then $AC^m = C^{m-\nu}$.*

Proof. $A = a(D) = (D - \lambda E)^\nu b(D)$. By Lemma 1 $b(D) : C^m \to C^m$ is an isomorphism. It remains to be shown that $(D - \lambda E)^\nu C^m = C^{m-\nu}$. But the matrix of the operator $D - \lambda E$ in the basis

$$e_k = \frac{t^k}{k!}e^{\lambda t}, \quad 0 \leq k < m,$$

is a nilpotent Jordan block, i.e., this operator acts on the basis as the shift $0 \leftarrow\!\!| \, e_0 \leftarrow\!\!| \, e_1 \leftarrow\!\!| \cdots \leftarrow\!\!| \, e_{m-1}$. The operator $(D - \lambda E)^\nu$ acts as a shift by ν places and therefore maps C^m onto $C^{m-\nu}$. □

Corollary. *Let λ be a root of the characteristic equation $a(\lambda) = 0$ of multiplicity ν. Let $f \in C^k$ be a quasi-polynomial of degree less than k with exponent λ. Then Eq.* (3) *has a solution $\varphi \in C^{k+\nu}$ in the form of a quasi-polynomial with exponent λ of degree less than $k + \nu$.*

For the proof it suffices to set $m = k + \nu$ in Lemma 2. □

Proof of the theorem. Consider the set Σ of all sums of quasi-polynomials. This is an infinite-dimensional subspace of the space F. By the preceding corollary the image $A(\Sigma)$ of the operator $A = a(D) : \Sigma \to \Sigma$ contains all quasi-polynomials. Being a vector space, $A(\Sigma)$ coincides with Σ. Therefore Eq. (3) has a particular solution that is a sum of quasi-polynomials. It remains only to add the general solution of the homogeneous equation. The latter is a sum of quasi-polynomials according to § 25.

The theorem is now proved. □

Remark 1. *If $f = e^{\lambda t}p_{<k}(t)$, then there exists a particular solution of Eq.* (3) *of the form $\varphi = t^\nu e^{\lambda t} q_{<k}(t)$.*

Indeed, by Lemma 2 there exists a particular solution in the form of a quasi-polynomial of degree less than $k + \nu$; but the terms of degree less than ν satisfy the homogeneous equation (cf. the corollary of Sect. 2), so that they can be discarded.

Remark 2. *Suppose Eq.* (3) *is real. If the λ are real, then the solution can be sought in the form of a real quasi-polynomial, while if $\lambda = \alpha \pm i\omega$, the solution* *can be sought in the form* $e^{\alpha t}(p(t)\cos\omega t + q(t)\sin\omega t)$. *Here the sine can appear in the solution even in the case when only cosines appear on the right-hand side.*

Problem 1. In what form can one write particular solutions of the following 13 equations:

$$1,2)\ \ddot{x} \pm x = t^2; \quad 3,4)\ \ddot{x} \pm x = e^{2t}; \quad 5,6)\ \ddot{x} \pm x = te^{-t};$$
$$7,8)\ \ddot{x} \pm x = t^3 \sin t; \quad 9,10)\ \ddot{x} \pm x = te^t \cos t;$$
$$11,12)\ \ddot{x} \pm 2ix = t^2 e^t \sin t; \quad 13)\ x^{(\mathrm{iv})} + 4x = t^2 e^t \cos t?$$

6. The Method of Complex Amplitudes

In the case of complex roots it is usually simpler to carry out the computations as follows.

Suppose Eq. (3) is real and the function $f(t)$ is represented as the real part of a complex-valued function $f(t) = \operatorname{Re} F(t)$. Let Φ be a complex-valued solution of the equation $a(D)\Phi = F$. Then, taking the real part, we verify that $a(D)\varphi = f$, where $\varphi = \operatorname{Re}\Phi$ (since $a = \operatorname{Re} a$).

Thus *to find the solutions of a linear inhomogeneous equation with right-hand side* f, *it suffices to regard* f *as the real part of a complex-valued function* F, *solve the equation with right-hand side* F, *and take the real part of the solution.*

Example 1. Let $f(t) = \cos\omega t = \operatorname{Re} e^{i\omega t}$. The degree of the quasi-polynomial $F(t) = e^{i\omega t}$ is 0, so that a solution can be sought in the form $Ct^\nu e^{i\omega t}$, where C is a complex constant (called the *complex amplitude*) and ν is the multiplicity of the root $i\omega$. Finally

$$\varphi(t) = \operatorname{Re}(Ct^\nu e^{i\omega t}).$$

If $C = re^{i\theta}$, then

$$\varphi(t) = rt^\nu \cos(\omega t + \theta).$$

Thus the complex amplitude C contains information about both the amplitude (r) and the phase (θ) of the real solution.

Example 2. Consider the behavior of the pendulum (Fig. 184) (or any other linear oscillating system, for example a loaded spring or an electrical oscillator) under the action of an external periodic force:

$$\ddot{x} + \omega^2 x = f(t), \quad f(t) = \cos\nu t = \operatorname{Re} e^{i\nu t}.$$

The characteristic equation $\lambda^2 + \omega^2 = 0$ has roots $\lambda = \pm i\omega$. If $\nu^2 \neq \omega^2$, then a particular solution should be sought in the form $\Phi = Ce^{i\nu t}$. Substituting into the equation, we find

$$C = \frac{1}{\omega^2 - \nu^2}. \qquad (4)$$

The quantity C can be written in trigonometric form: $C = r(\nu)e^{i\theta(\nu)}$.

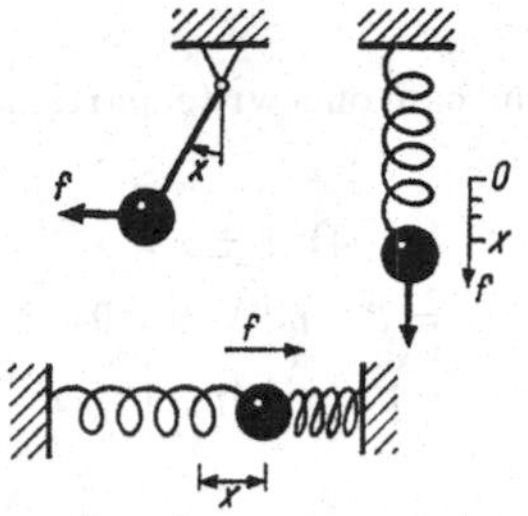

Fig. 184. A system oscillating under the action of the external force $f(t) = \cos \nu t$

According to formula (4) the amplitude r and the phase θ have the values indicated in Fig. 185[26]. The real part of Φ is $r\cos(\nu t + \theta)$. Thus the general solution of the inhomogeneous equation has the form

$$x = r\cos(\nu t + \theta) + C_1 \cos(\omega t + \theta_1).$$

Consequently *the oscillation of the pendulum under the influence of an external force consists of the "forced oscillation"* $r\cos(\nu t + \theta)$ *with the same frequency as the external force and the "free oscillation" with the natural frequency* ω.

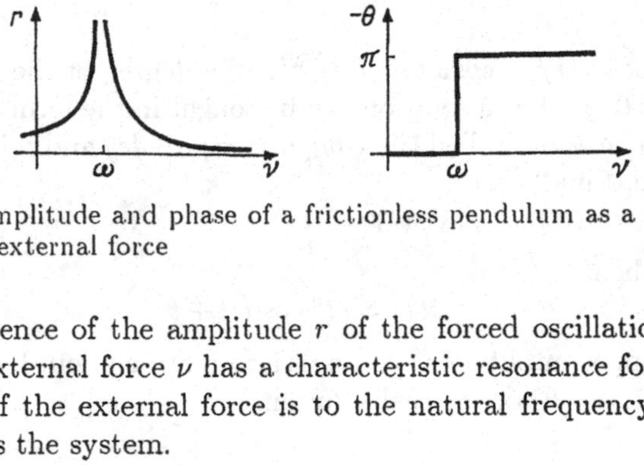

Fig. 185. The amplitude and phase of a frictionless pendulum as a function of the frequency of the external force

The dependence of the amplitude r of the forced oscillation on the frequency of the external force ν has a characteristic resonance form: the closer the frequency of the external force is to the natural frequency ω, the more strongly it rocks the system.

This phenomenon of resonance, which is observed when the frequency of the external force coincides with the natural frequency of an oscillating system, has great importance in applications. For example, in designing all kinds of devices it is necessary to take care that the natural frequency of the device not be close to the frequency of the external forces it will undergo. Otherwise even a small force, acting over an extended period of time, can rock the system and destroy it.

[26] The grounds for choosing $\theta = -\pi$ (not $+\pi$) for $\nu > \omega$ are shown by Example 3 below.

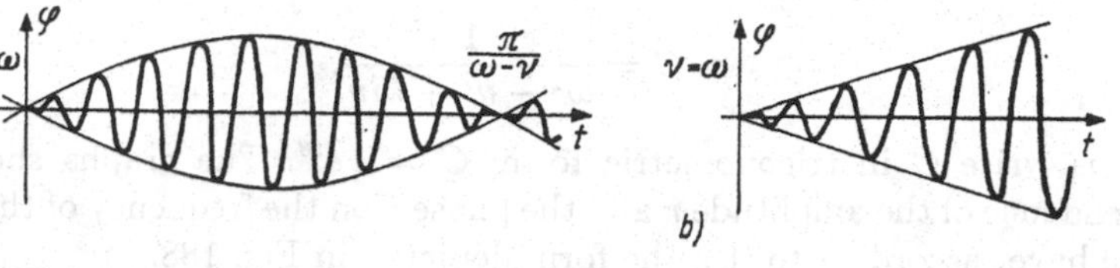

Fig. 186. The sum of two harmonics with frequencies close to each other (beats) and its limit in the case of resonance (reinforcement)

The phase of the forced oscillations θ has a jump of $-\pi$ when ν crosses the resonance value ω. For ν close to ω "beats" are observed (Fig. 186): the amplitude of the oscillations of the pendulum alternately increases (when the relation between the phases of the pendulum and the external force is such that the external force reinforces the pendulum, adding energy to it) and decreases (when the relation changes in such a way that the external force slows down the pendulum).

The closer the frequencies ν and ω are to each other the more slowly the relation between the phases, and *a fortiori* the period of the beats, changes. As $\nu \to \omega$ the period of the beats tends to infinity.

In the case of resonance ($\nu = \omega$) the relation of the phases is constant and the forced oscillations can grow to infinity (Fig. 186).

Indeed, by the general rule when $\nu = \omega$ we seek a solution of the form $x = \operatorname{Re} Cte^{i\omega t}$. Substituting into the equation, we find $C = 1/(2i\omega)$, whence $x = \frac{t}{2\omega}\sin\omega t$ (Fig. 186). The forced oscillations grow without bound.

Example 3. Consider the pendulum with friction $\ddot{x} + k\dot{x} + \omega^2 x = f(t)$. The characteristic equation $\lambda^2 + k\lambda + \omega^2 = 0$ has the following roots (Fig. 187):

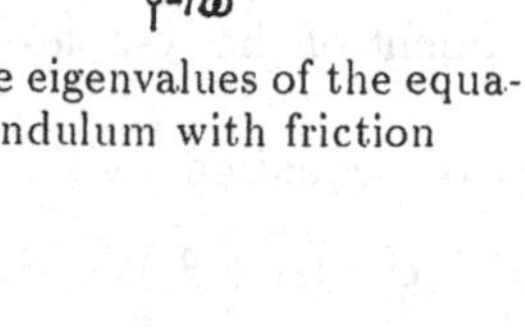

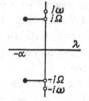

Fig. 187. The eigenvalues of the equation of the pendulum with friction

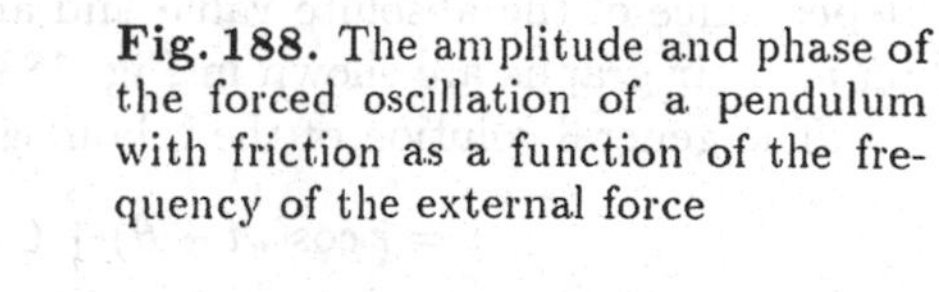

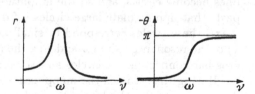

Fig. 188. The amplitude and phase of the forced oscillation of a pendulum with friction as a function of the frequency of the external force

$\lambda_{1,2} = -\alpha \pm i\Omega$, where $\alpha = \dfrac{k}{2}$ and $\Omega = \sqrt{\omega^2 - \dfrac{k^2}{4}}$. Let us assume that the coefficient of friction k is positive and small ($k^2 < 4\omega^2$). Consider a harmonic external force $f(t) = \cos\nu t = \operatorname{Re} e^{i\nu t}$. If the coefficient of friction k is nonzero, then $i\nu$ cannot be a root of the characteristic equation (since $\lambda_{1,2}$ have nonzero real parts). Therefore the solution must be sought in the form $x = \operatorname{Re} Ce^{i\nu t}$.

Substituting into the equation, we find

$$C = \frac{1}{\omega^2 - \nu^2 + ki\nu}. \tag{5}$$

Let us write C in trigonometric form: $C = re^{i\theta}$. The graphs showing the dependence of the amplitude r and the phase θ on the frequency of the external force have, according to (5), the form depicted in Fig. 188.

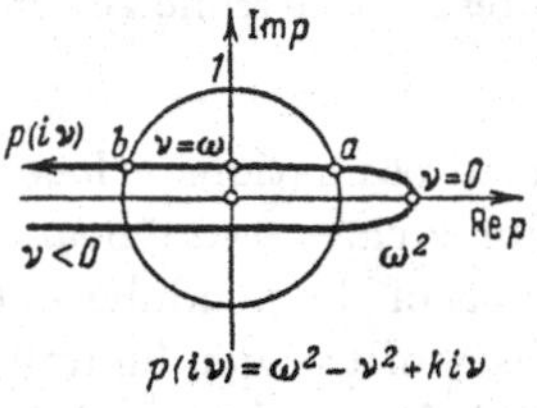

Fig. 189. The values of the characteristic polynomial on the imaginary axis

These graphs are constructed as follows. Consider the denominator of the fraction (5), i.e., the value of the characteristic polynomial p on the imaginary axis. The image of the mapping $\nu \mapsto p(i\nu) = \omega^2 - \nu^2 + ki\nu$ is called a *Mikhailov curve*. It can be seen from (5) that this curve (for our equation), is the parabola depicted in Fig. 189. In the case when the friction is small the parabola is "close" to a ray of the real axis traversed twice.

It is now easy to construct the image of the mapping $\nu \mapsto C(\nu) = 1/p(i\nu)$. This curve is called the *amplitude-phase characteristic*. To construct it, it suffices to perform an inversion and reflection in the real axis with the Mikhailov curve. The portion of the Mikhailov curve near the origin is almost indistinguishable from a pair of line segments and corresponds to neighborhoods of the points ω and $-\omega$ of the ν-axis with radii of order k. Under the inversion lines become circles, and so the amplitude-phase characteristic contains two parts that approximate large circles (of diameter $1/(k\omega)$) (Fig. 190). On the ν-axis these circles correspond to small neighborhoods (with radius of order k) of the resonance values ω and $-\omega$: the rest of the ν-axis corresponds to the crossbar connecting the circles and the terminal arcs.

Having thus studied the mapping $\nu \mapsto C(\nu)$, we can now easily study the dependence of the absolute value and argument of the complex amplitude C on ν: their graphs are shown in Fig. 188.

The general solution of the inhomogeneous equation, i.e.,

$$x = r\cos(\nu t + \theta) + C_1 e^{-\alpha t}\cos(\Omega t + \theta_1),$$

is obtained by adding the general solution of homogeneous equation (i.e., $C_1 e^{-\alpha t}\cos(\Omega t + \theta_1)$) to a particular solution of the inhomogeneous equation.

As $t \to +\infty$ this term tends to 0, so that only the forced oscillation $x = r\cos(\nu t + \theta)$ remains.

Let us compare the behavior of the pendulum with the coefficient of friction zero (Fig. 185) and positive (Fig. 188).

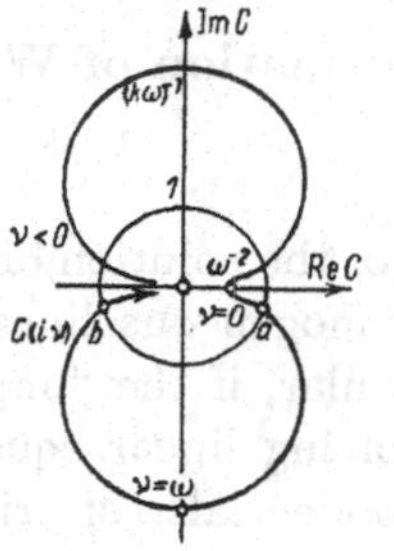

Fig. 190. The dependence of the complex amplitude on the frequency of the external force

We see that *the influence of a small friction on the resonance prevents the amplitude of the oscillations from becoming infinite under resonance and causes it to increase only to a definite finite value inversely proportional to the coefficient of friction.*

Indeed, the function $r(\nu)$ expressing the dependence of the amplitude of the steady-state oscillations on the frequency of the external force has a sharply defined maximum near $\nu = \omega$ (Fig. 188). It can be seen from formula (5) that the height of this maximum increases like $1/(k\omega)$ as k decreases.

From the "physical" point of view the finiteness of the amplitude of the steady-state forced oscillations when the coefficient of friction is nonzero is easy to predict by calculating the energy balance. With large amplitudes the energy lost to friction is larger than the energy communicated to the pendulum by the external force. Therefore the amplitude will decrease until a mode becomes established in which the loss of energy to friction is balanced by the work of the external force. The size of the amplitude of the steady-state oscillations increases in inverse proportion to the coefficient of friction as the latter tends to zero.

The shift in the phase θ is always negative: *the forced oscillation lags behind the impelling force.*

Problem 1. Prove that every solution of an inhomogeneous linear system of equations with constant coefficients and right-hand side in the form of a quasi-polynomial with vector coefficients

$$f = \sum_l e^{\lambda_l t} \sum_k c_{k,l} t^k$$

is a sum of quasi-polynomials with vector coefficients.

Problem 2. Prove that every solution of an inhomogeneous linear recursive equation with right-hand side in the form of a sum of quasi-polynomials

$$x_n + a_1 x_{n-1} + \cdots + a_k x_{n-k} = f(n)$$

is a sum of quasi-polynomials. Find a formula for the general term of the sequence 0, 2, 7, 18, 41, 88, ... ($x_n = 2x_{n-1} + n$).

7. Application to the Calculation of Weakly Nonlinear Oscillations

In studying the dependence of the solution of the equation on the parameters it is necessary to solve inhomogeneous linear equations – the equations of variations (cf. § 3). In particular, if the "unperturbed" system is linear, the problem often reduces to solving linear equations with the right-hand side in the form of a sum of exponentials (or trigonometric functions) or quasi-polynomials.

Problem 1. Study the dependence of the period of oscillation of a pendulum described by the equation $\ddot{x} = -\sin x$ on the amplitude A, regarding this last quantity as small.

Answer. $T = 2\pi(1 + A^2/16 + O(A^4))$.

For example, with an angle of deviation of 30° the period is 2% larger than the period for small oscillations.

Solution. Consider the solution of the pendulum equation with initial condition $x(0) = A$, $\dot{x}(0) = 0$ as a function of A.

By the theorem on differentiable dependence of the solution on the initial conditions this function is smooth. Let us expand it in a Taylor series in A about $A = 0$:

$$x = Ax_1(t) + A^2x_2(t) + A^3x_3(t) + O(A^4).$$

Then

$$\dot{x} = A\dot{x}^1 + A^2\dot{x}_2 + A^3\dot{x}_3 + O(A^4),$$
$$\ddot{x} = A\ddot{x}_1 + A^2\ddot{x}_2 + A^3\ddot{x}_3 + O(A^4),$$
$$\sin x = Ax_1 + A^2x_2 + A^3(x_3 - x_1^3/6) + O(A^4).$$

The equation $\ddot{x} = -\sin x$ holds for any A. From this we find equations for x_1, x_2, and x_3:

$$\ddot{x}_1 = -x_1, \quad \ddot{x}_2 = -x_2, \quad \ddot{x}_3 = -x_3 + x_1^3/6. \tag{6}$$

The initial condition $x(0) = A$, $\dot{x}(0) = 0$ holds for any A. From this we find initial conditions for Eq. (6):

$$x_1(0) = 1, \quad x_2(0) = x_3(0) = \dot{x}_1(0) = \dot{x}_2(0) = \dot{x}_3(0) = 0. \tag{7}$$

Solving Eq. (6) with conditions (7), we find $x_1 = \cos t$, $x_2 = 0$, and for x_3, we obtain the equation

$$\ddot{x}_3 + x_3 = (\cos^3 t)/6, \quad x_3(0) = \dot{x}_3(0) = 0.$$

Solving this equation (say by the method of complex amplitudes), we find

$$x_3 = \alpha(\cos t - \cos 3t) + \beta t \sin t,$$

where $\alpha = 1/192$ and $\beta = 1/16$.

Thus the influence of the nonlinearity ($\sin x \neq x$) on the oscillations of the pendulum reduces[27] to the addition of a term $A^3x_3 + O(A^4)$:

$$x = A\cos t + A^3[\alpha(\cos t - \cos 3t) + \beta t \sin t] + O(A^4).$$

The period of oscillation T is found as the maximum point for $x(t)$ near 2π for small A. This point is found from the condition $\dot{x}(T) = 0$, i.e.,

$$A\{-\sin t + A^2[(\beta - \alpha)\sin T + 3\alpha \sin 3T + \beta T \cos T] + O(A^3)\} = 0.$$

Let us solve this equation approximately for small A. We set $T = 2\pi + u$. For u we obtain the equation

$$\sin u = A^2[2\pi\beta + O(u)] + O(A^3).$$

By the implicit function theorem $u = 2\pi\beta A^2 + O(A^3)$, i.e., $T = 2\pi(1 + A^2/16 + o(A^2))$. Since T is an even function of A, we see that $o(A^2) = O(A^4)$.

Problem 2. Study the dependence of the period of oscillation on the amplitude A for the equation

$$\ddot{x} + \omega^2 x + ax^2 + bx^3 = 0.$$

Answer.

$$T = \frac{2\pi}{\omega}\left[1 + \left(\frac{5a^2}{12\omega^4} - \frac{3b}{8\omega^2}\right)A^2 + o(A^2)\right].$$

Problem 3. Obtain the same results from the explicit formula for the period (§ 12, Sect. 7).

§ 27. Nonautonomous Linear Equations

The portion of the theory of linear equations that is independent of translation-invariance carries over easily to linear equations and systems with nonconstant coefficients.

1. Definition

The equation

$$\dot{\boldsymbol{x}} = A(t)\boldsymbol{x}, \quad \boldsymbol{x} \in \boldsymbol{R}^n, \quad A(t) : \boldsymbol{R}^n \to \boldsymbol{R}^n, \tag{1}$$

[27] It is useful here to recall the bucket with a hole (cf. the caution in § 7, Sect. 5): the occurrence of the "secular" term $t \sin t$ in the formula for x_3 does not imply anything about the behavior of the pendulum as $t \to \infty$. Our approximation is good only on a finite interval of time; for large t the term $O(A^4)$ becomes large. And indeed, the true solution of the equation for the oscillations of a pendulum remains bounded (by the quantity A) for all t, as can be seen from the law of conservation of energy.

where t belongs to an interval I of the real axis, will be called a *linear (homogeneous) equation with variable coefficients*[28].

The solutions of Eq. (1) are depicted geometrically by integral curves in the strip $I \times \boldsymbol{R}^n$ of the extended phase space (Fig. 191). As usual, we shall assume that the function $A(t)$ is smooth[29].

Example 1. Consider the pendulum equation $\ddot{x} = -\omega^2 x$. The frequency ω is determined by the length of the pendulum. The oscillations of a pendulum of variable length are described by the analogous equation $\ddot{x} = -\omega^2(t)x$. This equation may be written in the form (1):

$$\begin{cases} \dot{x}_1 = x_2, \\ \dot{x}_2 = -\omega^2(t)x_1, \end{cases} \quad A(t) = \begin{pmatrix} 0 & 1 \\ -\omega^2(t) & 0 \end{pmatrix}.$$

An example of a pendulum of variable length is provided by a swing: by varying the position of the center of gravity a person on a swing can cause a periodic change in the value of the parameter ω (Fig. 192).

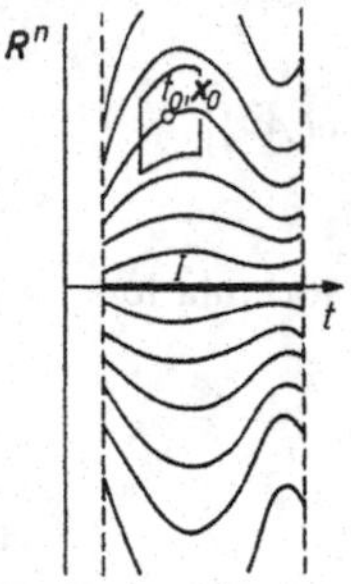

Fig. 191. The integral curves of a linear equation

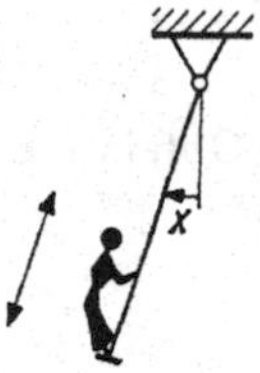

Fig. 192. A swing

2. The Existence of Solutions

One solution of Eq. (1) can be seen immediately: the zero solution. By the general theorems of Chapt. 2, for any initial conditions $(t_0, \boldsymbol{x}_0)$ in $I \times \boldsymbol{R}^n$ there exists a solution defined in some neighborhood of the point t_0. For a nonlinear equation it may be impossible to extend this solution to the entire interval I (Fig. 193). A peculiarity of linear equations is that the solution cannot go to infinity in a finite time.

Theorem. *Every solution of Eq.* (1) *can be extended to the entire interval.*

[28] We assume that the coefficients are real-valued. The complex case is completely analogous.

[29] It would suffice to assume that the function $A(t)$ is continuous (cf. below, Sect. 6 of § 32).

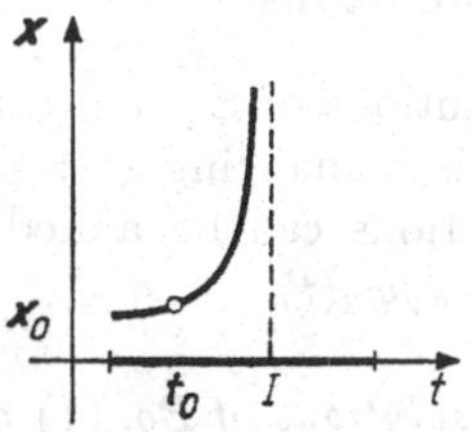

Fig. 193. A solution of the equation $\dot{x} = x^2$ that cannot be extended

The reason is that for a linear equation

$$\|\dot{\boldsymbol{x}}\| \leq C\|\boldsymbol{x}\|,$$

and so the solution grows no faster than e^{Ct}.

A precise proof can be carried out, for example, as follows. Let $[a, b]$ be a compact interval in I. Then on the interval $[a, b]$ the norm[30] of the operator $A(t)$ is bounded:

$$\|A(t)\| < C = C(a, b).$$

We shall prove the following *a priori estimate*:

If a solution φ *is defined on the interval* $[t_0, t]$ $(a \leq t_0 \leq t \leq b)$ (Fig. 194), *then*

$$\|\varphi(t)\| \leq e^{C(t-t_0)}\|\varphi(t_0)\|. \tag{2}$$

For the zero solution this is obvious. If $\varphi(t_0) \neq 0$, then by the uniqueness theorem $\varphi(\tau)$ does not vanish amywhere. Let us set $r(\tau) = \|\varphi(\tau)\|$. The function $L(\tau) = \ln r^2$ is defined for $t_0 \leq \tau \leq t$.

By hypothesis $\dot{L} \leq 2r\dot{r}/r^2 \leq 2C$. Therefore $L(t) \leq L(t_0) + 2C(t - t_0)$, which proves the a priori estimate (2).

Now suppose $\|\boldsymbol{x}_0\|^2 = B > 0$. Consider a compact subset of the extended phase space (Fig. 195)

$$F = \{t, \boldsymbol{x} : a \leq t \leq b, \|\boldsymbol{x}\|^2 \leq 2Be^{2C(b-a)}\}.$$

By the extension theorem the solution with initial condition $\varphi(t_0) = \boldsymbol{x}_0$ can be extended forward to the boundary of the cylinder F. The boundary of the cylinder F consists of the top and bottom ($t = a$, $t = b$) and the lateral surface ($\|\boldsymbol{x}\|^2 = 2Be^{2C(b-a)}$). The solution cannot exit through the lateral surface since, by the a priori estimate, $\|\varphi(t)\| \leq Be^{2C(b-a)}$. Thus the solution can be continued to the right to $t = b$. The proof of the extension to the left to a is similar.

Since a and b are arbitrary, the theorem is proved.

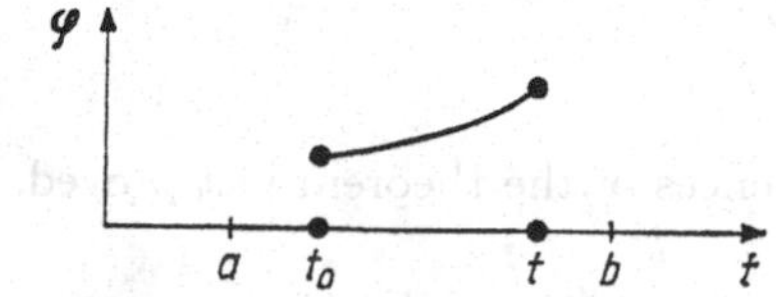

Fig. 194. The a priori estimate of the growth of the solution on $[a, b]$

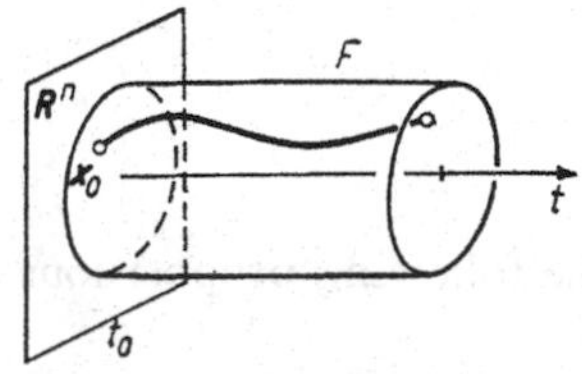

Fig. 195. The extension of the solution to $t = b$

[30] We are assuming that some Euclidean metric has been chosen in $\boldsymbol{R}^n$.

3. The Vector Space of Solutions

Consider the set X of all solutions of Eq. (1) that are defined on the entire interval I. Since a solution is a mapping $\varphi : I \to \boldsymbol{R}^n$ with values in the vector phase space $\boldsymbol{R}^n$, solutions can be added and multiplied by scalars: $(c_1\varphi_1 + c_2\varphi_2)(t) = c_1\varphi_1(t) + c_2\varphi_2(t)$.

Theorem. *The set X of all solutions of Eq.* (1) *defined on the interval I is a vector space.*

Proof. This is obvious:

$$\frac{d}{dt}(c_1\varphi_1 + c_2\varphi_2) = c_1\dot{\varphi}_1 + c_2\dot{\varphi}_2 = c_1A\varphi_1 + c_2A\varphi_2 = A(c_1\varphi_1 + c_2\varphi_2).$$

Theorem. *The vector space X of solutions of a linear equation is isomorphic to the phase space $\boldsymbol{R}^n$ of the equation.*

Proof. Let $t \in I$. Consider the mapping

$$B_t : X \to \boldsymbol{R}^n, \quad B_t\varphi = \varphi(t),$$

which assigns to each solution φ its value at the instant t.

The mapping B_t is linear (since the value of the sum of two solutions is the sum of their values). Its image is the entire phase space $\boldsymbol{R}^n$, since by the existence theorem, for every $\boldsymbol{x}_0 \in \boldsymbol{R}^n$ there exists a solution φ with the initial condition $\varphi(t_0) = \boldsymbol{x}_0$. The kernel of the mapping B_t is $\mathbf{o}$ since the solution with zero initial condition $\varphi(t_0) = \mathbf{o}$ is identically zero by the uniqueness theorem.

Thus B_t is an *isomorphism of X onto $\boldsymbol{R}^n$*. This is the fundamental result of the theory of linear equations.

Definition. A *fundamental system of solutions* of Eq. (1) is a basis of the vector space X of solutions.

Problem 1. Find a fundamental system of solutions of Eq. (1), where

$$A = \begin{pmatrix} 0 & 1 \\ -1 & 0 \end{pmatrix}.$$

The following propositions are consequences of the theorem just proved.

Corollary 1. *Eq.* (1) *has a fundamental system of n solutions $\varphi_1, \ldots, \varphi_n$.*

Corollary 2. *Every solution of Eq.* (1) *is a linear combination of the solutions of a fundamental system.*

Corollary 3. *Any set of $n+1$ solutions of Eq. (1) is linearly dependent.*

Corollary 4. *The mappings over the time from t_0 to t_1 corresponding to Eq. (1) (Fig. 196)*

$$g_{t_0}^{t_1} = B_{t_1} B_{t_0}^{-1} : \boldsymbol{R}^n \to \boldsymbol{R}^n$$

are linear isomorphisms.

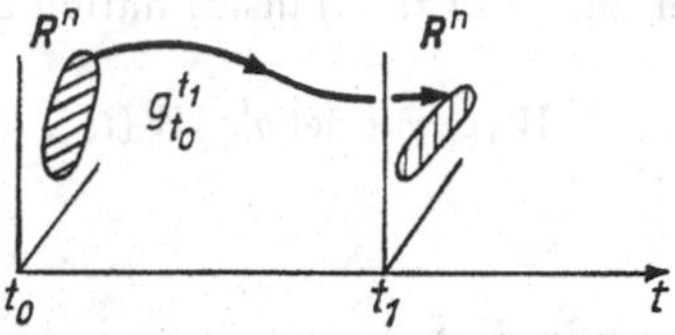

Fig. 196. The linear transformation of the phase space that is realized by the solutions of a linear equation over the time from t_0 to t_1

4. The Wronskian Determinant

Let $e_1, \ldots, e_n$ be a basis in the phase space $\boldsymbol{R}^n$. The choice of a basis fixes a unit of volume and an orientation in $\boldsymbol{R}^n$. Therefore every parallelepiped in the phase space has a definite volume.

Consider a system of n vector-valued functions $\varphi_k : I \to \boldsymbol{R}^n$, $(k = 1, \ldots, n)$.

Definition. The *Wronskian* of the system of vector-valued functions φ_k is the numerical function $W : I \to \boldsymbol{R}$ whose value at the point t equals the (oriented) volume of the parallelepiped spanned by the vectors $\varphi_1(t), \ldots, \varphi_n(t) \in \boldsymbol{R}^n$,

$$W(t) = \begin{vmatrix} \varphi_{1,1}(t) & \ldots & \varphi_{n,1}(t) \\ \ldots & \ldots & \ldots \\ \varphi_{1,n}(t) & \ldots & \varphi_{n,n}(t) \end{vmatrix},$$

where

$$\varphi_k(t) = \varphi_{k,1}(t) e_1 + \cdots + \varphi_{k,n}(t) e_n.$$

In particular, let φ_k be solutions of Eq. (1). Their images in the isomorphism B_t constructed above are vectors of the phase space $\varphi_k(t) \in \boldsymbol{R}^n$. They are linearly independent if and only if the Wronskian equals 0 at the point t. Hence:

Corollary 5. *A system of solutions $\varphi_1, \ldots, \varphi_n$ of Eq. (1) is fundamental if and only if its Wronskian is nonzero at some point t.*

Corollary 6. *If the Wronskian of a system of solutions of Eq.* (1) *vanishes at one point, then it vanishes identically for all t.*

Problem 1. Can the Wronskian of a system of linearly independent vector-valued functions vanish identically?

Answer. Yes.

Problem 2. Prove that the Wronskian of a system of fundamental solutions is proportional to the determinant of the transformation over the time from t_0 to t_1:

$$W(t) = \left(\det g_{t_0}^{t_1}\right) W(t_0).$$

Hint. The solution is given in Sect. 6.

5. The Case of a Single Equation

We consider a single nth-order equation

$$x^{(n)} + a_1 x^{(n-1)} + \cdots + a_n x = 0 \tag{3}$$

with, in general, variable coefficients $a_k = a_k(t)$, $t \in I$.

Some second-order equations with variable coefficients are encountered so frequently that they have special names and their solutions have been studied and tabulated in as much detail as the sine and cosine (cf., for example, Jahnke, Emde, and Losch *Tables of Functions*, McGraw-Hill, New York, 1960).

Example 1. Bessel's equation: $\ddot{x} + \frac{1}{t}\dot{x} + \left(1 - \frac{\nu^2}{t^2}\right)x = 0.$

Example 2. The hypergeometric equation of Gauss:

$$\ddot{x} + \frac{(\alpha+\beta+1)t - \gamma}{t(t-1)}\dot{x} + \frac{\alpha\beta}{t(t-1)}x = 0.$$

Example 3. The Mathieu equation: $\ddot{x} + (a + b\cos t)x = 0.$

We could have written Eq. (3) as a system of n first-order equations and applied the previous reasoning.

It is possible, however, to study the solution space X of Eq. (3) directly. It is a vector space of functions $\varphi : I \to \boldsymbol{R}$. It is naturally isomorphic to the solution space of the equivalent system of n equations. The isomorphism is given by assigning to the function φ the vector-valued function $\boldsymbol{\varphi} = (\varphi, \dot{\varphi}, \ldots, \varphi^{(n-1)})$ of derivatives of φ. Thus:

Corollary 7. *The solution space X of Eq.* (3) *is isomorphic to the phase space $\boldsymbol{R}^n$ of Eq.* (3), *an isomorphism being given by assigning to each solution $\varphi \in X$ the set of values of the derivatives at some point t_0:*

$$\varphi \mapsto (\varphi(t_0), \dot{\varphi}(t_0), \dots, \varphi^{(n-1)}(t_0)).$$

Definition. A basis of the vector space X is called a *fundamental system of solutions* of Eq. (3).

Problem 1. Exhibit a fundamental system of solutions of Eq. (3) in the case when the coefficients a_k are constant. (For example, for $\ddot{x} + ax = 0$.)

Answer. $\{t^r e^{\lambda t}\}$, where $0 \le r < \nu$, if λ is a root of the characteristic equation of multiplicity ν. In the case of complex roots ($\lambda = \alpha \pm i\omega$) the factor $e^{\lambda t}$ must be replaced by $e^{\alpha t} \cos \omega t$ and $e^{\alpha t} \sin \omega t$. In particular, for $\ddot{x} + ax = 0$

$$\begin{array}{llll} \cos \omega t \quad \text{and} \quad \sin \omega t & & & \text{if } a = \omega^2 > 0; \\ \cosh \alpha t \quad \text{and} \quad \sinh \alpha t & \text{or } e^{\alpha t} \text{ and } e^{-\alpha t} & & \text{if } a = -\alpha^2 < 0; \\ 1 \quad \text{and} \quad t & & & \text{if } a = 0. \end{array}$$

Definition. The *Wronskian of the system of functions* $\varphi_k : I \to \boldsymbol{R}$, $1 \le k \le n$, is the numerical function $W : I \to \boldsymbol{R}$ whose value at the point t equals

$$W(t) = \begin{vmatrix} \varphi_1(t) & \dots & \varphi_n(t) \\ \dot{\varphi}_1(t) & \dots & \dot{\varphi}_n(t) \\ \dots & \dots & \dots \\ \varphi_1^{(n-1)}(t) & \dots & \varphi_n^{(n-1)}(t) \end{vmatrix}.$$

In other words it is the Wronskian of the system of vector-valued functions $\boldsymbol{\varphi}_k : I \to \boldsymbol{R}^n$ obtained from φ_k in the usual way:

$$\boldsymbol{\varphi}_k = (\varphi_k(t), \dot{\varphi}_k(t), \dots, \varphi_k^{(n-1)}(t)), \quad k = 1, \dots, n.$$

Everything that was said about the Wronskian of a system of vector-valued solutions of Eq. (1) carries over without change to the Wronskian of a system of solutions of Eq. (3). In particular, we have the following proposition.

Corollary 8. *If the Wronskian of a system of solutions of Eq.* (3) *vanishes at even one point, then it vanishes identically for all t.*

Problem 2. Suppose the Wronskian of two functions vanishes at the point t_0. Does it follow that it vanishes identically?

Answer. No, unless the functions are analytic.

Corollary 9. *If the Wronskian of a system of solutions of Eq. (3) vanishes at even one point, then these solutions are linearly dependent.*

Problem 3. Suppose the Wronskian of two functions is identically zero. Does it follow that these functions are linearly dependent?

Corollary 10. *A system of n solutions of Eq. (3) is fundamental if and only if their Wronskian is nonzero at at least one point.*

Example 4. Consider the system of functions $e^{\lambda_1 t}, \ldots, e^{\lambda_n t}$. These functions form a fundamental system of solutions of a linear equation of the form (3) (which one?). Therefore they are linearly independent. Hence their Wronskian is nonzero. But this determinant is

$$W = \begin{vmatrix} e^{\lambda_1 t} & \ldots & e^{\lambda_n t} \\ \lambda_1 e^{\lambda_1 t} & \ldots & \lambda_n e^{\lambda_n t} \\ \ldots & \ldots & \ldots \\ \lambda_1^{(n-1)} e^{\lambda_1 t} & \ldots & \lambda_n^{(n-1)} e^{\lambda_N t} \end{vmatrix} = e^{(\lambda_1 + \cdots + \lambda_n)t} \begin{vmatrix} 1 & \ldots & 1 \\ \lambda_1 & \ldots & \lambda_n \\ \ldots & \ldots & \ldots \\ \lambda_1^{(n-1)} & \ldots & \lambda_n^{(n-1)} \end{vmatrix}.$$

Corollary 11. *The Vandermonde determinant*

$$\begin{vmatrix} 1 & \ldots & 1 \\ \lambda_1 & \ldots & \lambda_n \\ \ldots & \ldots & \ldots \\ \lambda_1^{(n-1)} & \ldots & \lambda_n^{(n-1)} \end{vmatrix}$$

is nonzero if the λ_k are pairwise distinct.

Example 5. Consider the pendulum equation $\ddot{x} + \omega^2 x = 0$. A fundamental system of solutions is: $(\cos \omega t, \sin \omega t)$. The Wronskian of this system $W = \begin{vmatrix} \cos \omega t & \sin \omega t \\ -\omega \sin \omega t & \omega \cos \omega t \end{vmatrix} = \omega$ is constant. This is not surprising, since the phase flow of the pendulum equation preserves area (cf. § 16, Sect. 4).

Let us now study how the volume of figures in the phase space changes in the general case under the action of the transformations $g_{t_0}^t$ over the time from t_0 to t.

6. Liouville's Theorem

The Wronskian of the solutions of Eq. (1) is a solution of the differential equation

$$\dot{W} = aW, \quad \textit{where} \quad a(t) = \operatorname{tr} A(t) \ \textit{(the trace of } A(t)\textit{)}. \tag{4}$$

Corollary.

$$W(t) = e^{\int_{t_0}^{t} a(\tau)\,d\tau} W(t_0), \quad \det g_{t_0}^{t} = e^{\int_{t_0}^{t} a(\tau)\,d\tau}. \tag{5}$$

Indeed Eq. (4) is easy to solve:

$$\frac{dW}{W} = a\,dt, \quad \ln W - \ln W_0 = \int_{t_0}^{t} a(\tau)\,d\tau.$$

Incidentally it can be seen once again from formula (5) that the Wronskian of the solutions is either identically zero or does not vanish at any point.

Problem 1. Find the volume of the image of the unit cube $0 \le x_i \le 1$ under the action of the transformation over unit time of the phase flow of the system

$$\dot{x}_1 = 2x_1 - x_2 - x_3, \quad \dot{x}_2 = x_1 + x_2 + x_3, \quad \dot{x}_3 = x_1 - x_2 - x_3.$$

Solution. $\operatorname{tr} A = 2$, so that $W(t) = e^{2t} W(0) = e^{2t}$.

A short proof of Liouville's Theorem. If the coefficients are constant, then Eq. (4) is Liouville's formula from § 16. "Freezing" the coefficients $A(t)$ (setting them equal to their values at some instant τ), we verify Eq. (4) for any τ. □

A long proof. Consider the linear transformation of the phase space $g_\tau^{\tau+\Delta} : \boldsymbol{R}^n \to \boldsymbol{R}^n$ (Fig. 197) over the small time interval from τ to $\tau + \Delta$. This transformation maps the value of each solution φ of Eq. (1) at the instant τ to its value at the instant $\tau + \Delta$. According to Eq. (1)

$$\varphi(\tau + \Delta) = \varphi(\tau) + A(\tau)\varphi(\tau)\Delta + o(\Delta),$$

i.e., $g_\tau^{\tau+\Delta} = E + \Delta A(\tau) + o(\Delta)$.

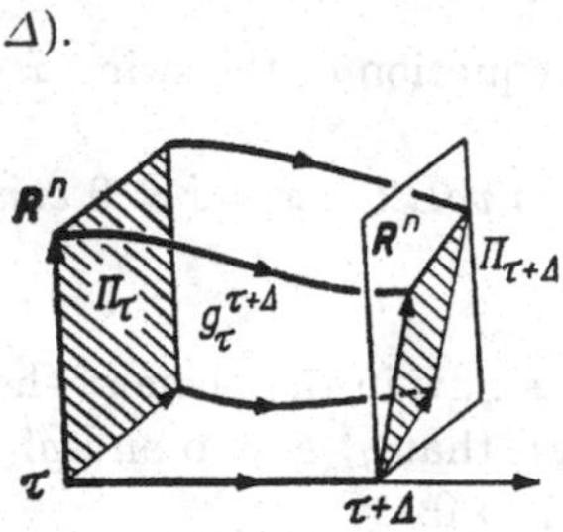

Fig. 197. The action of the phase flow on the parallelepiped spanned by a fundamental system of solutions

Consequently according to § 16 the coefficient of dilation of volumes under the transformation $g_\tau^{\tau+\Delta}$ is $\det g_\tau^{\tau+\Delta} = 1 + \Delta a + o(\Delta)$, where $a = \operatorname{tr} A$.

But $W(\tau)$ is the volume of the parallelepiped Π_τ spanned by the values of our system of solutions at the instant τ. The transformation $g_\tau^{\tau+\Delta}$ maps these values

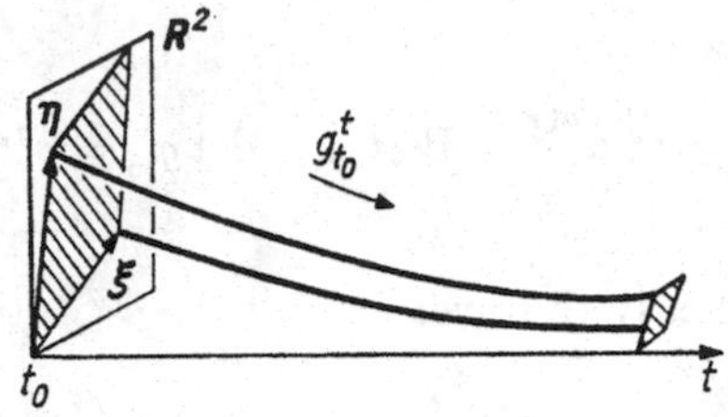

Fig. 198. The phase flow of an asymptotically stable linear system

into the set of values of the same system of solutions at the instant $\tau + \Delta$. The parallelepiped $\Pi_{\tau+\Delta}$ spanned by these new values has volume $W(\tau + \Delta)$. Thus

$$W(\tau + \Delta) = \left(\det g_\tau^{\tau+\Delta}\right) W(\tau) = [1 + a(\tau)\Delta + o(\Delta)]W(\tau),$$

whence $\dot{W} = aW$, which was to be proved. □

Corollary. *The Wronskian of a system of solutions of Eq.* (3) *equals*

$$W(t) = e^{-\int_{t_0}^t a_1(\tau)\,d\tau} W(t_0).$$

The negative sign appears because in writing Eq. (3) in the form of the system (1) the term $a_1 x^{(n-1)}$ must be transposed to the right-hand side. The only nonzero element of the diagonal of the matrix of the resulting system

$$\begin{pmatrix} 0 & 1 & & \\ & & \ddots & \\ & & & 1 \\ -a_n & & & -a_1 \end{pmatrix}$$

is $-a_1$.

Example 1. Consider the equation of the swing $\ddot{x} + f(t)x = 0$.

Theorem. *The equilibrium position $x = \dot{x} = 0$ cannot be asymptotically stable for any f.*

Proof. Consider some basis $\boldsymbol{\xi}, \boldsymbol{\eta}$ in the plane of the initial conditions $\boldsymbol{R}^2$ (Fig. 198). Stability would mean that $g_{t_0}^t \boldsymbol{\xi} \to \mathbf{o}$ and $g_{t_0}^t \boldsymbol{\eta} \to \mathbf{o}$. Then for a suitable fundamental system $W(t) \to 0$.

The equation is equivalent to the system

$$\dot{x}_1 = x_2, \quad \dot{x}_2 = -f(t)x_1$$

with matrix $\begin{pmatrix} 0 & 1 \\ -f & 0 \end{pmatrix}$. Since $\operatorname{tr} A = 0$, it follows that $W(t) = \text{const}$, contradicting the relation $W \to 0$. □

Problem 2. Consider the swing with friction $\ddot{x} + \alpha(t)\dot{x} + \omega^2(t)x = 0$. Show that asymptotic stability is impossible if the coefficient of friction is negative ($\alpha(t) < 0\ \forall t$).

Is it true that for a positive coefficient of friction the equilibrium position $(0, 0)$ is always stable?

Remark. The *divergence* of a vector field $\boldsymbol{v}$ in Euclidean space $\boldsymbol{R}^n$ with Cartesian coordinates x_i is the function $\operatorname{div} \boldsymbol{v} = \sum_{i=1}^{n} \frac{\partial v_i}{\partial x_i}$.

In particular *for a linear vector field $\boldsymbol{v}(\boldsymbol{x}) = A\boldsymbol{x}$ the divergence is the trace of the operator A*:

$$\operatorname{div} A\boldsymbol{x} = \operatorname{tr} A.$$

The divergence of a vector field determines the rate of dilation of volumes under the corresponding phase flow.

Let D be the domain of definition of the (not necessarily linear) equation $\dot{\boldsymbol{x}} = \boldsymbol{v}(\boldsymbol{x})$ in Euclidean space. We denote by $D(t)$ the image of the region D under the action of the phase flow and by $V(t)$ the volume of the region $D(t)$.

Problem *3. Prove the following theorem.

Liouville's Theorem.

$$\frac{dV}{dt} = \int_{D(t)} \operatorname{div} \boldsymbol{v}\, dx \quad \text{(Fig. 199)}.$$

Corollary 1. *If* $\operatorname{div} \boldsymbol{v} = 0$*, then the phase flow preserves the volume of any region.*

Such a phase flow can be pictured as the flow of an incompressible "phase fluid" in the phase space.

Corollary 2. *The phase flow of the Hamilton equations*

$$\dot{p}_k = -\frac{\partial H}{\partial q_k}, \quad \dot{q}_k = \frac{\partial H}{\partial p_k}, \qquad k = 1, \ldots, n,$$

preserves volumes.

Proof. $\operatorname{div} \boldsymbol{v} = \sum \frac{\partial^2 H}{\partial q_k \partial p_k} - \frac{\partial^2 H}{\partial p_k \partial q_k} \equiv 0.$ □

This fact plays a fundamental role in statistical physics.

7. Sturm's Theorems on the Zeros of Solutions of Second-order Equations

The solutions of second-order linear equations possess certain peculiar oscillation properties. Sturm spoke of "the theorems whose names I have the honor to bear."

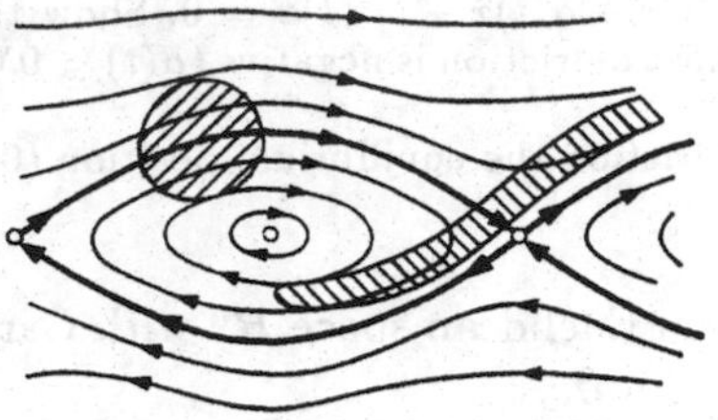

Fig. 199. The phase flow of a vector field of divergence 0 preserves areas

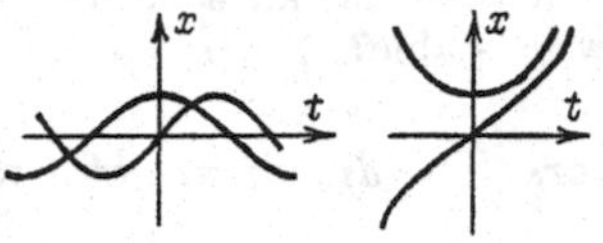

Fig. 200. The solutions of the equation $\ddot{x} \pm \omega^2 x = 0$

Consider the equations with constant coefficients (Fig. 200):

$$\ddot{x} + \omega^2 x = 0, \quad \ddot{x} - k^2 x = 0.$$

The solutions of the first equation have infinitely many zeros. The distance between two successive zeros of any nonzero solution of this equation is π/ω. Each nonzero solution of the second equation has at most one zero. In both cases between any two zeros of a solution that is not identically zero there is a zero of any other solution.

Sturm's theorems show that analogous phenomena hold for the equations with variable coefficients

$$\ddot{x} + q(t)x = 0 \tag{6}$$

(the more general equation $\ddot{x} + p(t)\dot{x} + q(t)x = 0$ is easily reduced to the form (6)).

Consider the phase plane for Eq. (6) with coordinates $(x,\ y = \dot{x})$. The phase curves of a nonautonomous equation may intersect. Nevertheless it is possible to obtain some information about these curves for a second-order equation. This information is the basis of Sturm's theorem.

Proposition 1. *The phase curves of Eq.* (6) *intersect the ray* $x = 0$, $y > 0$ *in points where* x *is increasing and the ray* $x = 0$, $y < 0$ *in points where* x *is decreasing.*

Proof. Write Eq. (6) in the form of a system

$$\dot{x} = y, \quad \dot{y} = -q(t)x.$$

On the line $x = 0$ the phase velocity vector has the component $(y, 0)$ for any q (Fig. 201), which proves Prop. 1. □

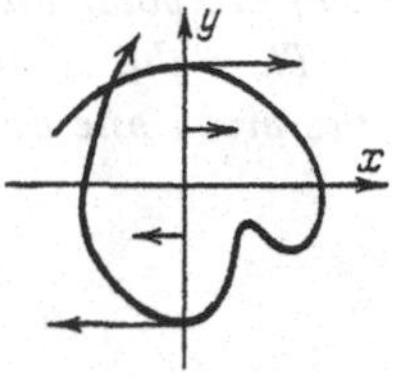

Fig. 201. The phase plane of the equation $\ddot{x} + q(t)x = 0$

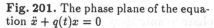

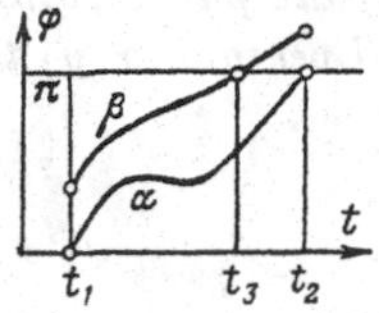

Fig. 202. Proof of the theorem on zeros

We remark that for $y \neq 0$ the phase velocity vector is nonzero on the axis $x = 0$. Therefore the zeros of any (not identically zero) solution of Eq. (1) are isolated and on any segment of the t-axis there are only finitely many such zeros.

The following proposition is an immediate consequence of Prop. 1.

Proposition 2. *For any two successive intersections of a phase curve with the line $x = 0$ one occurs with $y > 0$, the other with $y < 0$.*

We denote by φ the polar angle measured from the positive direction of the y-axis in the direction of the positive x-axis. The following result is a consequence of Prop. 2.

Proposition 3. *Between any two successive intersections of a phase curve with the axis $x = 0$ the quantity φ increases by π along the phase curve.*

The following result is an obvious consequence of this proposition.

Theorem. *On the interval between two successive zeros of any solution of Eq. (1) there is a zero of any other solution.*

Proof. Consider the polar angle φ along the two solutions (Fig. 202), $\varphi = \alpha(t)$, $\varphi = \beta(t)$. Suppose the zeros of the first solution correspond to $t = t_1$ and $t = t_2$. Assume that for the first solution we have $y > 0$ for $t = t_1$ (if such is not the case, reverse the sign of the first solution). Then we may assume $\alpha(t_1) = 0$. By Prop. 3 we have $\alpha(t_2) = \pi$. We may assume that $0 \leq \beta(t_1) \leq \pi$ (if such is not the case, reverse the sign of the second solution).

If the solutions are linearly dependent, their zeros coincide and there is nothing to prove. If they are linearly independent, then the vectors corresponding to them in the phase plane are also linearly independent at each instant of time. Consequently, in this case we have $\beta(t) \neq \alpha(t)$ for all t.

Thus $\beta(t_1) < \pi = \alpha(t_2) < \beta(t_2)$. Consequently on the interval $[t_1, t_2]$ there exists t_3 for which $\beta(t_3) = \pi$; this is a zero of the second solution. □

A second consideration that lies at the basis of Sturm's theorems is that the angular velocity of the motion of a phase point of Eq. (6) about the origin can be computed explicitly.

Proposition 4. *Denote by $\dot\varphi$ the rate of variation of the polar angle φ during a motion of the phase point $(x(t), y(t))$ of Eq. (1). Then the value of $\dot\varphi$ is the same for all vectors (x, y) that are collinear with the given one and is equal to*

$$\dot\varphi = \frac{q(t)x^2 + y^2}{x^2 + y^2}.$$

Proof. If $\boldsymbol{r}$ is the radius-vector of the phase point, then the double of the angular momentum per unit mass is $[\boldsymbol{r}, \dot{\boldsymbol{r}}]$ and also equal to $-r^2\dot\varphi$ (the plane is oriented by the coordinates (x, y), but the angle φ is measured from the y-axis toward the x-axis). Therefore

$$\dot\varphi = -\frac{[\boldsymbol{r}, \dot{\boldsymbol{r}}]}{r^2} = -\frac{\begin{vmatrix} x & y \\ y & -qx \end{vmatrix}}{x^2 + y^2},$$

which was to be proved. □

It follows from Prop. 4 that *at equal values of the polar angle $\varphi \neq k\pi$ the larger the coefficient q in the equation, the faster the radius-vector of a phase point revolves.*

The following result is an easy consequence of this fact.

The Sturm Comparison Theorem. *Consider two equations of the form* (6)

$$\ddot x + q(t)x = 0, \quad \ddot x + Q(t)x = 0,$$

and assume that $Q \geq q$. Then on the interval between any two successive zeros of any solution of the first equation (the one with the smaller coefficient q) there is a zero of any solution of the second equation.

Proof. Assume to begin with that Q is strictly larger than q for all t. Denote by $\varphi = \alpha(t)$ the polar angle along the first solution and by $\varphi = A(t)$ the polar angle along the second solution. As above, we may assume that $\alpha(t_1) = 0$, $\alpha(t_2) = \pi$, and $0 < A(t_1) < \pi$. At the initial instant t_1 we have $A(t_1) > \alpha(t_1)$. Subsequently, for $t_1 < t < t_2$ $A(t)$ will remain larger than $\alpha(t)$. Indeed, if the function α overtook A for the first time at some instant τ, then at that instant the values of α and A would be the same and would be different from $k\pi$. But then at the instant when α overtook A the radius-vector of the overtaking point would be revolving more slowly ($\dot A(\tau) > \dot\alpha(\tau)$ according to Prop. 4, since $Q > q$) and there could be no overtaking. Thus $A(t_2) > \alpha(t_2) = \pi$. But $A(t_1) < \pi$. Hence there exists an instant $t_3 \in [t_1, t_2]$ where $A(t_3) = \pi$. This is a zero of the second solution.

In the case $Q \geq q$ the proof is obtained by a limiting passage from the case $Q > q$. □

Corollary. *The distance between any two successive zeros of Eq.* (6)

a) *is not larger than π/ω if $q(t) \geq \omega^2$ for all t,*

b) *is not less than π/Ω if $q(t) \leq \Omega^2$ for all t.*

In particular, if $q(t) \le 0$ for all t, then no solution of Eq. (6) except the identically zero solution can have two zeros.

Proof. a) Suppose an interval of length π/ω can be inserted strictly between two successive zeros. For each interval of length π/ω one can choose a solution of the equation $\ddot{x}+\omega^2 x = 0$ whose zeros will be the endpoints of this interval. On this interval the solution of Eq. (6), which has coefficient q not less than ω^2, will have no zeros.

This contradicts the comparison theorem. Hence it is impossible to insert an interval of length π/ω between the zeros of a solution of Eq. (6).

b) The proof is similar, using a comparison with the equation $\ddot{x}+\Omega^2 x = 0$. □

The study of the characteristic oscillations of solid media (a clamped string) leads to the following *Sturm-Liouville problem.*

Find the solutions of the equation

$$\ddot{x} + (q(t) + \lambda)x = 0, \tag{7}$$

that vanish at the endpoints of a given interval $0 \le t \le l$.

The values of the *spectral parameter* λ at which such solutions (not identically zero) exist are called the *eigenvalues* and the solutions themselves are called *eigenfunctions.*

Problem 1. Find the eigenfunctions and eigenvalues in the case $q \equiv 0$.

Answer. The eigenfunctions are $\sin\sqrt{\lambda_k}t$ (Fig. 203), and the eigenvalues are $\lambda = k^2\lambda_1$, $\lambda_1 = (\pi/l)^2$.

Fig. 203. The characteristic oscillations of a string

Solution. $\ddot{x} + \lambda x = 0$, $x(0) = x(l) = 0 \Rightarrow \lambda > 0 \Rightarrow x = a\cos\sqrt{\lambda}l + b\sin\sqrt{\lambda}l$; $x(0) = 0 \Rightarrow a = 0 \Rightarrow \sqrt{\lambda}l = k\pi \Rightarrow \lambda_k = k^2(\pi/l)^2$.

Theorem. *For any function q that is smooth on the interval $[0,l]$ the Sturm-Liouville problem has an infinite set of eigenvalues; the corresponding eigenfunctions may have an arbitrarily large number of zeros on the interval.*

Proof. Consider the solution of Eq. (7) with initial condition $x(0) = 0$, $\dot{x}(0) = 1$. We denote by $\varphi = \alpha(t,\lambda)$ the value of the polar angle along the phase curve for this equation; let $\alpha(0,\lambda) \equiv 0$. The function α is continuous. Consider the value $\alpha(l,\lambda)$ as a function of λ. As $\lambda \to +\infty$ the quantity $\alpha(l,\lambda)$ tends to infinity. Indeed, suppose $q+\lambda > \omega^2$. If ω is sufficiently large, then the interval π/ω will fit into the interval $[0, l]$ an arbitrarily large number of times, say k times. Thus the number of zeros of the solution of the equation on this interval with λ this large will be at least k (by the comparison theorem). Consequently $\alpha(l, \lambda) \geq \pi k$ (Prop. 3). Thus $\alpha(l, \lambda) \to \infty$ as $\lambda \to +\infty$. Hence there exists an infinite set of eigenvalues λ_k for which $\alpha(t, \lambda_k) = \pi k$. The theorem is now proved. □

Problem 2. Prove that $\lim_{k\to\infty} \lambda_k/k^2 = (\pi/l)^2$.

Problem 3. Extend these results to equations of the form

$$(p\dot{x})^{\cdot} + qx = 0, \quad p(t) > 0 \ \forall t.$$

Hint. Consider the phase plane (x, y), where $y = p\dot{x}$.

§ 28. Linear Equations with Periodic Coefficients

The theory of linear equations with periodic coefficients explains how to rock a swing and why the upper, usually unstable, equilibrium position of a pendulum becomes stable if the point of suspension of the pendulum is making sufficiently rapid oscillations in the vertical direction.

1. The Mapping over a Period

Consider the differential equation

$$\dot{\boldsymbol{x}} = \boldsymbol{v}(t, \boldsymbol{x}), \quad \boldsymbol{v}(t+T, \boldsymbol{x}) = \boldsymbol{v}(t, \boldsymbol{x}), \quad \boldsymbol{x} \in \boldsymbol{R}^n, \tag{1}$$

whose right-hand side is a periodic function of time (Fig. 204).

Example 1. The motion of a pendulum with periodically varying parameters (for example, the motion of a swing) is described by a system of equations of the form (1):

$$\dot{x}_1 = x_2, \quad \dot{x}_2 = -\omega^2(t)x_1; \quad \omega(t+T) = \omega(t). \tag{2}$$

We shall assume that all the solutions of Eq. (1) can be extended indefinitely: this is known to be the case for the linear equations in which we are especially interested.

The periodicity of the right-hand side of the equation manifests itself in the special properties of the phase flow of Eq. (1).

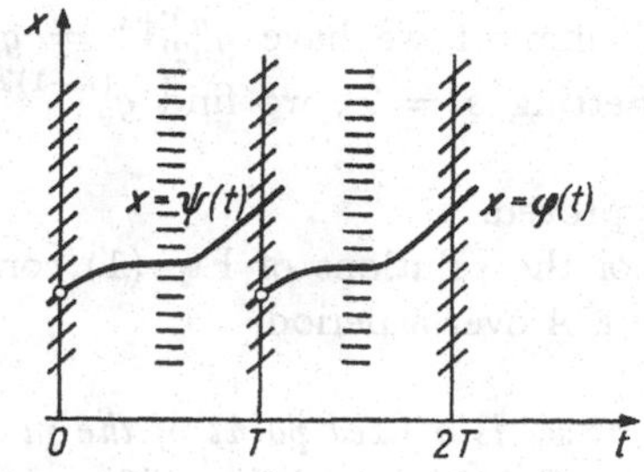

Fig. 204. The extended phase space of an equation with periodic coefficients

Lemma 1. *The transformation of the phase space over the time from t_1 to t_2, i.e., $g_{t_1}^{t_2} : \boldsymbol{R}^n \to \boldsymbol{R}^n$ remains unchanged when both the quantities t_1 and t_2 are simultaneously increased by the period T of the right-hand side of Eq.* (1).

Proof. We must prove that the translate $\psi(t) = \varphi(t+T)$ of the solution $\varphi(t)$ by the time T is a solution. But translating the extended phase space by T along the time axis maps the direction field of Eq. (1) into itself (Fig. 205). Therefore an integral curve of Eq. (1), when shifted by T, is tangent to the direction field at each point, and consequently remains an integral curve.

Thus, $g_{t_1+T}^{t_2+T} = g_{t_1}^{t_2}$, which was to be proved. □

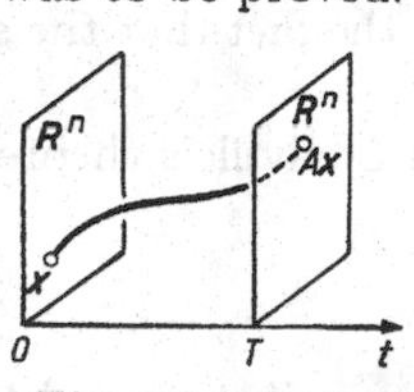

Fig. 205. The monodromy mapping

Consider in particular the transformation g_0^T realized by the phase flow over the time of one period T. This transformation will play an important role in what follows; we shall call it the *transformation over time* T and denote it (Fig. 205) by

$$A = g_0^T : \boldsymbol{R}^n \to \boldsymbol{R}^n.$$

Example 2. For the systems

$$\dot{x}_1 = x_2, \ \dot{x}_2 = -x_1 \quad \text{and} \quad \dot{x}_1 = x_1, \ \dot{x}_2 = -x_2,$$

which can be considered to be periodic with any period T, the mapping A is a rotation for the first system and a hyperbolic rotation for the second.

Lemma 2. *The transformations g_0^{nT} form a group $g_0^{nT} = A^n$. Moreover $g_0^{nT+s} = g_0^s g_0^{nT}$.*

Proof. According to Lemma 1 we have $g_{nT}^{nT+s} = g_0^s$. Therefore $g_0^{nT+s} = g_{nT}^{nT+s} g_0^{nT} = g_0^s g_0^{nT}$. Setting $s = T$, we find $g_0^{(n+1)T} = A g_0^{nT}$, whence by induction $g_0^{nT} = A^n$.

The lemma is now proved. □

To each property of the solutions of Eq. (1) corresponds an analogous property of the mapping A over a period.

Theorem. 1) *The point* $\boldsymbol{x}_0$ *is a fixed point of the mapping* A *(*$A\boldsymbol{x}_0 = \boldsymbol{x}_0$*) if and only if the solution with initial condition* $\boldsymbol{x}(0) = \boldsymbol{x}_0$ *is periodic with period* T.

2) *A periodic solution* $\boldsymbol{x}(t)$ *is Lyapunov-stable (resp. asymptotically stable) if and only if the fixed point* $\boldsymbol{x}_0$ *of the mapping* A *is Lyapunov stable (resp. asymptotically stable)*[31].

3) *If the system* (1) *is linear, i.e.,* $\boldsymbol{v}(t, \boldsymbol{x}) = \boldsymbol{V}(t)\boldsymbol{x}$ *is a linear function of* $\boldsymbol{x}$*, then the mapping* A *is linear.*

4) *If, in addition, the trace of the linear operator* $\boldsymbol{V}(t)$ *is zero, then the mapping* A *preserves volumes:* $\det A = 1$.

Proof. Assertions 1) and 2) follow from the relation $g_0^{T+s} = g_0^s A$ and the fact that the solution depends continuously on the initial conditions on the interval $[0, T]$.

Assertion 3) follows from the fact that the sum of two solutions of a linear system is again a solution.

Assertion 4) follows from Liouville's theorem. □

2. Stability Conditions

We shall now apply the theorem just proved to the mapping A of the phase plane of x_1 and x_2 into itself corresponding to the system (2). Since the system (2) is linear and the trace of the matrix on the right-hand side is zero, we obtain the following result.

Corollary. *The mapping* A *is linear and preserves areas (*$\det A = 1$*). A necessary and sufficient condition for the zero solution of the system of equations* (2) *to be stable is that the mapping* A *be stable.*

Problem 1. Prove that a rotation of the plane is a stable mapping and a hyperbolic rotation is unstable.

We shall now study in more detail the area-preserving linear mappings of a plane into itself.

---

[31] A fixed point $\boldsymbol{x}_0$ of the mapping A is called *Lyapunov stable* (resp. *asymptotically stable*) if for every $\varepsilon > 0$ there exists $\delta > 0$ such that $|\boldsymbol{x} - \boldsymbol{x}_0| < \delta$ implies $|A^n\boldsymbol{x} - A^n\boldsymbol{x}_0| < \varepsilon$ for all n, $0 < n < \infty$ simultaneously (resp. and also $A^n\boldsymbol{x} - A^n\boldsymbol{x}_0 \to \mathbf{0}$ as $n \to \infty$.

Theorem. *Let A be the matrix of an area-preserving linear transformation of a plane into itself* ($\det A = 1$)*. Then the mapping A is stable if $|\operatorname{tr} A| < 2$ and unstable if $|\operatorname{tr} A| > 2$.*

Proof. Let λ_1 and λ_2 be the eigenvalues of A. They satisfy the characteristic equation $\lambda^2 - \operatorname{tr} A\lambda + 1 = 0$ with real coefficients $\lambda_1 + \lambda_2 = \operatorname{tr} A$ and $\lambda_1\lambda_2 = \det A = 1$. The roots λ_1 and λ_2 of this real quadratic equation are real for $|\operatorname{tr} A| > 2$ and complex conjugates of each other if $|\operatorname{tr} A| < 2$ (Fig. 206). In the first case one of the eigenvalues is larger than 1 in absolute value and the other is smaller; the mapping A is a hyperbolic rotation and is unstable. In the second case the eigenvalues lie on the unit circle:

$$1 = \lambda_1\lambda_2 = \lambda_1\overline{\lambda}_1 = |\lambda_1|^2.$$

The mapping A is equivalent to a rotation through the angle α (where $\lambda_{1,2} = e^{i\pm\alpha}$), i.e., becomes a rotation under a suitable choice of a Euclidean structure on the plane (why?). Thus it is stable.

The theorem is now proved. □

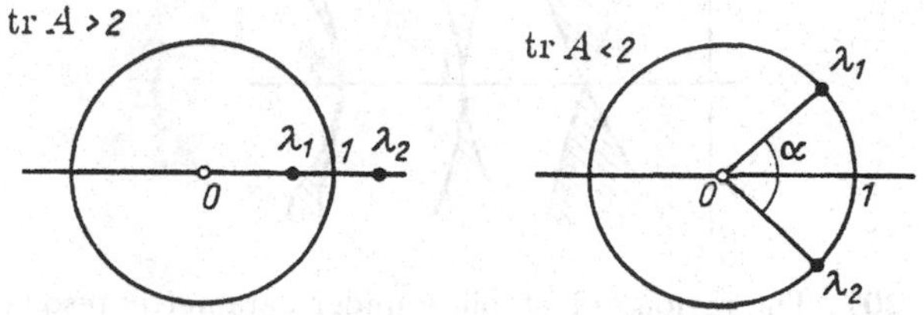

Fig. 206. The eigenvalues of a monodromy

Thus the whole question of the stability of the zero solution of the system (2) has been reduced to computing the trace of the matrix A. Unfortunately this trace can be computed explicitly only in special cases. It can always be found approximately by numerical integration of the equation on the interval $0 \le t \le T$. In the important case when $\omega(t)$ is nearly constant some simple general considerations are of use.

3. Strongly Stable Systems

Consider a linear system (1) with a two-dimensional phase space (i.e., with $n = 2$). Such a system is called *Hamiltonian* if the divergence of $\boldsymbol{v}$ is zero. For Hamiltonian systems, as shown above, the phase flow preserves area: $\det A = 1$.

Definition. The zero solution of a linear Hamiltonian system is *strongly stable* if it is stable and the zero solution of any nearby linear Hamiltonian system is also stable.

The following result is a consequence of the two preceding theorems.

Corollary. *If* $|\text{tr } A| < 2$, *then the zero solution is strongly stable.*

For if $|\text{tr } A| < 2$, then the condition $|\text{tr } A'| < 2$ also holds for the mapping A' corresponding to a sufficiently near system.

Let us apply this to a system with nearly constant coefficients. Consider, for example, the equation

$$\ddot{x} = -\omega^2(a + \varepsilon a(t))x, \quad \varepsilon \ll 1, \tag{3}$$

where $a(t + 2\pi) = a(t)$, for example $a(t) = \cos t$ (the pendulum whose frequency oscillates about ω with a small amplitude and period 2π)[32].

We shall represent each system (3) by a point in the plane of the parameters ε and ω (Fig. 207). Obviously the stable systems with $|\text{tr } A| < 2$ form an open set in the (ω, ε)-plane, as do the unstable systems with $|\text{tr } A| > 2$.

The boundary of stability is given by the equation $|\text{tr } A| = 2$.

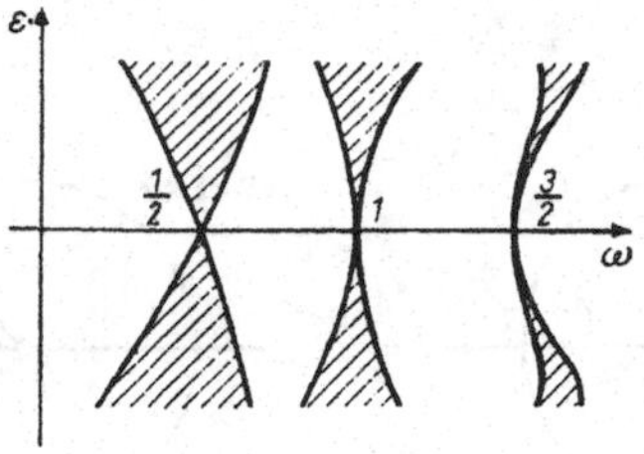

Fig. 207. The region of instability under parametric resonance

Theorem. *All points of the* ω*-axis except those with integer and half-integer coordinates,* $\omega = k/2$, $k = 0, 1, 2, \ldots$, *correspond to strongly stable systems* (3).

Thus the set of unstable systems can approach the ω-axis only at the points $\omega = k/2$. In other words, a swing can be started oscillating by small periodic changes in length only in the case when one period of the length change is close to an integer number of half-periods of the natural oscillations – a result familiar to everyone from experience.

The proof of the theorem just stated is based on the fact that when $\varepsilon = 0$, Eq. (3) has constant coefficients and can be solved explicitly.

Problem 1. For the system (3) with $\varepsilon = 0$ compute the matrix of the transformation A over the period $T = 2\pi$ in the basis $(x, \dot{x})$.

Solution. The general solution is

$$x = C_1 \cos \omega t + C_2 \sin \omega t.$$

[32] In the case $a(t) = \cos t$ Eq. (3) is called a *Mathieu equation.*

The particular solution with initial condition $x = 1$, $\dot{x} = 0$ is

$$x = \cos \omega t, \quad \dot{x} = -\omega \sin \omega t.$$

The particular solution with initial condition $x = 0$, $\dot{x} = 1$ is

$$x = (\sin \omega t)/\omega, \quad \dot{x} = \cos \omega t.$$

Answer. $A = \begin{pmatrix} \cos 2\pi\omega & \sin 2\pi\omega/\omega \\ -\omega \sin 2\pi\omega & \cos 2\pi\omega \end{pmatrix}$.

Therefore $|\mathrm{tr}\, A| = |2\cos 2\omega\pi| < 2$ if $\omega \neq k/2$, $k = 0, 1, \ldots$, and the theorem follows from the preceding corollary.

A closer analysis[33] shows that, in general, (and in particular when $a(t) = \cos t$) near the points $\omega = k/2$, $k = 1, 2, \ldots$, the region of instability (shaded in Fig. 207) actually does approach the ω-axis.

Thus when certain relations hold between the frequency of variation of the parameters and the natural frequency of the swing ($\omega \approx k/2$, $k = 1, 2, \ldots$), the lower equilibrium position of an ideal swing (3) is unstable, and it begins to oscillate under an arbitrarily small periodic variation in its length.

This phenomenon is known as *parametric resonance.* The characteristic property of parametric resonance is that it manifests itself most strongly in the case when the frequency of variation of the parameters ν (in Eq. (3) the frequency ν equals 1) is twice the natural frequency ω.

Remark. Parametric resonance is theoretically observed for an infinite set of ratios $\omega/\nu \approx k/2$, $k = 1, 2, \ldots$. In practice we usually observe only those cases when k is small ($k = 1, 2$ and rarely 3). What is happening is that

a) for large k the region of instability approaches the ω-axis by a narrow tongue and very rigid bounds are obtained for the resonance frequency ($\sim \varepsilon^k$ for $a(t) = \cos t$ in (3)).

b) the instability itself is expressed only weakly for large k, since the quantity $|\mathrm{tr}\, A| - 2$ is small and the eigenvalues are close to 1 for large k;

c) an arbitrarily small amount of friction leads to a minimum value ε_k of the amplitude which is required in order for parametric resonance to occur: for smaller ε the oscillations die out. As k increases, the quantity ε_k increases rapidly (Fig. 208).

We remark also that for Eq. (3) the quantity x increases without bound in the unstable case. In real systems the oscillations attain only a finite amplitude, since for large x the linearity of Eq. (3) itself ceases to hold and nonlinear effects must be taken into account.

[33] Cf., for example, Problem 1 of Sect. 4, which is explained below.

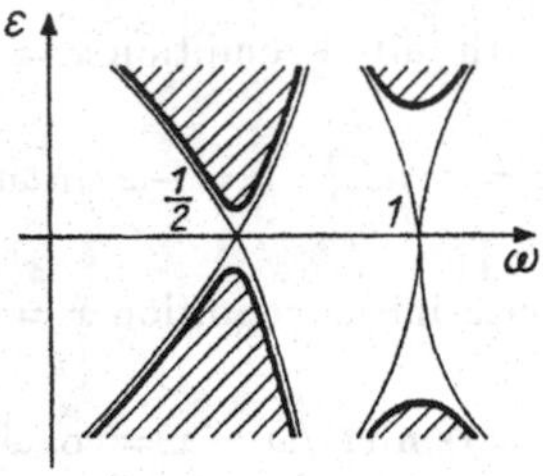

Fig. 208. The influence of a small amount of friction on the region of instability

4. Computations

Problem 1. Find the form of the regions of stability in the (ε, ω)-plane for the system described by the equation

$$\ddot{x} = -f(t)x, \quad f(t+2\pi) = f(t), \tag{4}$$
$$f(t) = \begin{cases} \omega + \varepsilon & \text{for } 0 \le t < \pi, \\ \omega - \varepsilon & \text{for } \pi \le t < 2\pi, \end{cases} \quad \varepsilon \ll 1.$$

Solution. It follows from the solution of the preceding problem (Prob. 1 of Sect. 3) that $A = A_2 A_1$, where

$$A_k = \begin{pmatrix} c_k & s_k/\omega_k \\ -\omega_k s_k & c_k \end{pmatrix}, \quad c_k = \cos \pi\omega_k, \quad s_k = \sin \pi\omega_k \quad \omega_{1,2} = \omega \pm \varepsilon.$$

Therefore the boundary of the zone of stability has the equation

$$|\operatorname{tr} A| = |2c_1c_2 - (\omega_1/\omega_2 + \omega_2/\omega_1)s_1s_2| = 2. \tag{5}$$

Since $\varepsilon \ll 1$, we have $\omega_1/\omega_2 = (\omega+\varepsilon)/(\omega-\varepsilon) \approx 1$.

We introduce the notation $\omega_1/\omega_2 + \omega_2/\omega_1 = 2(1+\Delta)$. Then, as one can easily compute, $\Delta = 2\varepsilon^2/\omega^2 + O(\varepsilon^4) \ll 1$. Using the relations $2c_1c_2 = \cos 2\pi\varepsilon + \cos 2\pi\omega$, $2s_1s_2 = \cos 2\pi\varepsilon - \cos 2\pi\omega$, we rewrite Eq. (5) in the form

$$-\Delta \cos 2\pi\varepsilon + (2+\Delta)\cos 2\pi\omega = \pm 2$$

or

$$\cos 2\pi\omega = (2 + \Delta\cos 2\pi\varepsilon)/(2+\Delta), \tag{6_1}$$
$$\cos 2\pi\omega = (-2 + \Delta\cos 2\pi\varepsilon)/(2+\Delta). \tag{6_2}$$

In the first case $\cos 2\pi\omega \approx 1$. Therefore we set $\omega = k + \alpha$, $|a| \ll 1$; $\cos 2\pi\omega = \cos 2\pi a = 1 - 2\pi^2 a^2 + O(a^4)$. We rewrite Eq. (6_1) in the form $\cos 2\pi\omega = 1 - \dfrac{\Delta}{2+\Delta}(1 - \cos 2\pi\varepsilon)$ or $2\pi^2 a^2 + O(a^4) = \Delta\pi^2\varepsilon^2 + O(\varepsilon^4)$. Substituting the value $\Delta = (2\varepsilon^2/\omega^2) + O(\varepsilon^4)$, we find $a = \pm\varepsilon^2/\omega^2 + o(\varepsilon^2)$, i.e., $\omega = k \pm \varepsilon^2/k^2 + o(\varepsilon^2)$ (Fig. 209).

Eq. (6_2) can be solved similarly; the result is

$$\omega = \varkappa \pm \varepsilon/(\pi\varkappa) + o(\varepsilon), \quad \varkappa = k + 1/2.$$

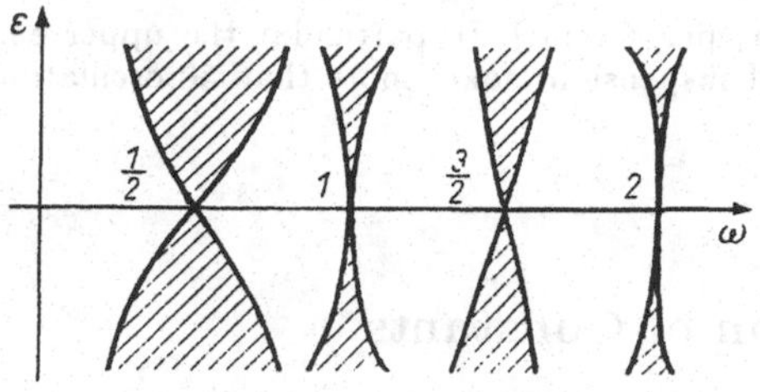

Fig. 209. The region of instability for Eq. (4)

Problem 2. Can the usually unstable upper equilibrium point of a pendulum become stable if the point of suspension oscillates in the vertical direction?

Solution. Suppose the length of the pendulum is l, the amplitude of the oscillations of the point of suspension is $a \ll l$, and the period of oscillation of the point of suspension is 2τ, where in the course of each half-period the acceleration of the point of suspension is constant and equal to $\pm c$ (then $c = 8a/\tau^2$). It turns out that for sufficiently rapid oscillations of the point of suspension ($\tau \ll 1$) the upper equilibrium point becomes stable. The equation of motion can be written in the form $\ddot{x} = (\omega^2 \pm \alpha^2)x$ (the sign reverses after time τ), where $\omega^2 = g/l$, $\alpha^2 = c/l$. If the oscillations of the point of suspension are sufficiently rapid, then $\alpha^2 > \omega^2$ ($\alpha^2 = 8a/(l\tau^2)$).

In analogy with the preceding problem, we have $A = A_2A_1$, where

$$A_1 = \begin{pmatrix} \cosh k\tau & k^{-1}\sinh k\tau \\ k\sinh k\tau & \cosh k\tau \end{pmatrix}, \qquad A_2 = \begin{pmatrix} \cos \Omega\tau & \Omega^{-1}\sin \Omega\tau \\ -\Omega\sin\Omega\tau & \cos\Omega\tau \end{pmatrix}.$$

$$k^2 = \alpha^2 + \omega^2, \qquad \Omega^2 = \alpha^2 - \omega^2.$$

The condition for stability $|\mathrm{tr}\, A| < 2$ therefore has the form

$$|2\cosh k\tau \cos\Omega\tau + (k/\Omega - \Omega/k)\sinh k\tau \sin\Omega\tau| < 2. \tag{7}$$

We shall show that this condition holds for sufficiently rapid oscillations of the point of suspension, i.e., when $c \geq g$. We introduce the dimensionless variables ε and μ given by

$$a/l = \varepsilon^2 \ll 1, \quad g/c = \mu^2 \ll 1.$$

Then $k\tau = 2\sqrt{2}\varepsilon\sqrt{1+\mu^2}$, $\Omega\tau = 2\sqrt{2}\varepsilon\sqrt{1-\mu^2}$, $k/\Omega - \Omega/k = 2\mu^2 + O(\mu^4)$.

Therefore the following expansions are valid up to $o(\varepsilon^4 + \mu^4)$ for small ε and μ:

$$\cosh k\tau = 1 + 4\varepsilon^2(1+\mu^2) + 2\varepsilon^4/3 + \cdots, \quad \cos\Omega\tau = 1 - 4\varepsilon^2(1-\mu^2) + 8\varepsilon^4/3 + \cdots,$$
$$(k/\Omega - \Omega/k)\sinh k\tau \sin\Omega\tau = 16\varepsilon^2\mu^2 + \cdots$$

Thus the stability condition (7) assumes the form

$$2(1 - 16\varepsilon^4 + 16\varepsilon^4/3 + 8\varepsilon^2\mu^2 + \cdots) + 16\varepsilon^2\mu^2 < 2.$$

Ignoring higher-order infinitesimals, we find $(2/3)16\varepsilon^4 \geq 32\mu^2\varepsilon^2$ or $\mu < \varepsilon/\sqrt{3}$, or $g/c < a/(3l)$. This condition can be rewritten in the form $N > \sqrt{3/32}\omega l/a \approx 0.3\omega l/a$, where $N = l/(2\tau)$ is the number of oscillations of the point of suspension in unit time. For example, if the length of the pendulum l is 20 cm and the amplitude of the oscillations of the point of suspension a equals 1 cm, then $N > 0.31\sqrt{980/20}\cdot$

$20 \approx 43$ (oscillations per second). In particular the upper equilibrium position is stable if the point of suspension makes more than 50 oscillations per second.

§ 29. Variation of Constants

In studying equations close to the "unperturbed" equations already studied the following device is frequently useful. Let c be a first integral of the "unperturbed" equation. Then c will no longer be a first integral for nearby "perturbed" equations. However, it is often possible to find out (precisely or approximately) how the values $c(\varphi(t))$ change with time, where φ is a solution of the "perturbed" equation. In particular, if the original equation is a homogeneous linear equation and the perturbed equation is inhomogeneous, this device leads to an explicit formula for the solution. Moreover, because the equation is linear, the perturbation is not required to be "small."

1. The Simplest Case

Consider the simplest linear inhomogeneous equation

$$\dot{\boldsymbol{x}} = \boldsymbol{f}(t), \quad \boldsymbol{x} \in \boldsymbol{R}^n, \quad t \in I, \tag{1}$$

corresponding to the simplest homogeneous equation

$$\dot{\boldsymbol{x}} = \mathbf{0}. \tag{2}$$

Equation (1) can be solved by quadrature:

$$\varphi(t) = \varphi(t_0) + \int_{t_0}^{t} \boldsymbol{f}(\tau)\, d\tau. \tag{3}$$

2. The General Case

Consider the linear inhomogeneous equation

$$\dot{\boldsymbol{x}} = A(t)\boldsymbol{x} + \boldsymbol{h}(t), \quad \boldsymbol{x} \in \boldsymbol{R}^n, \quad t \in I, \tag{4}$$

corresponding to the homogeneous equation

$$\dot{\boldsymbol{x}} = A(t)\boldsymbol{x}. \tag{5}$$

Assume that we know how to solve the homogeneous equation (5) and $\boldsymbol{x} = \varphi(t)$ is a solution of it. Let us take the initial conditions $\boldsymbol{c} = \varphi(t_0)$ as the

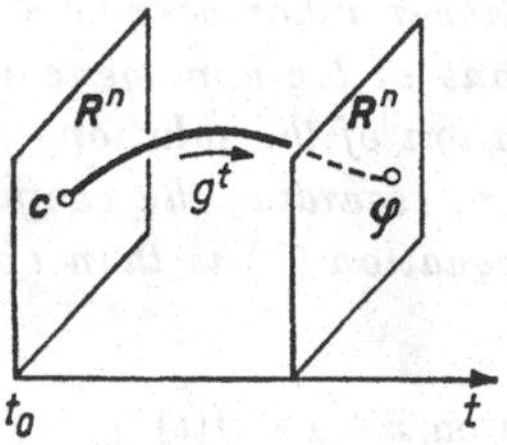

Fig. 210. The coordinates of the point c are first integrals of the homogeneous equation

coordinates (c, t) that rectify the integral curves of Eq. (5) in the extended phase space (Fig. 210).

In the new coordinates Eq. (5) assumes the very simple form (2). The passage to rectifying coordinates is carried out by a transformation that is linear in x. Therefore in the new coordinates the inhomogeneous equation (4) assumes the very simple form (1), and we can solve it.

3. Computations

We shall seek the solution of the inhomogeneous equation (4) in the form

$$\varphi(t) = g^t c(t), \quad c : I \to R^n, \tag{6}$$

where $g^t : R^n \to R^n$ is the linear operator of transformation over the time from t_0 to t for the homogeneous equation (5).

Differentiating on t, we find

$$\dot{\varphi} = \dot{g}^t c + g^t \dot{c} = A g^t c + g^t \dot{c} = A\varphi + g^t \dot{c}.$$

Substituting into Eq. (4), we find $g^t \dot{c} = h(t)$. Thus we have proved the following result.

Theorem. *Formula* (6) *gives the solution of Eq.* (4) *if and only if* c *satisfies the equation* $\dot{c} = f(t)$, *where* $f(t) = (g^t)^{-1} h(t)$.

This last equation has the very simple form (1). Applying formula (3), we obtain the following proposition.

Corollary. *The solution of the inhomogeneous equation* (4) *with the initial condition* $\varphi(t_0) = c$ *has the form*

$$\varphi(t) = g^t \left(c + \int_{t_0}^{t} (g^\tau)^{-1} h(\tau)\, d\tau \right).$$

Remark. In coordinate form the theorem just proved can be stated as follows:

In order to solve the linear inhomogeneous equation (4) knowing a fundamental system of solutions of the homogeneous equation (5), it suffices to substitute a linear combination of the solutions of a fundamental system into the inhomogeneous equation, regarding the coefficients as unknown functions of time. The very simple equation (1) is then obtained for determining these coefficients.

Problem 1. Solve the equation $\ddot{x} + x = f(t)$.

Solution. We form the homogeneous system of two equations:

$$\dot{x}_1 = x_2 \quad \dot{x}_2 = -x_1 .$$

A fundamental system of solutions of this system is known:

$$(x_1 = \cos t,\ x_2 = -\sin t); \quad (x_1 = \sin t,\ x_2 = \cos t).$$

By the general rule we seek a solution in the form

$$x_1 = c_1(t)\cos t + c_2(t)\sin t, \quad x_2 = -c_1(t)\sin t + c_2(t)\cos t.$$

To determine c_1 and c_2 we obtain the system

$$\dot{c}_1 \cos t + \dot{c}_2 \sin t = 0, \quad -\dot{c}_1 \sin t + \dot{c}_2 \cos t = f(t).$$

Consequently

$$\dot{c}_1 = -f(t)\sin t, \quad \dot{c}_2 = f(t)\cos t.$$

Answer. $x(t) = \left[x(0) - \int_0^t f(\tau)\sin\tau\, d\tau\right]\cos t + \left[\dot{x}(0) + \int_0^t f(\tau)\cos\tau\, d\tau\right]\sin t.$

Chapter 4. Proofs of the Main Theorems

In this chapter we prove theorems on the existence, uniqueness, continuity, and differentiability of solutions of ordinary differential equations, and also theorems on the rectification of a vector field and a direction field.

The proofs also contain a method for constructing approximate solutions.

§ 30. Contraction Mappings

The method studied below for finding a fixed point of a mapping of a metric space into itself is applied subsequently to construct solutions of differential equations.

1. Definition

Let $A : M \to M$ be a mapping of a metric space M (with metric ρ) into itself. The mapping M is called a *contraction* if there exists a constant λ, $0 < \lambda < 1$, such that

$$\rho(Ax, Ay) \leq \lambda \rho(x, y) \quad \forall x,\ y \in M. \tag{1}$$

Example 1. Let $A : \boldsymbol{R} \to \boldsymbol{R}$ be a real-valued function of a real variable (Fig. 211). If the derivative of A is merely less than 1 in absolute value, the mapping A may fail to be a contraction; but it is a contraction if $|A'| \leq \lambda < 1$.

Example 2. Let $A : \boldsymbol{R}^n \to \boldsymbol{R}^n$ be a linear operator. If all the eigenvalues of A lie strictly inside the unit disk, then there exists a Euclidean metric (a Lyapunov function, cf. § 22) on $\boldsymbol{R}^n$ such that A is a contraction.

Problem 1. Which of the following mappings of the real line (with the usual metric) into itself is a contraction?

$$1)\ y = \sin x; \quad 2)\ \sqrt{x^2+1}; \quad 3)\ \arctan x.$$

Problem 2. Can the sign $\leq$ in inequality (1) be replaced by $<$?

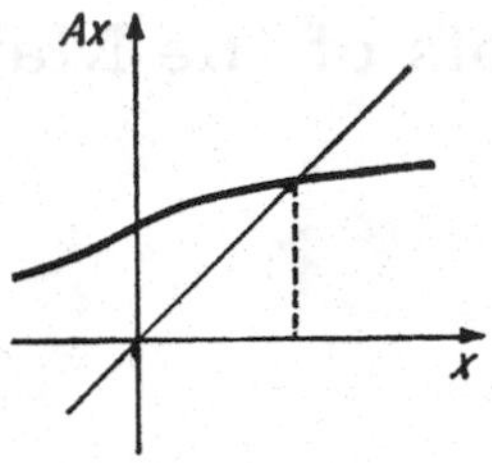

Fig. 211. The fixed point of a contraction mapping

2. The Contraction Mapping Theorem

A point $x \in M$ is called a *fixed point* of a mapping $A : M \to M$ if $Ax = x$.

Let $A : M \to M$ be a contraction mapping of a complete metric space M into itself. Then A has one and only one fixed point. For any point x in M the sequence of images of the point x under applications of A (Fig. 212)

$$x, \ Ax, \ A^2x, \ A^3x, \ldots$$

converges to the fixed point.

Proof. Let $\rho(x, Ax) = d$. Then

$$\rho(A^n x, A^{n+1}x) \leq \lambda^n d.$$

The series $\sum_{n=0}^{\infty} \lambda^n$ converges. Therefore the sequence $A^n x, = 0, 1, 2, \ldots$, is a Cauchy sequence. The space M is complete. Therefore the following limit exists: $X = \lim_{n \to \infty} A^n x$.

We shall show that X is a fixed point of A. We remark that every contraction mapping is continuous (one can take $\delta = \varepsilon$). Therefore

$$AX = A \lim_{n \to \infty} A^n x = \lim_{n \to \infty} A^{n+1} x = X.$$

We shall show that every fixed point Y coincides with X. Indeed

$$\rho(X, Y) = \rho(AX, AY) \leq \lambda \rho(X, Y), \quad \lambda < 1 \Rightarrow \rho(X, Y) = 0.$$

□

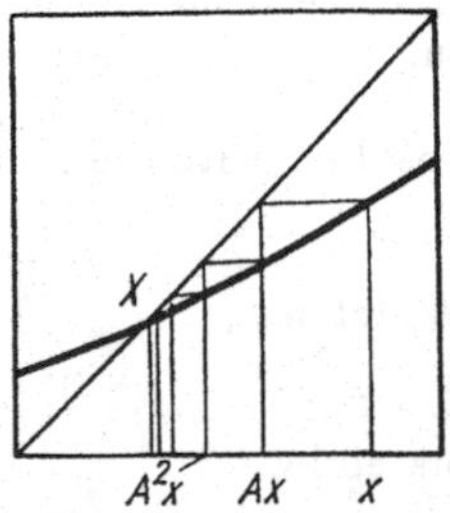

Fig. 212. The sequence of images of the point x under iteration of the contraction mapping A

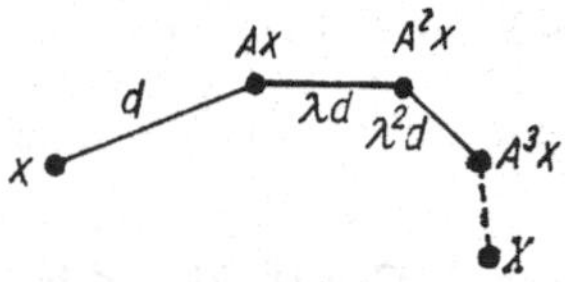

Fig. 213. An estimate of the accuracy of the approximation of x to the fixed point X

3. Remark

The points x, Ax, $A^2x, \ldots$ are called *successive approximations* to X. Let x be an approximation to the fixed point X of a contraction mapping A. The accuracy of this approximation is easily estimated in terms of the distance between the points x and Ax:

$$\rho(x, X) \le \frac{d}{1-\lambda},$$

for $d + \lambda d + \lambda^2 d + \cdots = \dfrac{d}{1-\lambda}$ (Fig. 213).

§ 31. Proof of the Theorems on Existence and Continuous Dependence on the Initial Conditions

In this section we construct a contraction mapping of a complete metric space whose fixed point defines the solution of a given differential equation.

1. The Successive Approximations of Picard

Consider the differential equation $\dot{\boldsymbol{x}} = \boldsymbol{v}(t, \boldsymbol{x})$, defined by the vector field $\boldsymbol{v}$ in some domain of the extended phase space $\boldsymbol{R}^{n+1}$ (Fig. 214).

We define the *Picard mapping* to be the mapping A that takes the function $\varphi : t \mapsto \boldsymbol{x}$ to the function $A\varphi : t \mapsto \boldsymbol{x}$, where

$$(A\varphi)(t) = \boldsymbol{x}_0 + \int_{t_0}^{t} \boldsymbol{v}(\tau, \varphi(\tau))\, d\tau.$$

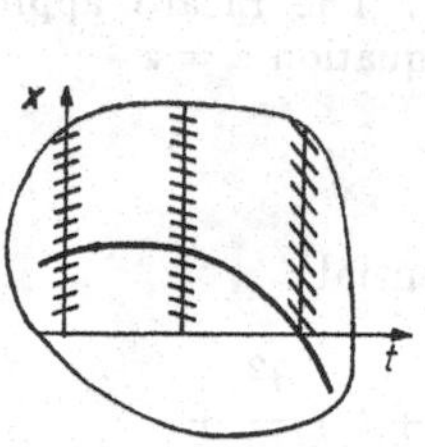

Fig. 214. An integral curve of the equation $\dot{\boldsymbol{x}} = \boldsymbol{v}(t, \boldsymbol{x})$

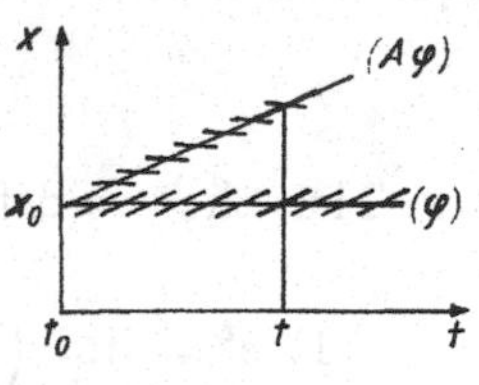

Fig. 215. The Picard mapping A

Geometrically, passing from φ to $A\varphi$ (Fig. 215) means constructing with respect to a curve (φ) a new curve $(A\varphi)$ whose tangent for each t is parallel to a given direction field, only not on the curve $(A\varphi)$ itself – for then $A\varphi$ would be a solution – but at the corresponding point of the curve (φ). We have

$$\left.\begin{array}{c} \varphi \textit{ is a solution} \\ \textit{with the initial condition} \\ \varphi(t_0) = \boldsymbol{x}_0 \end{array}\right. \Leftrightarrow (\varphi = A\varphi).$$

Motivated by the contraction mapping theorem, we consider the sequence of *Picard approximations* φ, $A\varphi$, $A^2\varphi, \ldots$ (starting, say, with $\varphi = \boldsymbol{x}_0$).

Example 1. $\dot{\boldsymbol{x}} = \boldsymbol{f}(t)$ (Fig. 216). $(A\varphi)(t) = \boldsymbol{x}_0 + \int_{t_0}^{t} \boldsymbol{f}(\tau)\, d\tau$. In this case the first step already leads to the exact solution.

Example 2. $\dot{\boldsymbol{x}} = \boldsymbol{x}$, $t_0 = 0$ (Fig. 217). The convergence of the approximations in this case can be observed directly. At the point t

$$\begin{aligned} \varphi &= 1, \\ A\varphi &= 1 + \int_0^t d\tau = 1 + t, \\ A^2\varphi &= a + \int_0^t (1+\tau)\, d\tau = 1 + t + t^2/2, \\ &\ldots\ldots\ \ldots \\ A^n\varphi &= 1 + t + t^2/2 + \cdots + t^n/n!, \\ \lim_{n\to\infty} A^n\varphi &= e^t. \end{aligned}$$

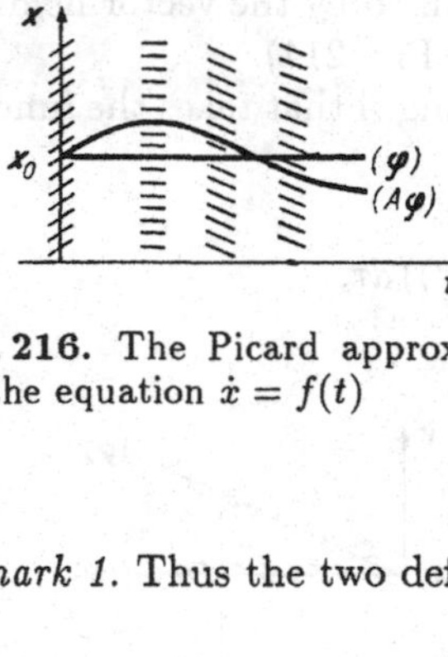

Fig. 216. The Picard approximation for the equation $\dot{x} = f(t)$

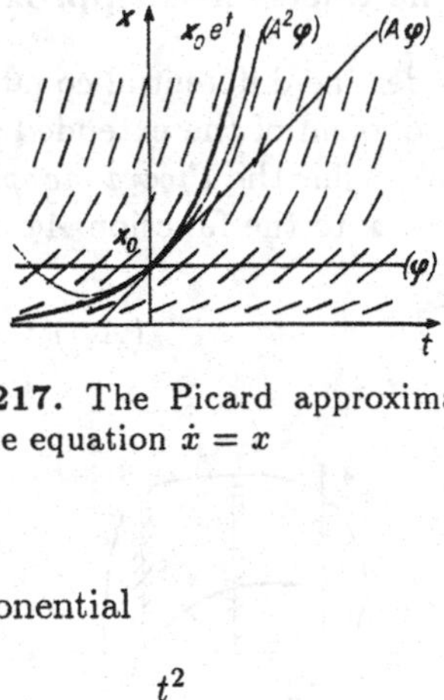

Fig. 217. The Picard approximation for the equation $\dot{x} = x$

Remark 1. Thus the two definitions of the exponential

$$1)\ e^t = \lim_{n\to\infty}\left(1 + \frac{t}{n}\right)^n, \quad 2)\ e^t = 1 + t + \frac{t^2}{2!} + \cdots$$

correspond to two methods of approximating the solutions of the the very simple differential equation $\dot{x} = x$: the broken line method of Euler, and the method of successive approximations of Picard. Historically the original definition of the exponential was simple:

3) e^t is the solution of the equation $\dot{x} = x$ with initial condition $x(0) = 1$.

Remark 2. Similarly one can prove that the approximations converge for the equation $\dot{x} = kx$. The reason for the convergence of the successive approximations in the general case is that the equation $\dot{x} = kx$ is the "worst case":

the successive approximations for any equation converge at least as fast as for some equation of the form $\dot{x} = kx$.

To prove convergence of the successive approximations we shall construct a complete metric space in which the Picard mapping is a contraction. We begin by recalling some facts from analysis.

2. Preliminary Estimates

1) *The norm.* We shall denote the norm of the vector $\boldsymbol{x}$ in the Euclidean space $\boldsymbol{R}^n$ by $|\boldsymbol{x}| = \sqrt{(\boldsymbol{x}, \boldsymbol{x})}$. The space $\boldsymbol{R}^n$ with the metric $\rho(\boldsymbol{x}, \boldsymbol{y}) = |\boldsymbol{x} - \boldsymbol{y}|$ is a complete metric space.

We note two important inequalities: the triangle inequality

$$|\boldsymbol{x} + \boldsymbol{y}| \leq |\boldsymbol{x}| + |\boldsymbol{y}|$$

and the Schwarz inequality[1]

$$|(\boldsymbol{x}, \boldsymbol{y})| \leq |\boldsymbol{x}|\,|\boldsymbol{y}|.$$

2) *The vector-valued integral.* Let $\boldsymbol{f} : [a, b] \to \boldsymbol{R}^n$ be a vector-valued function with values in $\boldsymbol{R}^n$, continuous on $[a, b]$. The vector-valued integral

$$\boldsymbol{I} = \int_a^b \boldsymbol{f}(t)\, dt \in \boldsymbol{R}^n$$

is defined in the usual way (using integral sums).

Lemma.

$$\left| \int_a^b \boldsymbol{f}(t)\, dt \right| \leq \left| \int_a^b |\boldsymbol{f}(t)|\, dt \right|. \tag{1}$$

Proof. Compare the integral sums using the triangle inequality: $\left| \sum \boldsymbol{f}(t_i)\Delta_i \right| \leq \sum |\boldsymbol{f}(t_i)|\,|\Delta_i|$, which was to be proved. □

3) *The norm of a linear operator.* Let $A : \boldsymbol{R}^m \to \boldsymbol{R}^n$ be a linear operator from one Euclidean space into another. We shall denote its norm by $|A| = \sup\limits_{\boldsymbol{x} \in \boldsymbol{R}^n \setminus 0} \dfrac{|A\boldsymbol{x}|}{|\boldsymbol{x}|}$. Then

[1] We recall the proofs of these inequalities. We pass a two-dimensional plane through the vectors $\boldsymbol{x}$ and $\boldsymbol{y}$ of Euclidean space. This plane inherits the Euclidean structure from $\boldsymbol{R}^n$. On the Euclidean plane both of these inequalities are known from elementary geometry. They are thereby proved for any Euclidean space as well, for example in $\boldsymbol{R}^n$. In particular we have proved without any calculation that

$$\left| \sum_{i=1}^n x_i y_i \right|^2 \leq \sum_{i=1}^n x_i^2 \sum_{i=1}^n y_i^2, \text{ and } \left| \int_a^b fg\, dt \right|^2 \leq \int_a^b f^2\, dt \int_a^b g^2\, dt.$$

$$|A+B| \leq |A| + |B|, \quad |AB| \leq |A|\,|B|. \tag{2}$$

The set of linear operators from $\boldsymbol{R}^m$ to $\boldsymbol{R}^n$ becomes a complete metric space if we set $\rho(A, B) = |A - B|$.

3. The Lipschitz Condition

Let $A : M_1 \to M_2$ be a mapping of the metric space M_1 (with metric ρ_1) into the metric space M_2 (with metric ρ_2), and let L be a positive real number.

Definition. The mapping A *satisfies a Lipschitz condition with constant* L (written $A \in \operatorname{Lip} L$) if it increases the distance between any two points of M_1 by a factor of at most L (Fig. 218):

$$\rho_2(Ax, Ay) \leq L\rho_1(x, y) \quad \forall x,\, y \in M_1.$$

The mapping A *satisfies a Lipschitz condition* if there exists a constant L such that $A \in \operatorname{Lip} L$.

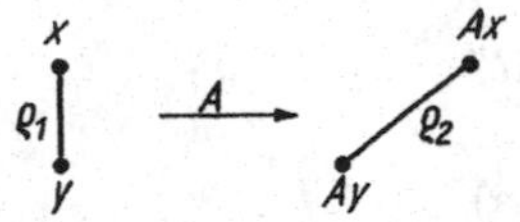

Fig. 218. The Lipschitz condition $\rho_2 \leq L\rho_1$

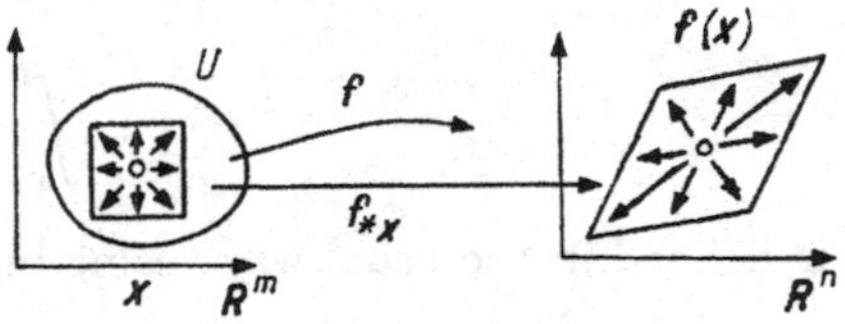

Fig. 219. The derivative of the mapping f

Problem 1. Do the following mappings satisfy a Lipschitz condition? (The metric is always the Euclidean metric.)

$$1)\ y = x^2,\ x \in \boldsymbol{R}; \quad 2)\ y = \sqrt{x},\ x > 0; \quad 3)\ y = \sqrt{x_1^2 + x_2^2},\ (x_1, x_2) \in \boldsymbol{R}^2;$$

$$4)\ y = \sqrt{x_1^2 - x_2^2},\ x_1^2 \geq x_2^2; \quad 5)\ y = \begin{cases} x \ln x & 0 < x \leq 1, \\ 0 & x = 0; \end{cases}$$

$$6)\ y = x^2,\ x \in \boldsymbol{C},\ |x| \leq 1.$$

Problem 2. Prove that

contraction $\Rightarrow$ *Lipschitz condition* $\Rightarrow$ *continuity.*

4. Differentiability and the Lipschitz Condition

Let $\boldsymbol{f} : U \to \boldsymbol{R}^n$ be a smooth mapping (of class C^r, $r \geq 1$) of the domain U of the Euclidean space $\boldsymbol{R}^m$ into the Euclidean space $\boldsymbol{R}^n$ (Fig. 219). The tangent space to the Euclidean space at each point has its own natural Euclidean structure. Therefore the derivative of f at a point $x \in U \subset R^m$

$$f_{*x} : T_x R^m \to T_{f(x)} R^n$$

is a linear operator from one Euclidean space into another. The following theorem is obvious.

Theorem. *A continuously differentiable mapping f satisfies a Lipschitz condition on each convex compact subset V of the domain U with constant L equal to the supremum of the derivative f on V:*

$$L = \sup_{x \in V} |f_{*x}|.$$

Proof. Join the points x and $y \in V$ with a line segment (Fig. 220): $z(t) = x + t(y - x)$, $0 \le t \le 1$. By the Barrow formula

$$f(y) - f(x) = \int_0^1 \frac{d}{dt}(f(z(\tau)))\, d\tau = \int_0^1 f_{*z(\tau)} \dot{z}(\tau)\, d\tau.$$

From formulas (1) and (2) of Sect. 2 and from the fact that $\dot{z} = y - x$, we have

$$\left| \int_0^1 f_{*z(\tau)} \dot{z}(\tau)\, d\tau \right| \le \int_0^1 L|y - x|\, d\tau = L|y - x|,$$

which was to be proved. □

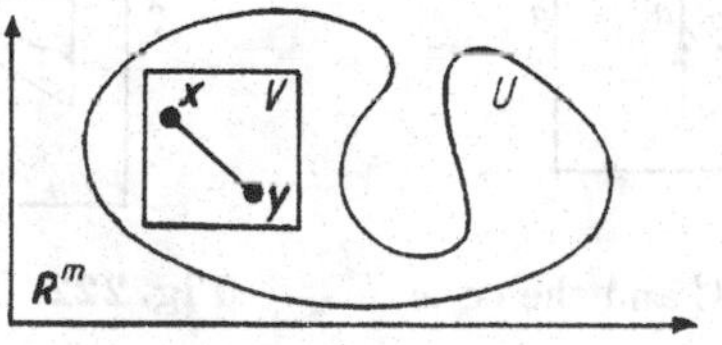

Fig. 220. Continuous differentiability implies a Lipschitz condition

Remark. The supremum of the norm of the derivative $|f_*|$ on V is attained. Indeed, by hypothesis $f \in C^1$, and so the derivative f_* is continuous. Consequently $|f_*|$ attains a maximum value L on the compact set V.

In approaching the proof of the convergence of the Picard approximations, we study them in a small neighborhood of a single point. To describe this neighborhood we use the following four numbers.

5. The Quantities C, L, a', b'

Let the right-hand side v of the differential equation

$$\dot{\boldsymbol{x}} = \boldsymbol{v}(t, \boldsymbol{x}) \tag{3}$$

be defined and differentiable (of class C^r, $r \geq 1$) in a domain U of the extended phase space: $U \subset \boldsymbol{R}^1 \times \boldsymbol{R}^n$. We fix a Euclidean structure in $\boldsymbol{R}^n$, and hence also in $T_{\boldsymbol{x}}\boldsymbol{R}^n$.

Consider any point $(t_0, \boldsymbol{x}_0) \in U$ (Fig. 221). The cylinder

$$\mathcal{C} = \{t, \boldsymbol{x} : |t - t_0| \leq a, \ |\boldsymbol{x} - \boldsymbol{x}_0| \leq b\}$$

lies in the domain U for sufficiently small a and b. We denote by C and L the suprema of the quantities $|\boldsymbol{v}|$ and $|\boldsymbol{v}_*|$ on this cylinder[2]. These suprema are attained, since the cylinder is compact: $|\boldsymbol{v}| \leq C$, $|\boldsymbol{v}_*| \leq L$.

Consider the cone K_0 with vertex $(t_0, \boldsymbol{x}_0)$, aperture C, and height a':

$$K_0 = \{t, \boldsymbol{x} : |t - t_0| \leq a', \ |\boldsymbol{x} - \boldsymbol{x}_0| \leq C|t - t_0|\}.$$

If the number a' is sufficiently small, this cone K_0 lies inside the cylinder $\mathcal{C}$. If the numbers $a' > 0$ and $b' > 0$ are sufficiently small, then the cylinder $\mathcal{C}$ also contains every cone $K_{\boldsymbol{x}}$ obtained from K_0 by a parallel translation of the vertex to the point $(t_0, \boldsymbol{x})$, where $|\boldsymbol{x} - \boldsymbol{x}_0| \leq b'$.

We shall assume that a' and b' are chosen so small that $K_{\boldsymbol{x}} \subset \mathcal{C}$. The solution φ of Eq. (3) with initial condition $\varphi(t_0) = \boldsymbol{x}$ will be sought in the form $\varphi(t) = \boldsymbol{x} + \boldsymbol{h}(t, \boldsymbol{x})$ (Fig. 222).

The corresponding integral curve lies inside the cone $K_{\boldsymbol{x}}$.

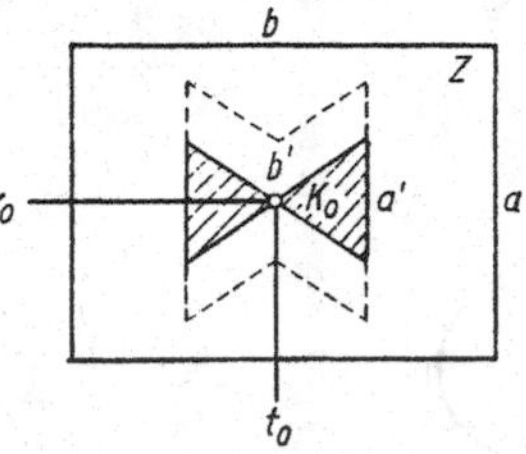

Fig. 221. The cylinder $\mathcal{C}$ and the cone K_0

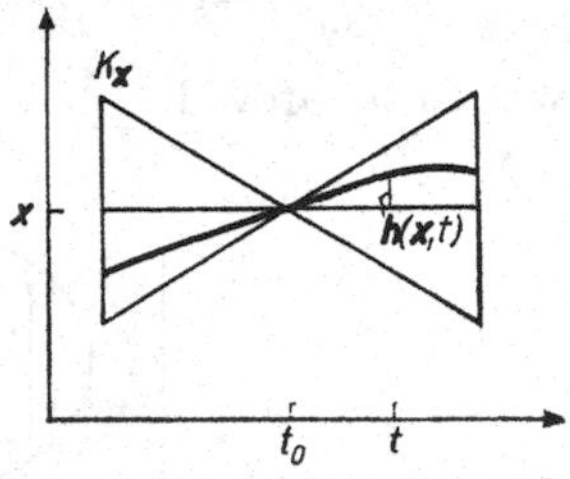

Fig. 222. The definition of $\boldsymbol{h}(t, \boldsymbol{x})$

6. The Metric Space M

Consider the set of all continuous mappings $\boldsymbol{h}$ of the cylinder $|\boldsymbol{x} - \boldsymbol{x}_0| \leq b'$, $|t - t_0| \leq a'$ into the Euclidean space $\boldsymbol{R}^n$. We denote by M the set of such mappings that satisfy the additional condition

$$|\boldsymbol{h}(t, \boldsymbol{x})| \leq C|t - t_0| \tag{4}$$

(in particular, $\boldsymbol{h}(t_0, \boldsymbol{x} = 0)$).

We introduce a metric ρ in M by setting

[2] Here and below the asterisk denotes the derivative (with respect to $\boldsymbol{x}$) for fixed t.

$$\rho(\boldsymbol{h}_1, \boldsymbol{h}_2) = \|\boldsymbol{h}_1 - \boldsymbol{h}_2\| = \max_{\substack{|\boldsymbol{x}-\boldsymbol{x}_0|\le b' \\ |t-t_0|\le a'}} |\boldsymbol{h}_1(t,\boldsymbol{x}) - \boldsymbol{h}_2(t,\boldsymbol{x})|.$$

Theorem. *The set M, endowed with the metric ρ, is a complete metric space.*

Proof. A uniformly convergent sequence of continuous functions converges to a continuous function. If the functions of the sequence satisfied inequality (4), then the limiting function also satisfies inequality (4) with the same constant C. □

We remark that the space M depends on three positive numbers: a', b', and C.

7. The Contraction Mapping $A : M \to M$

We define a mapping $A : M \to M$ by setting[3]

$$(A\boldsymbol{h})(t,\boldsymbol{x}) = \int_{t_0}^{t} \boldsymbol{v}(\tau, \boldsymbol{x} + \boldsymbol{h}(\tau,\boldsymbol{x}))\, d\tau. \tag{5}$$

Because of inequality (4) the point $(\tau, \boldsymbol{x} + \boldsymbol{h}(\tau,\boldsymbol{x}))$ belongs to the cone $K_{\boldsymbol{x}}$ and consequently to the domain of definition of the vector field $\boldsymbol{v}$.

Theorem. *If the quantity a' is sufficiently small, then formula* (5) *defines a contraction mapping of the space M into itself.*

Proof. 1. We shall show that *A maps M into itself.* The function $A\boldsymbol{h}$ is continuous, since the integral of a continuous function that depends continuously on a parameter is itself continuously dependent on the parameter and on the upper limit of integration. The function $A\boldsymbol{h}$ satisfies inequality (4), since

$$|(A\boldsymbol{h})(t,\boldsymbol{x})| \le \left| \int_{t_0}^{t} \boldsymbol{v}(\cdots)\, d\tau \right| \le \left| \int_{t_0}^{t} C\, dt \right| \le C|t - t_0|.$$

Thus $AM \subset M$.

2. We shall show that *the mapping A is a contraction*:

$$\|A\boldsymbol{h}_1 - A\boldsymbol{h}_2\| \le \lambda \|\boldsymbol{h}_1 - \boldsymbol{h}_2\|, \quad 0 < \lambda < 1.$$

To do this we estimate the value of $A\boldsymbol{h}_1 - A\boldsymbol{h}_2$ at the point $(t, \boldsymbol{x})$. We have (Fig. 223)

$$(A\boldsymbol{h}_1 - A\boldsymbol{h}_2)(t,\boldsymbol{x}) = \int_{t_0}^{t} (\boldsymbol{v}_1 - \boldsymbol{v}_2)\, d\tau,$$

[3] In comparing this mapping with the Picard mapping of Sect. 1, it should be kept in mind that we are now seeking a solution in the form $\boldsymbol{x} + \boldsymbol{h}$.

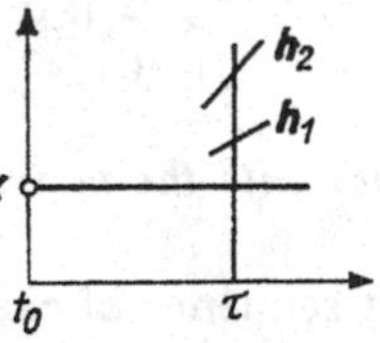

Fig. 223. Comparison of $\boldsymbol{v}_1$ and $\boldsymbol{v}_2$

where $\boldsymbol{v}_i(\tau) = \boldsymbol{v}(\tau, \boldsymbol{x} + \boldsymbol{h}_i(\tau, \boldsymbol{x}))$, $i = 1, 2$.

According to the theorem of Sect. 4 for a fixed τ the function $\boldsymbol{v}(\tau, \boldsymbol{x})$ satisfies a Lipschitz condition with constant L (on the second argument). Therefore

$$|\boldsymbol{v}_1(\tau) - \boldsymbol{v}_2(\tau)| \le L|\boldsymbol{h}_1(\tau, \boldsymbol{x}) - \boldsymbol{h}_2(\tau, \boldsymbol{x})| \le L\|\boldsymbol{h}_1 - \boldsymbol{h}_2\|.$$

According to the lemma of Sect. 2

$$|(A\boldsymbol{h}_1 - A\boldsymbol{h}_2)(t, \boldsymbol{x})| \le \left| \int_{t_0}^{t} L\|\boldsymbol{h}_2 - \boldsymbol{h}_2\| \, d\tau \right| \le La'\|\boldsymbol{h}_2 - \boldsymbol{h}_2\|.$$

For $La' < 1$ the mapping is a contraction.

The theorem is now proved. □

8. The Existence and Uniqueness Theorem

Corollary. *Suppose the right-hand side $\boldsymbol{v}$ of the differential equation (3) is continuously differentiable in a neighborhood of the point $(t_0, \boldsymbol{x}_0)$ of the extended phase space. Then there is a neighborhood of the point t_0 such that a solution of Eq. (3) is defined in this neighborhood with the initial condition $\boldsymbol{\varphi}(t_0) = \boldsymbol{x}$, where $\boldsymbol{x}$ is any point sufficiently close to $\boldsymbol{x}_0$; moreover this solution depends continuously on the initial point $\boldsymbol{x}$.*

Proof. The contraction mapping A, according to the theorem of § 30, has a fixed point $\boldsymbol{h} \in M$. Set $\boldsymbol{g}(t, \boldsymbol{x}) = \boldsymbol{x} + \boldsymbol{h}(t, \boldsymbol{x})$. Then

$$\boldsymbol{g}(t, \boldsymbol{x}) = \boldsymbol{x} + \int_{t_0}^{t} \boldsymbol{v}(\tau, \boldsymbol{g}(\tau, \boldsymbol{x})) \, d\tau, \qquad \frac{\partial \boldsymbol{g}(t, \boldsymbol{x})}{\partial t} = \boldsymbol{v}(t, \boldsymbol{g}(t, \boldsymbol{x})).$$

We see that $\boldsymbol{g}$ satisfies Eq. (3) for fixed $\boldsymbol{x}$ and for $t = t_0$ it satisfies the initial condition $\boldsymbol{g}(t, \boldsymbol{x}_0) = \boldsymbol{x}$). The function $\boldsymbol{g}$ is continuous, since $\boldsymbol{h} \in M$.

The corollary is now proved. □

Thus we have proved an existence theorem for Eq. (3) and produced a solution that depends continuously on the initial conditions.

Problem 1. Prove the uniqueness theorem.

Solution 1. Set $b' = 0$ in the definition of M. From the uniqueness of the fixed point of the contraction mapping $A : M \to M$ it follows that the solution (with initial condition $\varphi(t_0) = \boldsymbol{x}_0$) is unique.

Solution 2. Let φ_1 and φ_2 be two solutions with the same initial condition $\varphi_1(t_0) = \varphi_2(t_0)$ defined for $|t - t_0| < \alpha$. Let $0 < \alpha' < \alpha$. We set $\|\varphi\| = \max_{|t-t_0|\le\alpha'} |\varphi(t)|$. We have

$$\varphi_1(t) - \varphi_2(t) = \int_{t_0}^{t} \boldsymbol{v}(\tau, \varphi_1(\tau)) - \boldsymbol{v}(\tau, \varphi_2(\tau))\, d\tau.$$

For sufficiently small α' the points $(\tau, \varphi_1(\tau))$ and $(\tau, \varphi_2(\tau))$ lie inside the cylinder, where $\boldsymbol{v} \in \operatorname{Lip} L$. Therefore $\|\varphi_1 - \varphi_2\| \le L\alpha'\|\varphi_1 - \varphi_2\|$, whence $L\alpha' < 1$ implies $\|\varphi_1 - \varphi_2\| = 0$. Thus the solutions φ_1 and φ_2 coincide in some neighborhood of the point t_0.

The local uniqueness theorem is now proved.

9. Other Applications of Contraction Mappings

Problem 1. Prove the inverse function theorem.

Hint. It suffices to invert, in a neighborhood of the point $\mathbf{o} \in \boldsymbol{R}^n$, a C^1-mapping whose linear part equals 1, $\boldsymbol{y} = \boldsymbol{x} + \varphi(\boldsymbol{x})$, where $\varphi'(\mathbf{o}) = \mathbf{o}$ (the general case reduces to this one by a linear change of coordinates).

We seek the solution in the form $\boldsymbol{x} = \boldsymbol{y} + \psi(\boldsymbol{y})$. Then for ψ we obtain the equation $\psi(\boldsymbol{y}) = -\varphi(\boldsymbol{y} + \psi(\boldsymbol{y}))$.

Consequently the function ψ being sought is a fixed point of the mapping A defined by the formula

$$(A\psi)(\boldsymbol{y}) = -\varphi(\boldsymbol{y} + \psi(\boldsymbol{y})).$$

The mapping A is a contraction (in a suitable metric) because the derivative of the function φ is small in a neighborhood of the point $\mathbf{o}$ (by the condition $\varphi'(\mathbf{o}) = \mathbf{o}$).

Problem 2. Prove that the Euler broken line tends to a solution as the step size tends to zero.

Solution. Let $\boldsymbol{g}_\Delta = \boldsymbol{x} + \boldsymbol{h}_\Delta$ be the Euler broken line with step Δ and initial point $\boldsymbol{g}_\Delta(t, \boldsymbol{x}_0) = \boldsymbol{x}$ (Fig. 224). In other words, for $t \ne t_0 + k\Delta$

$$\frac{\partial \boldsymbol{g}_\Delta(t, \boldsymbol{x})}{\partial t} = \boldsymbol{v}(s(t), \boldsymbol{g}_\Delta(s(t), \boldsymbol{x})),$$

where $s(t) = t_0 + k\Delta$, k being the integer part of $(t - t_0)/\Delta$. The difference between the Euler broken line and the solution $\boldsymbol{g}$ can be estimated by the formula of Sect. 3 of § 30:

$$\|\boldsymbol{g}_\Delta - \boldsymbol{g}\| = \|\boldsymbol{h}_\Delta - \boldsymbol{h}\| \le (1 - \lambda)^{-1}\|A\boldsymbol{h}_\Delta - \boldsymbol{h}_\Delta\|.$$

But

$$(A\boldsymbol{h}_\Delta)(t, \boldsymbol{x}) = \int_{t_0}^{t} \boldsymbol{v}(\tau, \boldsymbol{g}_\Delta(\tau, \boldsymbol{x}))\, d\tau, \quad \boldsymbol{h}_\Delta(t, \boldsymbol{x}) = \int_{t_0}^{t} \boldsymbol{v}(s(\tau), \boldsymbol{g}_\Delta(s(\tau), \boldsymbol{x}))\, d\tau.$$

As $\Delta \to 0$ the difference between the two integrands tends to zero uniformly in τ for $|\tau| \le a'$ (because of the uniform continuity of $\boldsymbol{v}$). Therefore $\|A\boldsymbol{h}_\Delta - \boldsymbol{h}_\Delta\| \to 0$ as $\Delta \to 0$, and the Euler broken line tends to the solution.

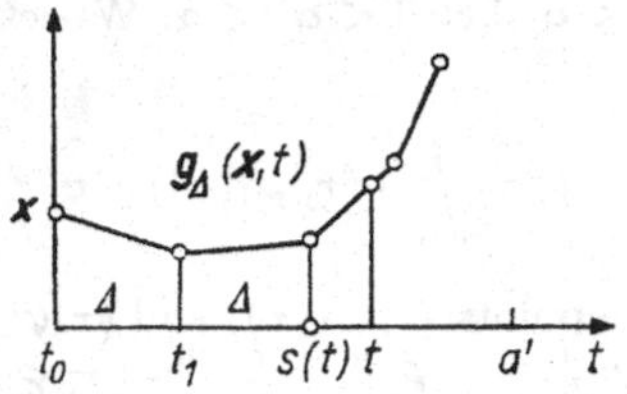

Fig. 224. The Euler broken line

Problem *3. Consider a diffeomorphism A of a neighborhood of the point $\mathbf{o}$ in $\boldsymbol{R}^n$ onto a neighborhood of the point $\mathbf{o}$ in $\boldsymbol{R}^n$ that maps $\mathbf{o}$ to $\mathbf{o}$. Assume that the linear part of A at $\mathbf{o}$ (i.e., the linear operator $A_{*\mathbf{o}} : \boldsymbol{R}^n \to \boldsymbol{R}^n$) does not have any eigenvalues of absolute value 1. Let the number of eigenvalues with $|\lambda| < 1$ be m_- and let the number with $|\lambda| > 1$ be m_+. Then $A_{*\mathbf{o}}$ has an invariant subspace $\boldsymbol{R}^{m_-}$ (an incoming space) and an invariant subspace $\boldsymbol{R}^{m_+}$ (an outgoing space), whose points tend to $\mathbf{o}$ under the application of $A_{*\mathbf{o}}^N$ as $N \to +\infty$ (for $\boldsymbol{R}^{m_-}$) or as $N \to -\infty$ (for $\boldsymbol{R}^{m_+}$) (Fig. 225).

Prove that *the original nonlinear operator A also has invariant submanifolds M^{m_-} and M^{m_+} in a neighborhood of the point $\mathbf{o}$* (incoming and outgoing manifolds) *that are tangent to the subspaces $\boldsymbol{R}^{m_-}$ and $\boldsymbol{R}^{m_+}$ at the point $\mathbf{o}$ ($A^N\boldsymbol{x} \to \mathbf{o}$ as $N \to +\infty$ on M^{m_-} and as $N \to -\infty$ for $x \in M^{m_+}$).*

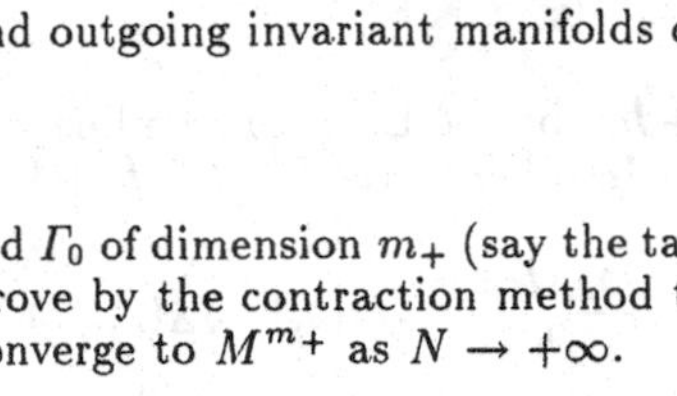

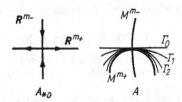

Fig. 225. The incoming and outgoing invariant manifolds of a diffeomorphism and of its linear part

Hint. Take any submanifold Γ_0 of dimension m_+ (say the tangent to $\boldsymbol{R}^{m_+}$ at $\mathbf{o}$) and apply powers of A to it. Prove by the contraction method that the approximations $\Gamma_N = A^N\Gamma_0$ so obtained converge to M^{m_+} as $N \to +\infty$.

Problem *4. Prove that there exist incoming and outgoing invariant manifolds for the nonlinear saddle $\dot{\boldsymbol{x}} = \boldsymbol{v}(\boldsymbol{x})$, $\boldsymbol{v}(\mathbf{o}) = \mathbf{o}$ (it is assumed that none of the eigenvalues of the operator $A = \boldsymbol{v}_*(\mathbf{o})$ lies on the imaginary axis).

§ 32. The Theorem on Differentiability

In this section we shall prove the rectification theorem.

1. The Equation of Variations

Associated with a differentiable mapping $f : U \to V$ is a linear operator on the tangent space at each point

$$f_{*x} : T_x U \to T_{f(x)} V.$$

In exactly the same way with a differential equation

$$\dot{x} = v(t, x), \quad x \in U \subset R^n, \tag{1}$$

there is associated a system of differential equations

$$\begin{cases} \dot{x} = v(t, x), & x \in U \subset R^n, \\ \dot{y} = v_*(t, x) y, & y \in T_x U, \end{cases} \tag{2}$$

called the *system of equations of variations* for Eq. (1) and *linear* with respect to the tangent vector y (Fig. 226).

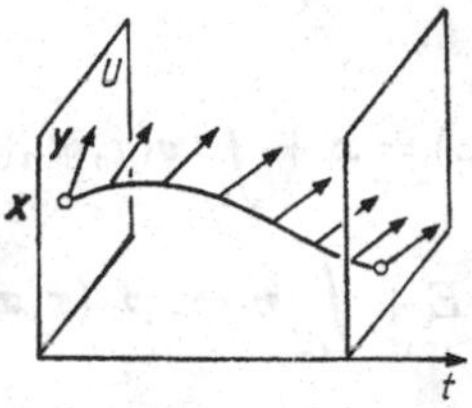

Fig. 226. The solution of the equation of variations with the initial condition (x, y)

The asterisk in formula (2) (and in subsequent formulas) denotes the derivative with respect to x for a fixed t. Thus $v_*(t, x)$ is a linear operator from R^n into R^n.

Along with the system (2) it is convenient to consider the system

$$\dot{x} = v(t, x), \quad x \in U \subset R^n, \quad \dot{z} = v_*(t, x) z, \quad z : R^n \to R^n. \tag{3}$$

The system (3) is obtained from (2) by replacing the unknown vector y with the unknown linear operator z. We shall also use the name *equation of variations* in reference to the system (3).

Remark. In general given a linear equation

$$\dot{y} = A(t) y, \quad y \in R^n, \tag{2'}$$

it is useful to consider the associated equation

$$\dot{z} = A(t)z, \quad z : \boldsymbol{R}^n \to \boldsymbol{R}^n, \tag{3'}$$

in the linear operator z.

Knowing a solution of one of Eqs. (2′) and (3′), it is easy to find a solution of the other (how?).

2. The Differentiability Theorem

Suppose the right-hand side $\boldsymbol{v}$ of Eq. (1) *is twice continuously differentiable in some neighborhood of the point $(t_0, \boldsymbol{x}_0)$. Then the solution $\boldsymbol{g}(t, \boldsymbol{x})$ of Eq.* (1) *with initial condition $\boldsymbol{g}(t_0, \boldsymbol{x}) = \boldsymbol{x}$ depends on the initial condition $\boldsymbol{x}$ in a continuously differentiable manner as $\boldsymbol{x}$ and t vary in some (perhaps smaller) neighborhood of the point $(t_0, \boldsymbol{x}_0)$:*

$$\boldsymbol{v} \in C^2 \Rightarrow \boldsymbol{g} \in C^1_{\boldsymbol{x}}$$

(it is of class C^1 in $\boldsymbol{x}$).

Proof. $\boldsymbol{v} \in C^2 \Rightarrow \boldsymbol{v}_* \in C^1$. Therefore the system of equations of variations (3) satisfies the hypotheses of § 31 and the sequence of Picard approximations converges uniformly to a solution of the system in a sufficiently small neighborhood of the point t_0. We choose initial conditions $\varphi_0 = \boldsymbol{x}$ (sufficiently close to $\boldsymbol{x}_0$), and $\psi_0 = E$. We denote the Picard approximations by φ_n (for $\boldsymbol{x}$) and by ψ_n (for z), i.e., we set

$$\varphi_{n+1}(t, \boldsymbol{x}) = \boldsymbol{x} + \int_{t_0}^{t} \boldsymbol{v}(\tau, \varphi_n(\tau, \boldsymbol{x}))\, d\tau, \tag{4}$$

$$\psi_{n+1}(t, \boldsymbol{x}) = E + \int_{t_0}^{t} \boldsymbol{v}_*(\tau, \varphi_n(\tau, \boldsymbol{x}))\psi_n(\tau, \boldsymbol{x})\, d\tau. \tag{5}$$

We remark that $\varphi_{0*} = \psi_0$. From the definitions (4) and (5) we conclude by induction on n that $\varphi_{n+1*} = \psi_{n+1}$. Therefore the sequence $\{\psi_n\}$ is the sequence of derivatives of the sequence $\{\varphi_n\}$. Both sequences (4) and (5) converge uniformly (being the sequences of Picard approximations of the system (3)) for $|t - t_0|$ sufficiently small. Thus the sequence $\{\varphi_n\}$ converges uniformly along with the derivatives on $\boldsymbol{x}$. Therefore the limiting function $\boldsymbol{g}(t, \boldsymbol{x}) = \lim_{n\to\infty} \varphi_n(t, \boldsymbol{x})$ is continuously differentiable with respect to $\boldsymbol{x}$, which was to be proved. □

Remark. We have simultaneously proved the following theorem.

Theorem. *The derivative $\boldsymbol{g}_*$ of a solution of Eq.* (1) *with respect to the initial condition $\boldsymbol{x}$ satisfies the equation of variations* (3) *with the initial condition $z(t_0) = E$:*

$$\frac{\partial}{\partial t}\boldsymbol{g}(t, \boldsymbol{x}) = \boldsymbol{v}(t, \boldsymbol{g}(t, \boldsymbol{x})), \quad \frac{\partial}{\partial t}\boldsymbol{g}_*(t, \boldsymbol{x}) = \boldsymbol{v}_*(t, \boldsymbol{g}(t, \boldsymbol{x}))\boldsymbol{g}_*(t, \boldsymbol{x}).$$

$$\boldsymbol{g}(t_0, \boldsymbol{x}) = \boldsymbol{x}, \quad \boldsymbol{g}_*(t_0, \boldsymbol{x}) = E.$$

This theorem explains the meaning of the equations of variations: they describe the action of the transformation over the time from t_0 to t on the vectors tangent to the phase space (Fig. 227).

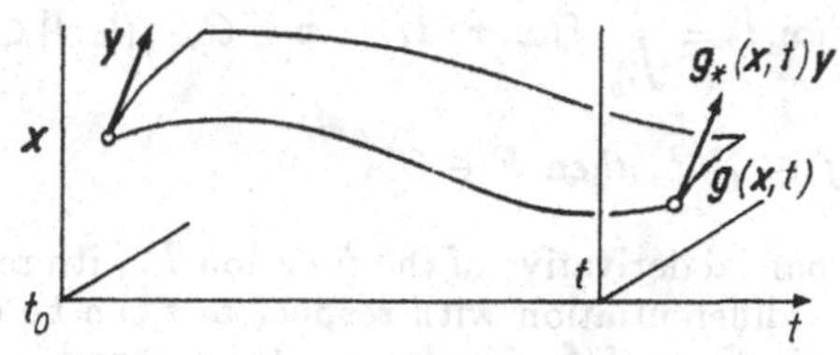

Fig. 227. The action of the transformation over the time from t_0 to t on a curve in the phase space and on a tangent vector to it

3. Higher Derivatives with Respect to $\boldsymbol{x}$

Let $r \geq 2$ be an integer.

Theorem T_r. *Suppose the right-hand side $\boldsymbol{v}$ of Eq. (1) is r times continuously differentiable in some neighborhood of the point $(t_0, \boldsymbol{x}_0)$. Then the solution $\boldsymbol{g}(t, \boldsymbol{x})$ of Eq. (1) with initial condition $\boldsymbol{g}(t_0, \boldsymbol{x}) = \boldsymbol{x}$ is $r-1$ times continuously differentiable as a function of the initial condition $\boldsymbol{x}$ when $\boldsymbol{x}$ and t vary in some (possibly smaller) neighborhood of the point $(t_0, \boldsymbol{x}_0)$:*

$$\boldsymbol{v} \in C^r \Rightarrow \boldsymbol{g} \in C^{r-1}_{\boldsymbol{x}}.$$

Proof. $\boldsymbol{v} \in C^r \Rightarrow \boldsymbol{v}_* \in C^{r-1}$. Hence the system of equations in variations (3) satisfies the hypotheses of Theorem T_{r-1}. Therefore Theorem T_r, $r > 2$ follows from Theorem T_{r-1}:

$$\boldsymbol{v} \in C^r \Rightarrow \boldsymbol{v}_* \in C^{r-1} \Rightarrow \boldsymbol{g}_* \in C^{r-2}_{\boldsymbol{x}} \Rightarrow \boldsymbol{g} \in C^{r-1}_{\boldsymbol{x}}.$$

But Theorem T_2 was proved in Sect. 2. Thus Theorem T_r is proved. □

4. Derivatives in $\boldsymbol{x}$ and t

Let $r \geq 2$ be an integer.

Theorem T'_r. *Under the hypotheses of Theorem T_r the solution $\boldsymbol{g}(t, \boldsymbol{x})$ is a differentiable function of class C^{r-1} in the variables $\boldsymbol{x}$ and t jointly: $\boldsymbol{v} \in C^r \Rightarrow \boldsymbol{g} \in C^{r-1}$.*

This theorem is an obvious corollary of the preceding theorem. Here is a formal proof.

Lemma. *Let $\boldsymbol{f}$ be a function (with values in $\boldsymbol{R}^n$) defined on the direct product of the domain G of the Euclidean space $\boldsymbol{R}^n$ and the interval I of the t-axis:*

$$\boldsymbol{f} : G \times I \to \boldsymbol{R}^n.$$

Form the integral

$$\boldsymbol{F}(\boldsymbol{x}, t) = \int_{t_0}^{t} \boldsymbol{f}(\boldsymbol{x}, \tau)\, d\tau, \quad \boldsymbol{x} \in G, \quad [t_0, t] \subset I.$$

If $\boldsymbol{f} \in C^r_{\boldsymbol{x}}$ *and* $\boldsymbol{f} \in C^{r-1}$, *then* $\boldsymbol{F} \in C^r$.

Indeed, any rth partial derivative of the function $\boldsymbol{F}$ with respect to the variables x_i and t containing a differentiation with respect to t can be expressed in terms of $\boldsymbol{f}$ and the partial derivatives of $\boldsymbol{f}$ of order less than r, and is therefore continuous: and every rth partial derivative with respect to the variables x_i is continuous by hypothesis.

Proof of the theorem. We have

$$\boldsymbol{g}(t, \boldsymbol{x}) = \boldsymbol{x} + \int_{t_0}^{t} (\boldsymbol{v}(\tau, \boldsymbol{g}(\tau, \boldsymbol{x}))\, d\tau.$$

We use the notation $\boldsymbol{f}(\tau, \boldsymbol{x}) = \boldsymbol{v}(\tau, \boldsymbol{g}(\tau, \boldsymbol{x}))$ and apply the lemma. We find that for $1 \le \rho \le r$

$$\boldsymbol{g} \in C^{\rho-1} \cap C^{\rho}_{\boldsymbol{x}} \Rightarrow \boldsymbol{g} \in C^{\rho}.$$

According to Theorem T_r we have $\boldsymbol{g} \in C^{\rho}_{\boldsymbol{x}}$ for $\rho < r$. We obtain successively

$$\boldsymbol{g} \in C^0 \Rightarrow \boldsymbol{g} \in C^1 \Rightarrow \cdots \Rightarrow \boldsymbol{g} \in C^{r-1}.$$

But according to § 31 we have $\boldsymbol{g} \in C^0$ (the solution depends continuously on $(\boldsymbol{x}, t)$).

Theorem T'_r is now proved. □

Problem 1. Prove that if the right-hand side of the differential equation (1) is infinitely differentiable, then the solution is also an infinitely differentiable function of the initial conditions:

$$\boldsymbol{v} \in C^{\infty} \Rightarrow \boldsymbol{g} \in C^{\infty}.$$

Remark. It can also be proved that if the right-hand side $\boldsymbol{v}$ is analytic (can be expanded in a convergent Taylor series in a neighborhood of each point), then the solution $\boldsymbol{g}$ is also an analytic function of $\boldsymbol{x}$ and t.

It is natural to regard differential equations with analytic right-hand sides both as functions of complex-valued unknowns and (what is especially important) as functions of complex-valued time. For this theory cf., for example, the book of V. V. Golubev, *Vorlesungen über Differentialgleichungen im Komplexen*, Hochschulbücher für Mathematik, Bd. 43, VEB Deutscher Verlag der Wissenschaften, Berlin, 1958.

5. The Rectification Theorem

This theorem is an obvious corollary of Theorem T'_r. Before proving it we recall two simple geometric propositions. Let L_1 and L_2 be two subspaces of a third vector space L (Fig. 228). The subspaces L_1 and L_2 are called *transversal* if their sum is the whole space L: $L_1 + L_2 = L$. For example, a line in $\boldsymbol{R}^3$ is transversal to a plane if it intersects it in a nonzero angle.

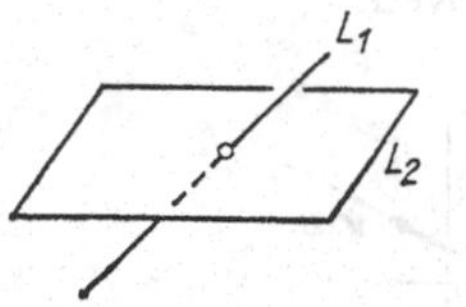

Fig. 228. The line L_1 is transversal to the plane L_2 in the space $\boldsymbol{R}^3$

Proposition 1. *For each k-dimensional subspace $\boldsymbol{R}^k$ in $\boldsymbol{R}^n$ there is a subspace of dimension $n-k$ transversal to it (and in fact there is one among the C_n^k coordinate planes of the space $\boldsymbol{R}^n$).*

The proof can be found in courses of linear algebra (the theorem on the rank of a matrix).

Proposition 2. *If a linear transformation $A : L \to M$ maps some pair of transversal subspaces into transversal subspaces, then its range is all of M.*

Proof. $AL = AL_1 + AL_2 = M$.

Proof of the rectification theorem: the nonautonomous case (cf. Chapt. 2, § 8, Sect. 1). Consider the mapping G of a domain of the direct product $\boldsymbol{R} \times \boldsymbol{R}^n$ into the extended phase space of the equation

$$\dot{\boldsymbol{x}} = \boldsymbol{v}(t, \boldsymbol{x}), \tag{1}$$

given by the formula $G(t, \boldsymbol{x}) = (t, \boldsymbol{g}(t, \boldsymbol{x}))$, where $\boldsymbol{g}(t, \boldsymbol{x})$ is a solution of Eq. (1) with the initial condition $\boldsymbol{g}(t_0, \boldsymbol{x}) = \boldsymbol{x}$.

We shall show that G is a rectifying diffeomorphism in a neighborhood of the point $(t_0, \boldsymbol{x}_0)$.

a) *The mapping G is differentiable* (of class C^{r-1} if $\boldsymbol{v} \in C^r$) by Theorem T_r'.

b) *The mapping G leaves t fixed*: $G(t, \boldsymbol{x}) = (t, \boldsymbol{g}(t, \boldsymbol{x}))$.

c) *The mapping G_* takes the standard vector field* $\mathbf{e}$ *($\dot{\boldsymbol{x}} = \mathbf{o}$, $t = 1$) into the given field*: $G_*\mathbf{e} = (1, \boldsymbol{v})$ (since $\boldsymbol{g}(t, \boldsymbol{x})$ is a solution of Eq. (1)).

d) *The mapping G is a diffeomorphism in a neighborhood of the point* $(t_0, \boldsymbol{x}_0)$.

Indeed, let us compute the restrictions of the linear operator $G_*|_{t_0, \boldsymbol{x}_0}$ to the transversal planes $\boldsymbol{R}^n$ and $\boldsymbol{R}^1$ (Fig. 229). We find:

$$G_*|_{\boldsymbol{R}^n : t = t_0} = E, \quad G_*|_{\boldsymbol{R}^1 : \boldsymbol{x} = \boldsymbol{x}_0} \mathbf{e} = \boldsymbol{v} + \mathbf{e}.$$

The plane $\boldsymbol{R}^n$ and the line with direction vector $\boldsymbol{v}+\boldsymbol{e}$ are transversal. Thus G_* is a linear transformation of $\boldsymbol{R}^{n+1}$ onto $\boldsymbol{R}^{n+1}$, and consequently an isomorphism (the Jacobian of G_* at the point $(t_0, \boldsymbol{x}_0)$ is nonzero). By the inverse function theorem G is a local diffeomorphism.

The theorem is now proved. □

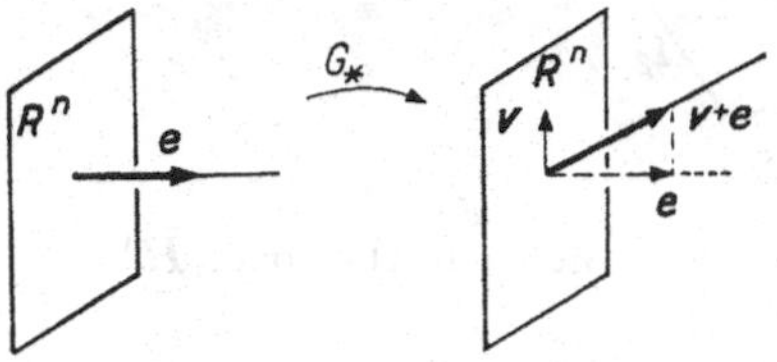

Fig. 229. The derivative of the mapping G at the point $(t_0, \boldsymbol{x}_0)$

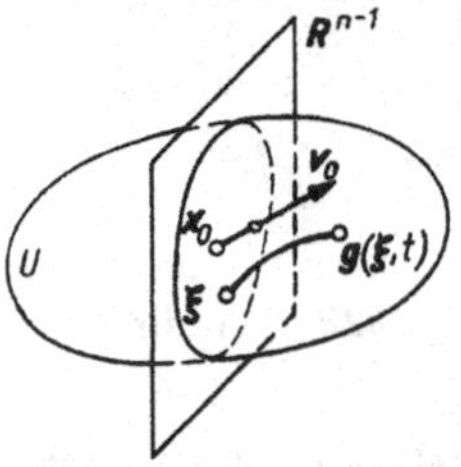

Fig. 230. The construction of a diffeomorphism that rectifies a vector field

Proof of the rectification theorem: the autonomous case (§ 7, Sect. 1). Consider the autonomous equation

$$\dot{\boldsymbol{x}} = \boldsymbol{v}(\boldsymbol{x}), \quad \boldsymbol{x} \in U \subset \boldsymbol{R}^n. \tag{6}$$

Suppose the phase velocity vector $\boldsymbol{v}_0$ is nonzero at the point $\boldsymbol{x}_0$ (Fig. 230). Then there exists an $(n-1)$-dimensional hyperplane $\boldsymbol{R}^{n-1} \subset \boldsymbol{R}^n$ passing through $\boldsymbol{x}_0$ and transversal to $\boldsymbol{v}_0$ (more precisely, the corresponding plane in the tangent space $T_{\boldsymbol{x}_0}U$ is transversal to the line $\boldsymbol{R}^1$ with direction $\boldsymbol{v}_0$).

We define a mapping G of the domain $\boldsymbol{R} \times \boldsymbol{R}^{n-1}$, where $\boldsymbol{R}^{n-1} = \{\boldsymbol{\xi}\}$, $\boldsymbol{R} = \{t\}$, into the domain $\boldsymbol{R}^n$ by the formula $G(t, \boldsymbol{\xi}) = \boldsymbol{g}(t, \boldsymbol{\xi})$, where $\boldsymbol{\xi}$ lies in $\boldsymbol{R}^{n-1}$ near $\boldsymbol{x}_0$ and $\boldsymbol{g}(t, \boldsymbol{\xi})$ is the value of the solution of Eq. (6) with the initial condition $\varphi(0) = \boldsymbol{\xi}$ at the instant t. We shall show that in a sufficiently small neighborhood of the point $(\boldsymbol{\xi} = \boldsymbol{x}_0, t = 0)$ the mapping G^{-1} is a rectifying diffeomorphism.

a) *The mapping G is differentiable* ($G \in C^{r-1}$ if $\boldsymbol{v} \in C^r$) by Theorem T'_r.

b) *The mapping G^{-1} is rectifying* since G_* maps the standard vector field $\boldsymbol{e}$ ($\dot{\boldsymbol{\xi}} = \mathbf{o}$, $\dot{t} = 1$) to $G_*\boldsymbol{e} = \boldsymbol{v}$, because $\boldsymbol{g}(t, \boldsymbol{\xi})$ satisfies Eq. (6).

c) *The mapping G is a local diffeomorphism.*

Indeed, let us calculate the linear operator $G_*|_{t_0, \boldsymbol{x}_0}$ on the transversal planes $\boldsymbol{R}^{n-1}$ and $\boldsymbol{R}^1$. We find $G_*|_{\boldsymbol{R}^{n-1}} = E$, $G_*|_{\boldsymbol{R}^1}\boldsymbol{e} = \boldsymbol{v}_0$.

Thus the operator $G_*|_{t_0, \boldsymbol{x}_0}$ maps the pair of transversal subspaces $\boldsymbol{R}^{n-1}$ and $\boldsymbol{R}^1 \subset \boldsymbol{R}^n$ onto a pair of transversal subspaces. Therefore $G_*|_{t_0, \boldsymbol{x}_0}$ is a linear transformation of $\boldsymbol{R}^n$ onto $\boldsymbol{R}^n$ and consequently an isomorphism. By the inverse function theorem G is a local diffeomorphism. The theorem is now proved. □

Remark. Since the theorem on differentiability is proved with the loss of one derivative ($\boldsymbol{v} \in C^r \Rightarrow \boldsymbol{g} \in C^{r-1}$), we can also guarantee only class C^{r-1} of smoothness for our rectifying diffeomorphisms. In reality the rectifying diffeomorphism constructed belongs to C^r; the proof is given below.

6. The Last Derivative

In the differentiability theorem (Sect. 2) we assumed the field $\boldsymbol{v}$ was twice continuously differentiable. In reality it would suffice to have only continuous first-order derivatives.

Theorem. *If the right-hand side $\boldsymbol{v}(t, \boldsymbol{x})$ of the differential equation $\dot{\boldsymbol{x}} = \boldsymbol{v}(t, \boldsymbol{x})$ is continuously differentiable, then the solution $\boldsymbol{g}(t, \boldsymbol{x})$ with initial condition $\boldsymbol{g}(t_0, \boldsymbol{x}) = \boldsymbol{x}$ is a continuously differentiable function of the initial conditions:*

$$\boldsymbol{v} \in C^1 \Rightarrow \boldsymbol{g} \in C^1_{\boldsymbol{x}}. \tag{7}$$

Corollaries.

1) $\boldsymbol{v} \in C^r \Rightarrow \boldsymbol{g} \in C^r$ *for* $r \geq 1$.

2) *The rectifying diffeomorphisms constructed in Sect. 5 are r times continuously differentiable if $\boldsymbol{v} \in C^r$.*

The corollaries are deduced from relation (7) by repeating verbatim the reasoning of Sects. 3, 4, and 5. The proof of relation (7) itself requires some clever tricks.

Proof of the theorem. We begin with the following remarks.

Lemma 1. *The solution of a linear equation $\dot{\boldsymbol{y}} = A(t)\boldsymbol{y}$ whose right-hand side depends continuously on t exists, is unique, is determined uniquely by the initial conditions $\varphi(t_0) = \boldsymbol{y}_0$, and is a continuous function of $\boldsymbol{y}_0$ and t.*

Indeed the proof of the existence, uniqueness and continuity theorems (§ 31) used only the differentiability with respect to $\boldsymbol{x}$ for a fixed t (actually only the Lipschitz condition on $\boldsymbol{x}$). Therefore the proof remains valid if the dependence on t is assumed to be merely continuous. The lemma is now proved. □

We remark that the solution is a linear function of $\boldsymbol{y}_0$ and a continuously differentiable function of t, hence belongs to the class C^1 jointly in $\boldsymbol{y}_0$ and t.

Lemma 2. *If the linear transformation A in Lemma 1 also depends on a parameter α in such a way that the function $A(t, \alpha)$ is continuous, then the solution will be a continuous function of $\boldsymbol{y}_0$, t, and α.*

Indeed, the solution can be constructed as the limit of a sequence of Picard approximations. Each approximation depends continuously on $\boldsymbol{y}_0$, t, and α. The sequence of approximations converges uniformly with respect to $\boldsymbol{y}_0$, t, and α in a sufficiently small neighborhood of each point $(\boldsymbol{y}_{0,0}, t_0, \alpha_0)$. Therefore the limit is a continuous function of $\boldsymbol{y}_0$, t, and α.

Lemma 2 is now proved. □

We now apply Lemma 2 to the equation of variations.

Lemma 3. *The system of equations of variations*

$$\dot{\boldsymbol{x}} = \boldsymbol{v}(t, \boldsymbol{x}), \quad \dot{\boldsymbol{y}} = \boldsymbol{v}_*(t, \boldsymbol{x})\boldsymbol{y}$$

has a solution that is uniquely determined by its initial data and depends continuously on them provided the field $\boldsymbol{v}$ is of class C^1.

Indeed, the first equation of the system has a solution by the existence theorem of § 31. This solution is uniquely determined by its initial conditions $(t_0, \boldsymbol{x}_0)$ and depends continuously on these conditions. Let us substitute this solution into the second equation. We then obtain an equation that is linear in $\boldsymbol{y}$. Its right-hand side depends continuously on t and on the initial condition $\boldsymbol{x}_0$ (regarded as a parameter) of the solution of the first equation under consideration. By Lemma 2 this linear equation has a solution that is determined by its initial data $\boldsymbol{y}_0$ and is a continuous function of t, $\boldsymbol{y}_0$ and the parameter $\boldsymbol{x}_0$

Lemma 3 is now proved. □

Thus *the equations of variations are solvable even in the case $\boldsymbol{v} \in C^1$*. We remark that in the case $\boldsymbol{v} \in C^2$ we proved that the derivative of the solution with respect to the initial data satisfies the equation of variations (3). We cannot assert this now: in fact we don't even know whether such a derivative exists.

To prove that the solution is differentiable with respect to the initial conditions we begin by studying a special case.

Lemma 4. *If the vector field $\boldsymbol{v}(t, \boldsymbol{x})$ of class C^1 vanishes at the point $\boldsymbol{x} = \mathbf{o}$, for all t together with its derivative $\boldsymbol{v}_*$, then the solution of the equation $\dot{\boldsymbol{x}} = \boldsymbol{v}(t, \boldsymbol{x})$ is differentiable with respect to the initial conditions at the point $\boldsymbol{x} = \mathbf{o}$.*

Indeed, by hypothesis $|\boldsymbol{v}(t, \boldsymbol{x})| = o(|\boldsymbol{x}|)$ in a neighborhood of the point $\boldsymbol{x} = \mathbf{o}$. Let us estimate the error in the approximation $\boldsymbol{x} = \boldsymbol{x}_0$ to the solution $\boldsymbol{x} = \boldsymbol{\varphi}(t)$ with the initial condition $\boldsymbol{\varphi}(t_0) = \boldsymbol{x}_0$ according to the formula of Sect. 3 of § 30. For sufficiently small $|\boldsymbol{x}_0|$ and $|t - t_0|$ we find

$$|\boldsymbol{\varphi} - \boldsymbol{x}_0| \leq (1 - \lambda)^{-1} \left| \int_{t_0}^{t} \boldsymbol{v}(\tau, \boldsymbol{x}_0)\, d\tau \right| \leq K \max_{t_0 \leq \tau \leq t} \boldsymbol{v}(\tau, \boldsymbol{x}_0),$$

where the constant K is independent of $\boldsymbol{x}_0$.

Thus $|\boldsymbol{\varphi} - \boldsymbol{x}_0| = o(|\boldsymbol{x}_0|)$, from which it follows that $\boldsymbol{\varphi}$ is differentiable with respect to $\boldsymbol{x}_0$ at zero, which was to be proved. □

We now reduce the general case to the special situation of Lemma 4: to do this it suffices to choose a suitable coordinate system in the extended phase space. First of all we can always assume that the solution under consideration is zero:

Lemma 5. *Suppose $\boldsymbol{x} = \boldsymbol{\varphi}(t)$ is a solution of the equation $\dot{\boldsymbol{x}} = \boldsymbol{v}(t, \boldsymbol{x})$ with right-hand side of class C^1 defined in a domain of the extended phase space $\boldsymbol{R} \times \boldsymbol{R}^n$. Then there exists a C^1-diffeomorphism of the extended phase space that preserves time $((t, \boldsymbol{x}) \mapsto (t, \boldsymbol{x}_1(t, \boldsymbol{x})))$ and maps the solution $\boldsymbol{\varphi}$ to $\boldsymbol{x}_1 \equiv \mathbf{o}$.*

Indeed it suffices to make the shift $\boldsymbol{x}_1 = \boldsymbol{x} - \boldsymbol{\varphi}(t)$, since $\boldsymbol{\varphi} \in C^1$.

Lemma 5 is now proved. □

In the coordinate system $(t, \boldsymbol{x}_1)$ the right-hand side of our equation is $\mathbf{o}$ at the point $\boldsymbol{x}_1 = \mathbf{o}$. We shall show that the derivative of the right-hand side with respect to $\boldsymbol{x}_1$ can also be made to vanish using a suitable change of coordinates that is linear in $\boldsymbol{x}$.

Lemma 6. *Under the hypotheses of Lemma 5 the coordinates $(t, \boldsymbol{x}_1)$ can be chosen so that the equation $\dot{\boldsymbol{x}} = \boldsymbol{v}(t, \boldsymbol{x})$ is equivalent to the equation $\dot{\boldsymbol{x}}_1 = \boldsymbol{v}_1(t, \boldsymbol{x}_1)$, where the field $\boldsymbol{v}_1$ equals $\mathbf{o}$ at the point $\boldsymbol{x}_1 = \mathbf{o}$ along with its derivative $\partial \boldsymbol{v}_1/\partial \boldsymbol{x}_1$. Moreover the function $\boldsymbol{x}_1(t, \boldsymbol{x})$ can be chosen to be linear (but not necessarily homogeneous) with respect to $\boldsymbol{x}$.*

According to Lemma 5 we may assume that $\boldsymbol{v}_1(t, \mathbf{o}) = \mathbf{o}$.
To prove Lemma 6 we begin by considering a special case:

Lemma 7. *The assertion of Lemma 6 is true for a linear equation $\dot{\boldsymbol{x}} = A(t)\boldsymbol{x}$.*

Indeed it suffices to take as $\boldsymbol{x}_1$ the value of the solution with initial condition $\varphi(t) = \boldsymbol{x}$ at a fixed instant t_0. According to Lemma 1 we have $\boldsymbol{x}_1 = B(t)\boldsymbol{x}$, where $B(t) : \boldsymbol{R}^n \to \boldsymbol{R}^n$ is a linear operator of class C^1 with respect to t. In the coordinates $(t, \boldsymbol{x}_1)$ our linear equation assumes the form $\dot{\boldsymbol{x}}_1 = \mathbf{o}$.
Lemma 7 is now proved. □

Proof of Lemma 6. We linearize the equation $\dot{\boldsymbol{x}} = \boldsymbol{v}(t, \boldsymbol{x})$ at zero, i.e., we form the equation of variations $\dot{\boldsymbol{x}} = A(t)\boldsymbol{x}$, where $A(t) = \boldsymbol{v}_*(t, \mathbf{o})$.

By hypothesis $\boldsymbol{v} \in C^1$, so that $A \in C^0$. By Lemma 7 we can choose C^1-coordinates $\boldsymbol{x}_1 = B(t)\boldsymbol{x}$ such that in the new coordinates the linearized equation assumes the form $\dot{\boldsymbol{x}}_1 = \mathbf{o}$. It is easy to verify that in this coordinate system the right-hand side of the original nonlinear equation will have zero as its linear part.

Indeed, let us introduce the notation $\boldsymbol{v} = A\boldsymbol{x} + \boldsymbol{R}$, (so that $\boldsymbol{R} = o(|\boldsymbol{x}|)$) and $\boldsymbol{x} = C\boldsymbol{x}_1$ (so that $C = B^{-1}$). The differential equation for $\boldsymbol{x}_1$ is obtained from $\dot{\boldsymbol{x}} = \boldsymbol{v}$ by substituting $\boldsymbol{x} = C\boldsymbol{x}_1$. We obtain

$$\dot{C}\boldsymbol{x}_1 + C\dot{\boldsymbol{x}}_1 = AC\boldsymbol{x}_1 + \boldsymbol{R}.$$

But by definition of C the first terms on the left and right (the linear terms in $\boldsymbol{x}_1$) are equal. Thus

$$\dot{\boldsymbol{x}}_1 = C^{-1}\boldsymbol{R}(t, C\boldsymbol{x}_1) = O(|\boldsymbol{x}_1|).$$

Lemma 6 is now proved. □

Combining Lemmas 6 and 4, we arrive at the following conclusion:

Lemma 8. *The solution of the differential equation $\dot{\boldsymbol{x}} = \boldsymbol{v}(t, \boldsymbol{x})$ with right-hand side of class C^1 depends differentiably on the initial condition. The derivative z of the solution with respect to the initial condition satisfies the system of equations of variations*

$$\dot{\boldsymbol{x}} = \boldsymbol{v}(t, \boldsymbol{x}), \quad \dot{\boldsymbol{z}} = \boldsymbol{v}_*(t, \boldsymbol{x})\boldsymbol{z}, \quad \boldsymbol{z}(t_0) = E : \boldsymbol{R}^n \to \boldsymbol{R}^n.$$

To prove Lemma 8 it suffices to write the equation in the system of coordinates of Lemma 6 and apply Lemma 4.

To prove the theorem it remains to verify the continuity of the derivative of the solution with respect to the initial condition. According to Lemma 8 this derivative exists and satisfies the system of equations of variations. It follows from Lemma 3 that the solutions of this system depend continuously on $\boldsymbol{x}_0$ and t.

Hence the theorem is proved. □

Chapter 5. Differential Equations on Manifolds

In this chapter we define differentiable manifolds and prove a theorem on the existence of the phase flow defined by a vector field on a manifold.

Many interesting and profound results have been obtained in the theory of differential equations on manifolds. There will not be time to discuss these in the present chapter, which is only a brief introduction to this area at the junction of analysis and topology.

§ 33. Differentiable Manifolds

The concept of a *differentiable* or *smooth* manifold plays a role in geometry and analysis that is as fundamental as the concepts of group and vector space in algebra.

1. Examples of Manifolds

Although the definition of a manifold will be given later, the following objects, for example, are manifolds (Fig. 231):

1. The vector space $\boldsymbol{R}^n$ or any domain (open subset) U of it.

2. The sphere S^n defined in the Euclidean space $\boldsymbol{R}^{n+1}$ by the equation $x_1^2 + \cdots + x_{n+1}^2 = 1$, in particular the circle S^1.

3. The torus $T^2 = S^1 \times S^1$ (cf. § 24).

4. The projective space $\boldsymbol{RP}^n = \{(x_0 : x_1 : \ldots : x_n)\}$. We recall that the points of this space are the lines passing through the origin in $\boldsymbol{R}^{n+1}$. Such a line is determined by any of its points (except the origin). The coordinates of this point $(x_1, \ldots, x_n)$ in $\boldsymbol{R}^{n+1}$ are called *homogeneous coordinates* of the corresponding point of the projective space.

This last example is particularly instructive. In studying the following definitions it is useful to keep in mind the affine coordinates in projective space (cf. Example 3 of Sect. 3 below).

2. Definitions

A *differentiable manifold* M is a set M together with a differentiable manifold structure on it.

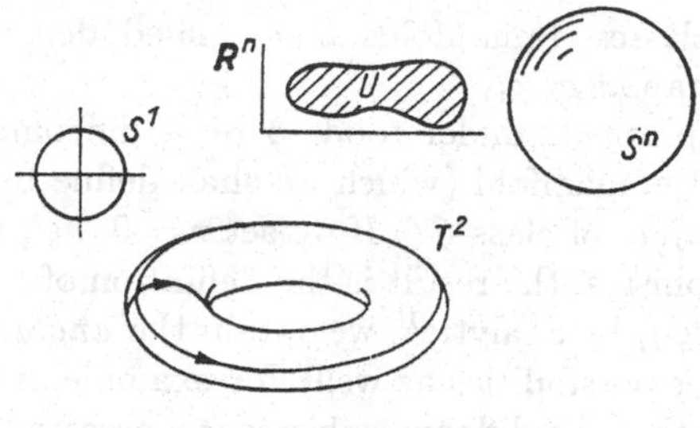

Fig. 231. Examples of manifolds

A *differentiable manifold structure* is introduced on the set M if an *atlas* consisting of *charts* that are *consistent* is prescribed.

Definition 1. A *chart* is a domain $U \subset \boldsymbol{R}^n$ together with a one-to-one mapping $\varphi : W \to U$ of a subset W of the manifold M onto U (Fig. 232). We call $\varphi(x)$ the *image* of the point $x \in W \subset M$ on the chart U.

Consider the charts (Fig. 233)

$$\varphi_i : W_i \to U_i \text{ and } \varphi_j : W_j \to U_j.$$

If the sets W_i and W_j intersect, then their intersection $W_i \cap W_j$ has an image on both charts:

$$U_{ij} = \varphi_i(W_i \cap W_j), \quad U_{ji} = \varphi_j(W_j \cap W_i).$$

The transition from one chart to the other is defined by a mapping of *subsets of vector spaces*

$$\varphi_{ij} : U_{ij} \to U_{ji}, \quad \varphi_{ij}(x) = \varphi_j(\varphi_i^{-1}(x)).$$

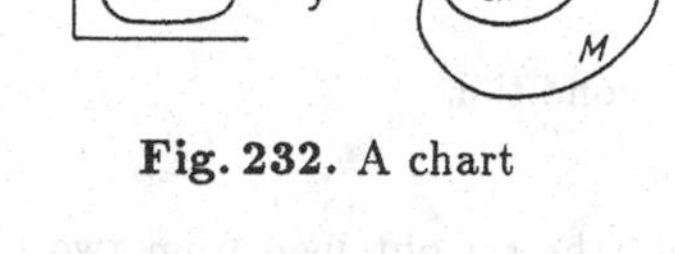

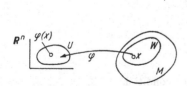

Fig. 232. A chart

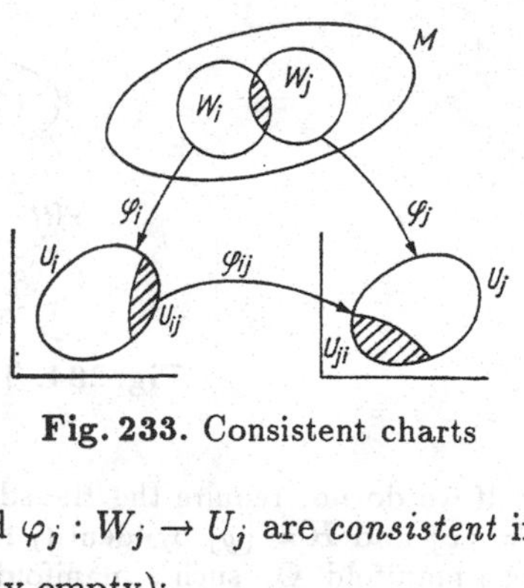

Fig. 233. Consistent charts

Definition 2. Two charts $\varphi_i : W_i \to U_i$ and $\varphi_j : W_j \to U_j$ are *consistent* if

1) the sets U_{ij} and U_{ji} are open (possibly empty);

2) the mappings φ_{ij} and φ_{ji} (which are defined if $W_i \cap W_j$ is nonempty) are diffeomorphisms of domains of $\boldsymbol{R}^n$.

Remark. Various classes of manifolds are obtained, depending on the smoothness class of the mappings φ_{ij}.

If a diffeomorphism is understood to be a diffeomorphism of class C^r, $1 \leq r \leq \infty$, then the manifold (which we shall define below) will be called a *differentiable manifold* of class C^r. If we set $r = 0$, i.e., require only that the φ_{ij} be homeomorphisms, the result is the definition of a *topological manifold.* If we require that φ_{ij} be analytic[1], we obtain the *analytic manifolds.*

There are other possibilities as well. For example, if we fix an orientation in $\boldsymbol{R}^n$ and require that the diffeomorphisms φ_{ij} preserve the orientation (that the Jacobian of φ_{ij} be positive at each point), we obtain the definition of an *oriented manifold.*

Definition 3. A collection of charts $\varphi_i : W_i \to U_i$ is an *atlas* on M if

1) any two charts are consistent;

2) any point $x \in M$ has an image on at least one chart.

Definition 4. Two atlases on M are *equivalent* if their union is again an atlas (i.e., if any chart of the first atlas is consistent with any chart of the second).

It is easy to see that Definition 4 really does give an equivalence relation.

Definition 5. A *differentiable manifold structure* on M is an equivalence class of atlases.

At this point we note two conditions that are frequently imposed on manifolds to avoid pathology.

1. *The Hausdorff condition*: Any two points x, $y \in M$ have disjoint neighborhoods (Fig. 234). That is, either there exist two charts $\varphi_i : W_i \to U_i$ and $\varphi_j : W_j \to U_j$ with W_i and W_j disjoint and containing x and y respectively, or there exists a chart in which both x and y have an image.

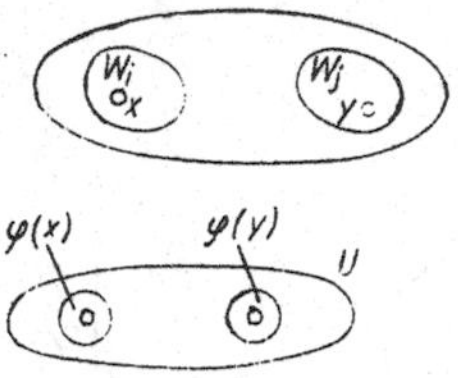

Fig. 234. The Hausdorff condition

If we do not require the Hausdorff condition, the set obtained from two lines $\boldsymbol{R} = \{x\}$ and $\boldsymbol{R} = \{y\}$ by identifying points with equal negative coordinates x and y is a manifold. On such a manifold the theorem about the unique continuation of

[1] A function is analytic if its Taylor series converges to it in a neighborhood of each point.

solutions of a differential equation is not true, although the local uniqueness theorem is true.

1. *Second-countability.* There exists an atlas of M consisting of at most countably many charts.

In what follows the word *manifold* denotes a differentiable manifold that satisfies the Hausdorff and second-countability conditions.

3. Examples of Atlases

1. The sphere S^2 defined by the equation $x_1^2 + x_2^2 + x_3^2 = 1$ in $\boldsymbol{R}^3$ can be equipped with an atlas of two charts, for example, in stereographic projection (Fig. 235). Here

$$W_1 = S^2 \setminus N, \quad U_1 = \boldsymbol{R}_1^2;$$
$$W_2 = S^2 \setminus S, \quad U_2 = \boldsymbol{R}_2^2.$$

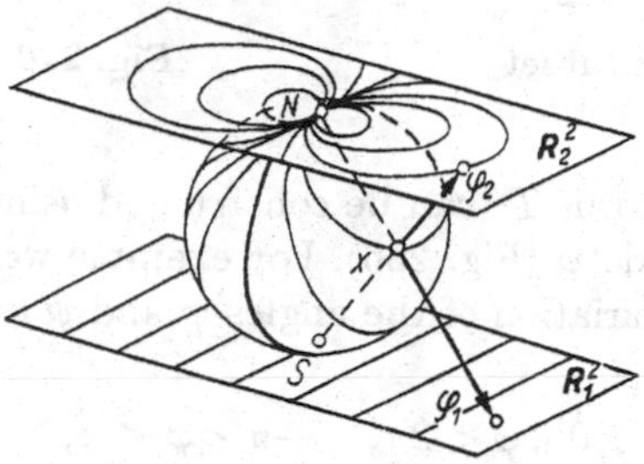

Fig. 235. An atlas of the sphere. A family of tangent circles on the sphere passing through the point N is represented on the lower chart by a family of parallel lines and on the upper chart by a family of tangent circles

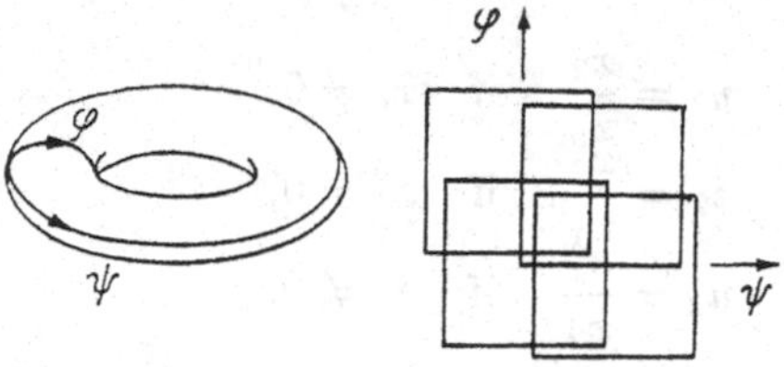

Fig. 236. An atlas of the torus

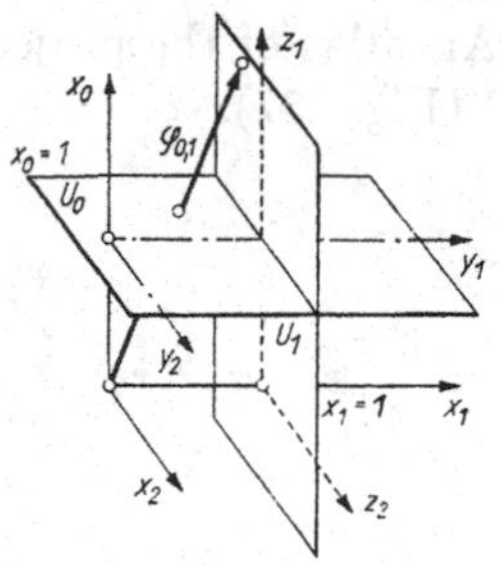

Fig. 237. Affine charts of the projective plane

Problem 1. Write down the formulas for the mappings $\varphi_{1,2}$ and verify that our two charts are consistent.

A differentiable structure on S^n can be defined similarly by an atlas of two charts.

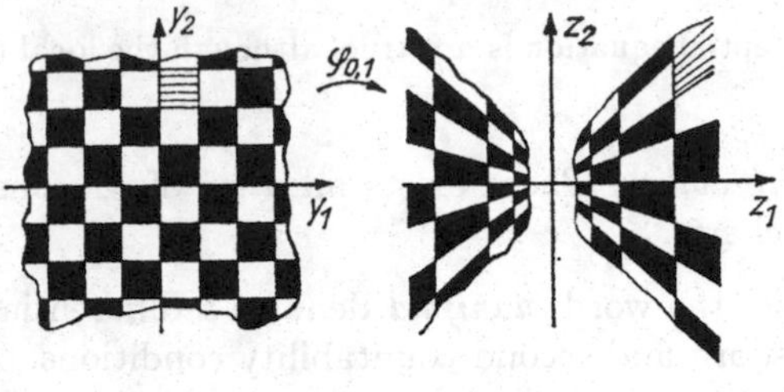

Fig. 238. The consistency of the charts of the projective plane

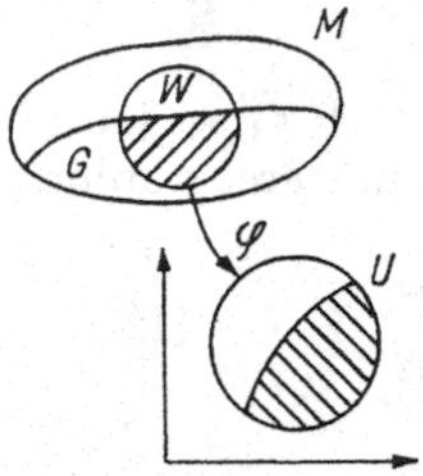

Fig. 239. An open subset

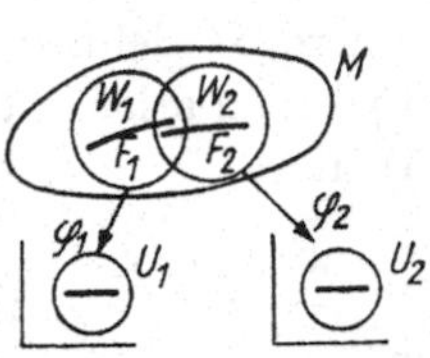

Fig. 240. A compact subset

2. An atlas of the torus T^2 can be constructed using angular coordinates: latitude φ and longitude ψ (Fig. 236). For example we can consider 4 charts corresponding to the variation of the angles φ and ψ in the intervals

$$0 < \varphi < 2\pi, \quad -\pi < \varphi < \pi,$$
$$0 < \psi < 2\pi, \quad -\pi < \psi < \pi.$$

3. An atlas of the projective plane $\boldsymbol{RP}^2$ can be made from three "affine charts" (Fig. 237):

$$x_0 : x_1 : x_2 \quad \begin{matrix} \overset{\varphi_0}{\nearrow} & y_1 = \dfrac{x_1}{x_0}, & y_2 = \dfrac{x_2}{x_0}, & \text{if } x_0 \neq 0, \\ \xrightarrow{\varphi_1} & z_1 = \dfrac{x_0}{x_1}, & z_2 = \dfrac{x_2}{x_1}, & \text{if } x_1 \neq 0, \\ \underset{\varphi_2}{\searrow} & u_1 = \dfrac{x_0}{x_2}, & u_2 = \dfrac{x_1}{x_2}, & \text{if } x_2 \neq 0. \end{matrix}$$

These charts are consistent. For example the consistency of φ_0 and φ_1 means that the mapping $\varphi_{0,1}$ of the domain $U_{0,1} = \{y_1, y_2 : y_1 \neq 0\}$ of the (y_1, y_2)-plane onto the domain $U_{1,0}$: $z_1 \neq 0$ of the (z_1, z_2)-plane given by the formulas $z_1 = y_1^{-1}$, $z_2 = y_2 y_1^{-1}$ is a diffeomorphism (Fig. 238).

Proof. $y_1 = z_1^{-1}$, $y_2 = z_2 z_1^{-1}$.

Similarly a differentiable structure in the projective space $\boldsymbol{RP}^n$ can be defined by an atlas of $n + 1$ affine charts.

4. Compactness

Definition. A subset G of the manifold M is *open* if its image $\varphi(W \cap G)$ on each chart $\varphi : W \to U$ is an open subset of the domain U of the vector space (Fig. 239).

Problem 1. Prove that the intersection of two open subsets and the union of any number of open subsets of a manifold are open.

Definition. A subset K of the manifold M is *compact* if every covering of it by open subsets contains a finite subcovering.

Problem 2. Prove that the sphere S^n is compact. Is the projective space $\boldsymbol{RP}^n$ compact?

Hint. The following theorem can be used for the solution.

Theorem. *Suppose a subset F of a manifold M* (Fig. 240) *is the union of a finite number of subsets F_i, each of which has a compact image on one of the charts $F_i \subset W_i$, $\varphi_i : W_i \to U_i$, i.e., $\varphi_i(F_i)$ is compact in $\boldsymbol{R}^n$.*

Then F is compact.

Proof. Let $\{G_j\}$ be an open covering of the set F. Then $\{\varphi_i(G_j \cap W_i)\}$ is an open covering of the compact set $\varphi_i(F_i)$ for each i. Choose a finite subcovering of it. Forcing j to range over this finite set of values, we obtain a finite number of the G_j that cover F. □

5. Connectedness and Dimension

Definition. A manifold M is *connected* (Fig. 241) if for any two of its points x and y there exists a finite chain of charts $\varphi_i : W_i \to U_i$ such that W_1 contains x, W_n contains y, $W_i \cap W_{i+1}$ is nonempty for each i, and U_i is connected[2].

If a manifold M is not connected, it decomposes in a natural way into connected components M_i.

Problem 1. Are the manifolds defined by the following equations connected in $\boldsymbol{R}^3$ (in $\boldsymbol{RP}^3$):

$$x^2 - y^2 - z^2 = C, \quad C \neq 0?$$

[2] That is, any two points of U_i can be joined by a broken line in $U_i \subset \boldsymbol{R}^n$.

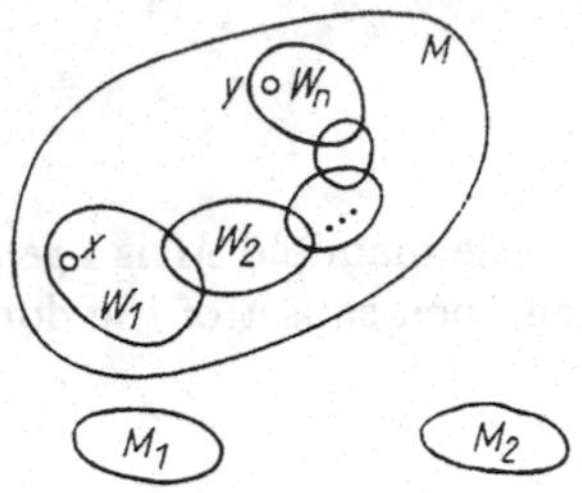

Fig. 241. A connected manifold M and a disconnected manifold $M_1 \cup M_2$

Problem 2. The set of all matrices of order n having nonzero determinant has a natural differentiable manifold structure (a domain in $\boldsymbol{R}^{n^2}$). How many connected components does this manifold have?

Theorem. *Let M be a connected manifold and $\varphi_i : W_i \to U_i$ its charts. Then the dimension of the vector space $\boldsymbol{R}^n$ in which U_i is a domain is the same for all charts.*

Proof. This follows from the fact that a diffeomorphism between domains of vector spaces is possible only when the spaces are of the same dimension, and the fact that any two domains W_i and W_j of a connected manifold M can be joined by a finite chain of pairwise intersecting domains. □

The number n defined in the theorem is called the *dimension* of the manifold M and is denoted $\dim M$.

For example, $\dim \boldsymbol{R}^n = \dim S^n = \dim T^n = \dim \boldsymbol{RP}^n = n$.

A disconnected manifold is called n-dimensional if all of its components have the same dimension n.

Problem 3. Equip the set $\mathrm{O}(n)$ of orthogonal matrices of order n with the structure of a differentiable manifold. Find its connected components and their dimension.

Answer. $\mathrm{O}(n) = \mathrm{SO}(n) \times \boldsymbol{Z}_2$, $\dim \mathrm{O}(n) = n(n-1)/2$.

6. Differentiable Mappings

Definition. A mapping $f : M_1 \to M_2$ of one C^r-manifold into another is said to be *differentiable* (of class C^r) if in local coordinates on M_1 and M_2 it is defined by a differentiable function (of class C^r).

In other words, let $\varphi_1 : W_1 \to U_1$ be a chart of M_1 representing the point $x \in W_1$ and $\varphi_2 : W_2 \to U_2$ a chart of M_2 representing the point $f(x) \in W_2$ (Fig. 242). Then the mapping of domains of Euclidean spaces $\varphi_2 \circ f \circ \varphi_1^{-1}$, which is defined in a neighborhood of the point $\varphi_1(x)$, must be differentiable of class C^r.

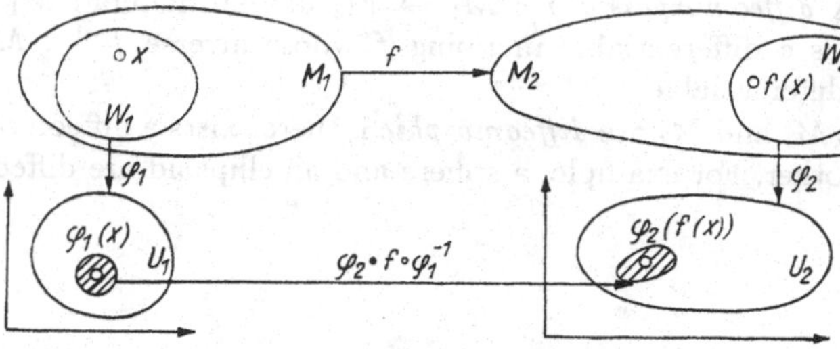

Fig. 242. A differentiable mapping

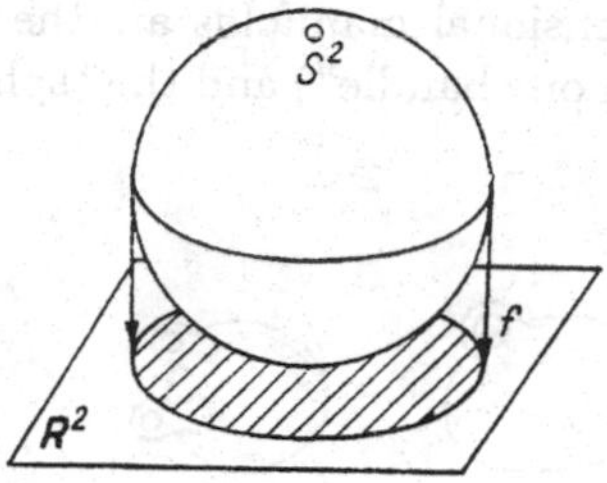

Fig. 243. Projection of the sphere on the plane yields a closed disk

Example 1. The projection of the sphere on the plane (Fig. 243) is a differentiable mapping $f : S^2 \to \boldsymbol{R}^2$.

We see that the image of a differentiable mapping is not necessarily a differentiable manifold.

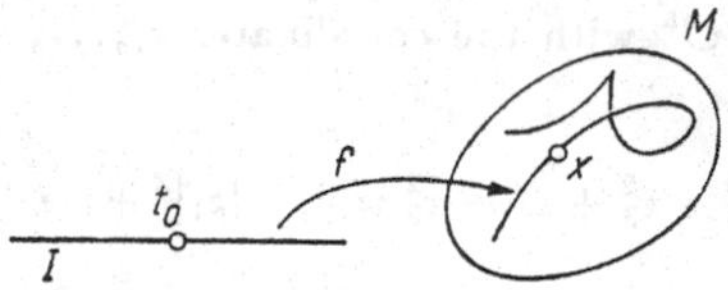

Fig. 244. A curve on a manifold M

Example 2. A *curve*[3] on the manifold M emanating from the point $x \in M$ at the instant t_0 is defined to be a differentiable mapping $f : I \to M$ from an interval I of the real t-axis containing the point t_0 into the manifold M with $f(t_0) = x$.

---

[3] Or a *parametrized curve*, since curves on M are sometimes defined as one-dimensional submanifolds of the manifold M (for the definition see Sect. 8 below). A parametrized curve may have points of self-intersection, cusps, and the like (Fig. 244).

Example 3. A *diffeomorphism* $f : M_1 \to M_2$ of the manifold M_1 onto the manifold M_2 is a differentiable mapping f whose inverse $f^{-1} : M_2 \to M_1$ exists and is differentiable.

Manifolds M_1 and M_2 are *diffeomorphic* if there exists a diffeomorphism of one onto the other. For example, a sphere and an ellipsoid are diffeomorphic.

7. Remark

It is easy to see that every connected one-dimensional manifold is diffeomorphic to a circle (if compact) or to the line (if not compact).

Examples of two-dimensional manifolds are the sphere, the torus (diffeomorphic to a "sphere with one handle") and the "sphere with n handles" (Fig. 245).

Fig. 245. Nondiffeomorphic two-dimensional manifolds

In courses of topology it is proved that every compact connected orientable two-dimensional manifold is diffeomorphic to a sphere with $n \geq 0$ handles. Little is known about three-dimensional manifolds. For example, it is unknown whether a compact simply-connected[4] three-dimensional manifold is diffeomorphic to the sphere S^3 (the *Poincaré conjecture*) or even homeomorphic to it.

In large dimensions the differential and topological classifications of manifolds diverge. For example there are exactly 28 smooth manifolds homeomorphic to the sphere S^7 but not diffeomorphic to one another. These are called the *Milnor spheres.*

A Milnor sphere in $\boldsymbol{C}^5$ with the coordinates $z_1, \ldots, z_5$ can be defined by the following two equations:

$$z_1^{6k-1} + z_2^3 + z_3^2 + z_4^2 + z_5^2 = 0, \quad |z_1|^2 + \cdots + |z_5|^2 = 1.$$

For $k = 1, 2, \ldots, 28$ we obtain the 28 Milnor spheres[5]. One of these 28 manifolds is diffeomorphic to the sphere S^7.

8. Submanifolds

The sphere in $\boldsymbol{R}^3$ defined by the equation $x^2 + y^2 + z^2 = 1$ provides an example of a subset of Euclidean space that inherits from it a natural differentiable

[4] A manifold is *simply connected* if every closed path in it can be continuously contracted to a point.

[5] Cf. E. Brieskorn, "Beispiele zur Differentialtopologie von Singularitäten," *Invent. Math.*, **2** (1966), 1–14.

manifold structure – the structure of a *submanifold of* $\boldsymbol{R}^3$. The general definition of a submanifold is the following.

Definition. A subset V of a manifold M (Fig. 246) is a *submanifold* if each point $x \in V$ has a neighborhood W in M and a chart $\varphi : W \to U$ such that $\varphi(W \cap V)$ is a domain of some affine subspace of the affine space $\boldsymbol{R}^n$ in which U lies.

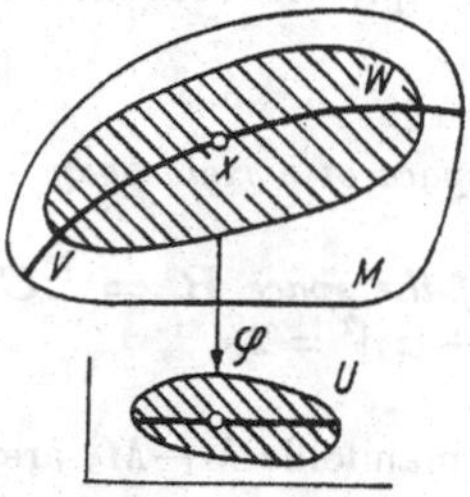

Fig. 246. A submanifold

A submanifold V itself has a natural manifold structure $W' = W \cap V$, $U' = \varphi(W')$).

The following fundamental fact is presented without proof and will not be used in what follows.

Theorem. *Every manifold M^n is diffeomorphic to a submanifold of a Euclidean space of sufficiently large dimension $\boldsymbol{R}^N$ (for example, it suffices to take $N > 2n$, where $n = \dim M^n$).*

Thus the abstract concept of a manifold actually encompasses a set of objects no larger than "n-dimensional surfaces in an N-dimensional space." The advantage of the abstract approach is that it immediately encompasses the cases when no imbedding in a Euclidean space is given in advance and introducing one would only lead to unnecessary complications (for example: the projective space $\boldsymbol{RP}^n$). The situation here is the same as with finite-dimensional vector spaces (they are all isomorphic to the coordinate space $\{(x_1, \ldots, x_n)\}$, but introducing coordinates often complicates matters).

9. An Example

In conclusion let us consider five interesting manifolds (Fig. 247).

$M_1 = \mathrm{SO}(3)$ is the *group of orthogonal matrices* of order three with determinant $+1$. Since a matrix has nine elements, M_1 is a subspace of the space $\boldsymbol{R}^9$. It is easy to see that this subset is indeed a submanifold.

$M_2 = T_1 S^2$ is the *set of unit tangent vectors to the sphere* S^2 in three-dimensional Euclidean space. The introduction of a differentiable manifold structure in this set is left to the reader (cf. § 34).

$M_3 = \boldsymbol{RP}^3$ is *three-dimensional projective space.*

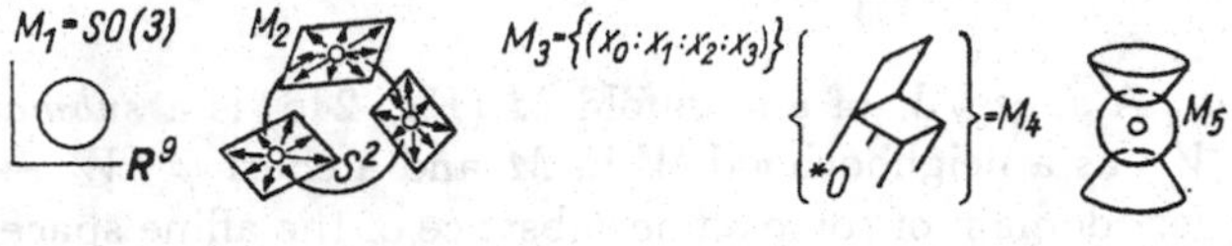

Fig. 247. Examples of three-dimensional manifolds

M_4 is the *configuration space of a rigid body* having a fixed point at the origin O.

M_5 is the *submanifold of the space* $\boldsymbol{R}^6 = {}^{\boldsymbol{R}}\boldsymbol{C}^3$ *defined by the equations* $z_1^2 + z_2^2 + z_3^2 = 0$ *and* $|z_1|^2 + |z_2|^2 + |z_3|^2 = 2$.

Problem * 1. Which of the manifolds M_1–M_5 are diffeomorphic?

§ 34. The Tangent Bundle. Vector Fields on a Manifold

With each smooth manifold M there is associated another manifold (of dimension twice as large) called the *tangent bundle*[6] of M and denoted TM. The tangent bundle enables us to carry over immediately to manifolds the whole theory of ordinary differential equations.

1. The Tangent Space

Let M be a smooth manifold. A *tangent vector* $\boldsymbol{\xi}$ *to* M *at the point* x is an equivalence class of curves emanating from x; two curves (Fig. 248)

$$\gamma_1 : I \to M, \quad \gamma_2 : I \to M$$

are equivalent if their images on any chart $\varphi\gamma_1 : I \to U$ and $\varphi\gamma_2 : I \to U$ are equivalent.

[6] The tangent bundle is a special case of a vector bundle; a still more general concept is that of a fibration. All these concepts are fundamental in topology and analysis, but we limit ourselves to just the tangent bundle, which is especially important for the theory of *ordinary* differential equations.

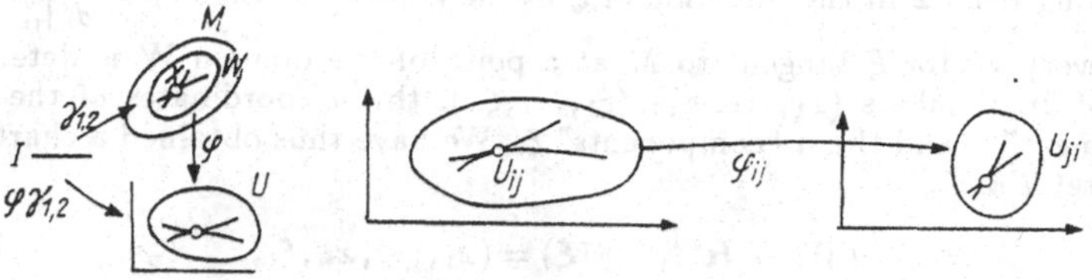

Fig. 248. A tangent vector

We remark that the concept of equivalence of curves is independent of the choice of the chart from the atlas (cf. § 5): equivalence on a chart φ_i implies equivalence on any other chart φ_j, since the transition mapping φ_{ij} from one chart to another is a diffeomorphism.

The set of vectors tangent to M at x has a vector-space structure independent of the choice of chart (cf. § 5). This vector space is called the *tangent space to M at x* and is denoted T_xM. Its dimension equals the dimension of M.

Example 1. Let M^n be a submanifold of an affine space $\boldsymbol{R}^N$ (Fig. 249) passing through x. Then T_xM^n can be thought of as an n-dimensional plane in $\boldsymbol{R}^N$ passing through x. In doing this one must remember, however, that *the tangent spaces to M at different points x and y are disjoint*: $T_xM \cap T_yM = \varnothing$.

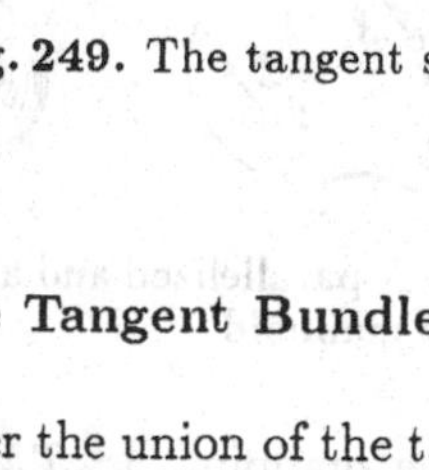

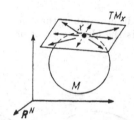

Fig. 249. The tangent space

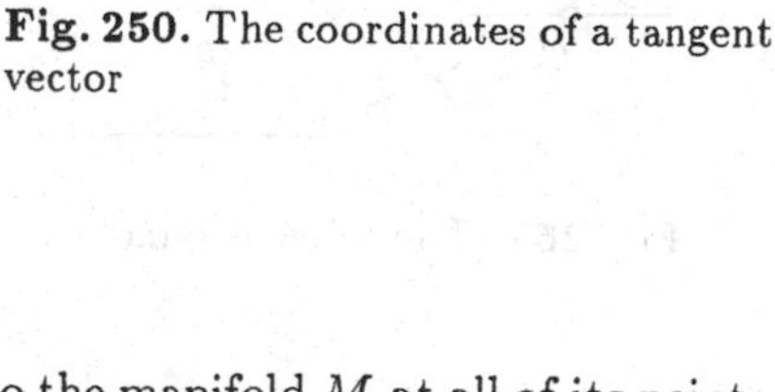

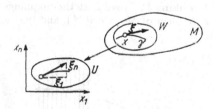

Fig. 250. The coordinates of a tangent vector

2. The Tangent Bundle

Consider the union of the tangent spaces to the manifold M at all of its points $TM = \bigcup_{x \in M} T_xM$.

The set TM has a natural smooth manifold structure.

Indeed, consider a chart on the manifold M, and let $(x_1, \dots, x_n) : W \to U \subset \boldsymbol{R}^n$ (Fig. 250) be local coordinates in a neighborhood W of the point x defining this chart. Every vector $\boldsymbol{\xi}$ tangent to M at a point $x \in W$ is determined by its set of components $(\xi_1, \dots, \xi_n)$ in the given coordinate system. To be specific, if $\gamma : I \to M$ is a curve emanating from x in the direction of $\boldsymbol{\xi}$ at the instant t_0, then $\xi_i = \left.\frac{d}{dt}\right|_{t=t_0} x_i(\gamma(t))$. Thus every vector $\boldsymbol{\xi}$ tangent to M at a point of the domain W is determined by a set of $2n$ numbers $(x_1, \dots, x_n)$, $(\xi_1, \dots, \xi_n)$, the n coordinates of the "point of attachment" x and the n "components" ξ_i. We have thus obtained a chart of a part of the set TM:

$$\psi : TW \to \boldsymbol{R}^{2n}, \quad \psi(\boldsymbol{\xi}) = (x_1, \dots, x_n, \xi_1, \dots, \xi_n).$$

The different charts for TM corresponding to different charts of the atlas of M are consistent (of class C^{r-1} if M is of class C^r). Indeed, let $(y_1, \dots, y_n)$ be another local coordinate system on M and let $(\eta_1, \dots, \eta_n)$ be the components of the vector in this system; then

$$y_i = y_i(x_1, \dots, x_n), \quad \eta_i = \sum_{j=1}^{n} \frac{\partial y_i}{\partial x_j} \xi_j \quad (i = 1, \dots, n),$$

are smooth functions of x_i and ξ_j.

Thus the set TM of all tangent vectors to M has received the structure of a smooth manifold of dimension $2n$.

Definition. The manifold TM is called the *tangent bundle*[7] of the manifold M.

There exist natural mappings $i : M \to TM$ (the *null section*) and $p : TM \to M$ (the *projection*): $i(x)$ is the zero vector of T_xM and $p(\boldsymbol{\xi})$ is the point x at which $\boldsymbol{\xi}$ is tangent to M (Fig. 251).

Problem 1. Prove that the mappings i and p are differentiable, that i is a diffeomorphism of M onto $i(M)$, and that $p \circ i : M \to M$ is the identity transformation.

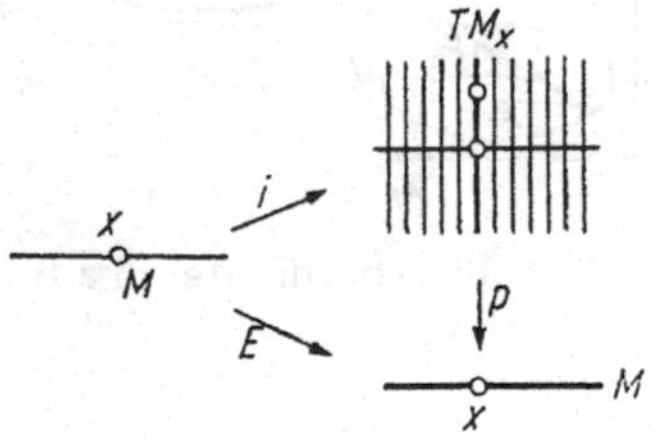

Fig. 251. The tangent bundle

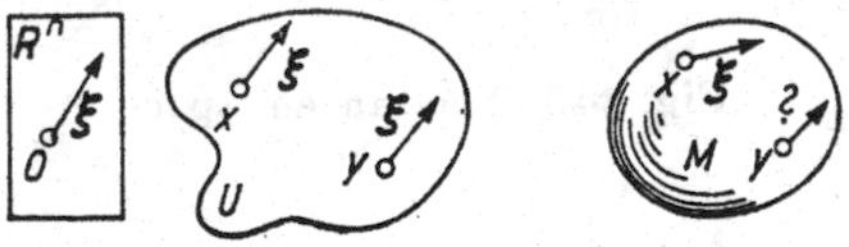

Fig. 252. A parallelized and a nonparallelizable manifold

The preimages of the points $x \in M$ under the mapping $p : TM \to M$ are called *fibers* of the bundle TM. Every fiber has the structure of a vector space. The manifold M is called the *base* of the bundle TM.

---

[7] We shall use this abbreviated name instead of the pedantic term *space of the tangent bundle*.

3. A Remark on Parallelizability

The tangent bundle of the affine space $\boldsymbol{R}^n$ or of a domain U of it has an additional direct-product structure: $TU = U \times \boldsymbol{R}^n$.

Indeed, a tangent vector to U can be defined by a pair $(x, \boldsymbol{\xi})$, where $x \in U$ and $\boldsymbol{\xi}$ is the vector of the space $\boldsymbol{R}^n$ for which the linear isomorphism with T_xU is shown (Fig. 252).

This can be otherwise expressed by saying that the affine space is *parallelizable*: equality is defined for tangent vectors to the domain U at different points of the space $\boldsymbol{R}^n$.

The tangent bundle of a manifold is by no means necessarily a product space, and in general it is impossible to give a sensible definition of equality of vectors attached at different points of a manifold M.

The situation here is the same as with the Möbius band (Fig. 253), which is a bundle with the circle as base and the line as a fiber, but is not the direct product of a circle and a line.

Definition. A manifold M is *parallelized* if a direct product structure has been introduced in its tangent bundle, i.e., a diffeomorphism $TM^n \cong M^n \times \boldsymbol{R}^n$ mapping T_xM linearly to $x \times \boldsymbol{R}^n$. A manifold is *parallelizable* if it can be parallelized.

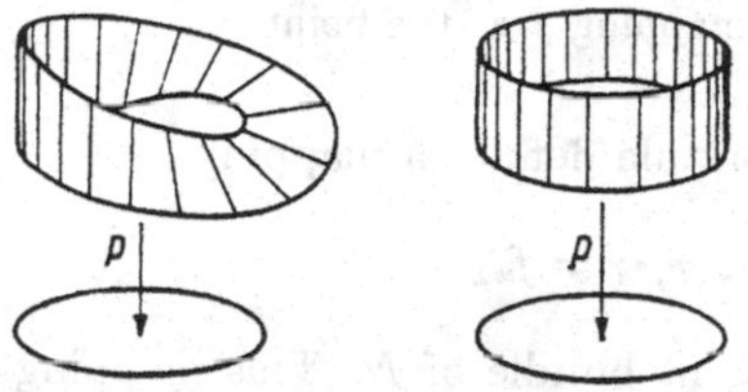

Fig. 253. A bundle that is not a direct product

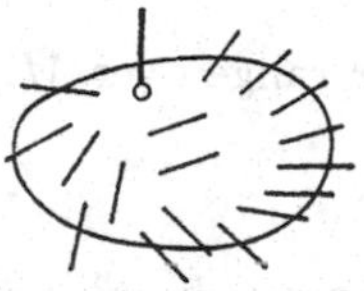

Fig. 254. The hedgehog theorem

Example 1. Any domain in Euclidean space is parallelized in the natural way.

Problem 1. Prove that the torus is parallelizable, but that the Möbius band is not.

Theorem *. *The only parallelizable spheres are S^1, S^3, and S^7. In particular, the two-sphere is not parallelizable:*

$$TS^2 \neq S^2 \times \boldsymbol{R}^2.$$

(It follows from this, for example, that a hedgehog cannot be combed: at least one quill is perpendicular to the surface (Fig. 254).)

The reader who has solved the problem at the end of § 33 will easily prove that S^2 is nonparallelizable (hint: $\boldsymbol{RP}^3 \not\cong S^2 \times S^1$). A parallelization of the circle S^1 is obvious. It is an instructive exercise to parallelize S^3 (hint: S^3 is a group, namely the group of quaternions of unit length). A complete proof of the theorem just stated requires a rather profound penetration of topology; it was attained comparatively recently.

Analysts are inclined to regard all bundles as direct products and all manifolds as parallelizable. Care should be taken to avoid this error.

4. The Tangent Mapping

Let $f : M \to N$ be a smooth mapping from the manifold M to the manifold N (Fig. 255). We denote by f_{*x} the induced mapping of the tangent spaces. It is defined as in § 6, and is a linear mapping of one vector space into another:

$$f_{*x} : T_xM \to T_{f(x)}N.$$

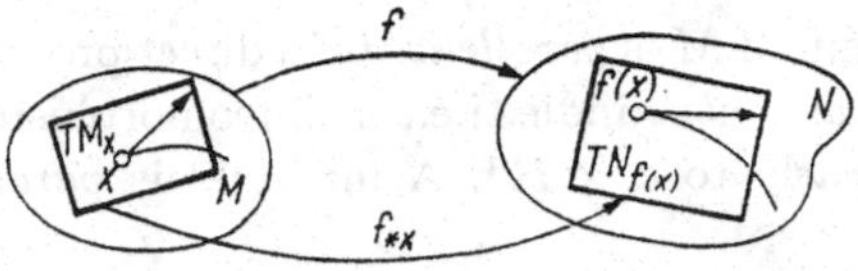

Fig. 255. The derivative of the mapping f at the point x

Let x range over M. The preceding formula defines a mapping

$$f_* : TM \to TN, \quad f_*|_{T_xM} = f_{*x},$$

of the tangent bundle of M into the tangent bundle of N. This mapping is differentiable (why?) and maps the fibers of TM linearly into the fibers of TN (Fig. 256).

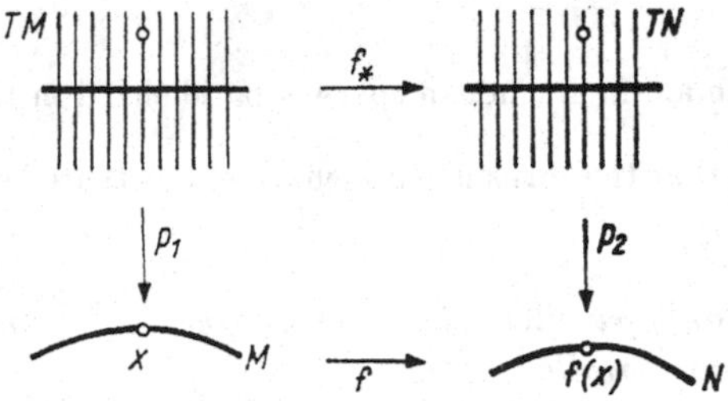

Fig. 256. The tangent mapping

The mapping f_* is the *tangent mapping to* f (the notation $Tf : TM \to TN$ is also used).

Problem 1. Let $f : M \to N$ and $g : N \to K$ be smooth mappings and $g \circ f : M \to K$ their composition. Prove that $(g \circ f)_* = (g_*) \circ (f_*)$, i.e., that

$$\begin{array}{ccccccc} & N & & & & TN & \\ {}^{f}\nearrow & & \searrow^{g} & \Rightarrow & {}^{f_*}\nearrow & & \searrow^{g_*} \\ M & \xrightarrow{g\circ f} & K & & TM & \xrightarrow{(g\circ f)_*} & TK \end{array}$$

Remark on terminology. In analysis this formula is called the rule for differentiating a composite function; in algebra it is called the (covariant) functorial property of passing to the tangent mapping.

5. Vector Fields

Let M be a smooth manifold (of class C^{r+1}) and TM its tangent bundle (Fig. 257).

Definition. A *vector field*[8] (of class C^r) $\boldsymbol{v}$ on M is a smooth mapping $\boldsymbol{v}: M \to TM$ (of class C^r) such that the mapping $p \circ \boldsymbol{v}: M \to M$ is the identity: the diagram

$$\begin{array}{ccc} & & TM \\ & {}^{\boldsymbol{v}}\nearrow & \\ M & & \downarrow p \\ & {}_{E}\searrow & \\ & & M \end{array}$$

is commutative, i.e., $p(\boldsymbol{v}(x)) = x$.

Remark. If M is a domain of the space $\boldsymbol{R}^n$ with coordinates $(x_1, \ldots, x_n)$, this definition coincides with the old one (§ 5).

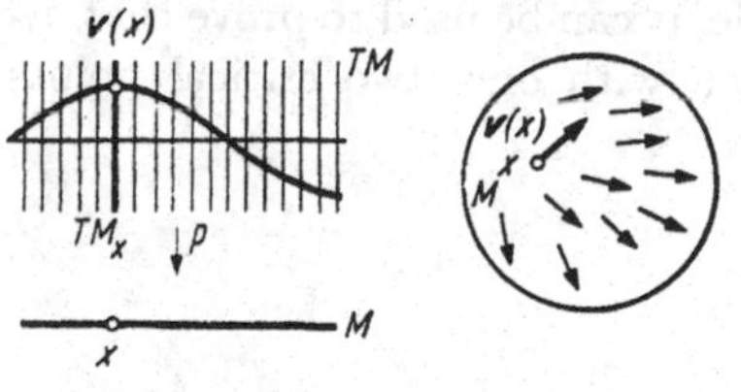

Fig. 257. A vector field

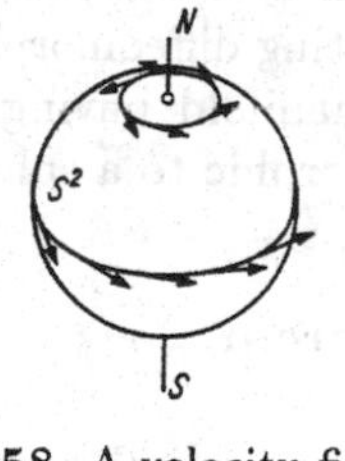

Fig. 258. A velocity field

No specific coordinate system occurs in the present definition, however.

Example. Consider the family g^t of rotations of the sphere S^2 about the SN-axis through the angle t (Fig. 258). Each point $x \in S^2$ of the sphere describes a curve (a parallel of latitude) in this rotation and has velocity

$$\boldsymbol{v}(x) = \left.\frac{d}{dt}\right|_{t=0} g^t x \in T_x S^2.$$

[8] The term *section of the tangent bundle* is also used.

We thus obtain a mapping $\boldsymbol{v} : S^2 \to TS^2$; it is obvious that $p\boldsymbol{v} = E$, i.e., $\boldsymbol{v}$ is a vector field on S^2.

In general if $g^t : M \to M$ is a one-parameter group of diffeomorphisms of the manifold M, there arises a phase velocity vector field on M, point-for-point as in § 5.

All the local theory of (nonlinear) ordinary differential equations carries over immediately to manifolds, since we took care in advance (in § 5) to keep the fundamental concepts independent of the coordinate system.

In particular, the basic local theorem on rectification of a vector field and the local theorems on existence, uniqueness, and continuity and differentiability with respect to the initial conditions all carry over to manifolds. The specifics of the manifold manifest themselves only in the study of nonlocal questions. The simplest of the latter are the questions of continuation of solutions and the existence of a phase flow with a given phase velocity vector field.

§ 35. The Phase Flow Defined by a Vector Field

The theorem proved below is the simplest theorem of the qualitative theory of differential equations: it gives conditions under which it makes sense to ask about the behavior of the solutions of a differential equation on an infinite time interval.

It follows in particular from this theorem that the solution is globally continuous and differentiable with respect to the initial data (i.e., on any finite time interval). This theorem is also useful as a technical method of constructing diffeomorphisms. For example, it can be used to prove that every closed manifold having a smooth function with only two critical points is homeomorphic to a sphere.

1. Theorem

Let M be a smooth manifold (of class C^r, $r \geq 2$) (Fig. 259) *and let $\boldsymbol{v} : M \to TM$ be a vector field. Suppose the vector $\boldsymbol{v}(x)$ is different from zero only in a compact part K of the manifold M. Then there exists a one-parameter group of diffeomorphisms $g^t : M \to M$ for which the field $\boldsymbol{v}$ is the phase velocity vector field:*

$$\frac{d}{dt} g^t x = \boldsymbol{v}(g^t x). \qquad (1)$$

Corollary 1. *Every vector field $\boldsymbol{v}$ on a compact manifold M is the phase velocity vector field of some one-parameter group of diffeomorphisms.*

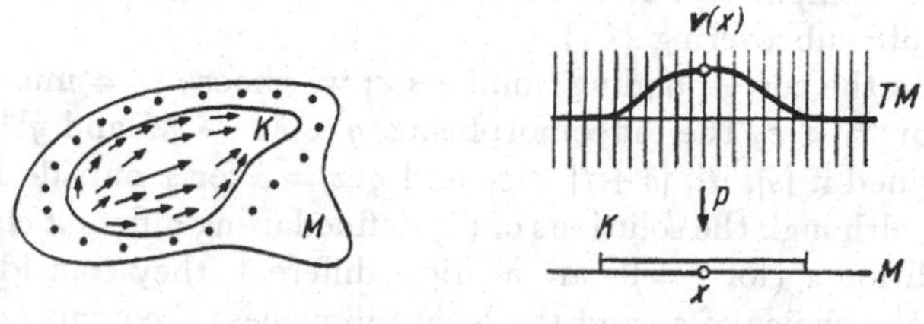

Fig. 259. A vector field vanishing outside a compact set K

In particular, under the hypotheses of the theorem or the hypotheses of Corollary 1, the following corollary holds.

Corollary 2. *Every solution of the differential equation*

$$\dot{x} = v(x), \quad x \in M, \tag{2}$$

can be extended forward and backward indefinitely. When this is done, the solution $g^t x$ at the instant t depends smoothly on t and on the initial condition x.

Remark. The hypothesis of compactness cannot be eliminated.

Example 1. $M = \mathbf{R}$, $\dot{x} = x^2$ (cf. § 1, Sect. 7): the solution cannot be extended indefinitely.

Example 2. $M = \{x : 0 < x < 1\}$, $\dot{x} = 1$.

Let us now proceed to the proof.

2. Construction of the Diffeomorphisms g^t for Small t

For each point $x \in M$ there exists an open neighborhood $U \subset M$ and a number $\varepsilon > 0$ such that for each point $y \in U$ and any t with $|t| < \varepsilon$ the solution $g^t y$ of Eq. (2) with initial condition y (at $t = 0$) exists, is unique, depends differentiably on t and y, and satisfies the condition

$$g^{t+s} y = g^t g^s y. \tag{3}$$

if $|s| < \varepsilon$, $|t| < \varepsilon$, and $|s + t| < \varepsilon$.

Indeed the point x has an image on some chart, and for equations in a domain of an affine space our assertion has been proved (cf. Chapt. 2 and Chapt. 4)[9].

[9] The proof of uniqueness requires a small additional argument: it must be verified that uniqueness of a solution with given initial conditions on each fixed chart implies its uniqueness on the manifold. On a non-Hausdorff manifold uniqueness

Thus the compact set K is covered with open neighborhoods U. We can choose a finite subcovering $\{U_i\}$.

Let ε_i be the corresponding numbers ε; we choose $\varepsilon_0 = \min \varepsilon_i > 0$.

Then for $|t| < \varepsilon_0$ the diffeomorphisms $g^t : M \to M$ and $g^{t+s} = g^t g^s$ are *globally* defined if $|s|$, $|t|$, $|s+t| < \varepsilon_0$ and $g^t x = x$ for x outside K.

Indeed, although the solutions of (2) defined using different charts with the initial condition x (for $t = 0$) are a priori different, they coincide for $|t| < \varepsilon_0$ because of the choice of ε_0 and the local uniqueness theorem.

Furthermore, by the local differentiability theorem the point $g^t x$ depends differentiably on t and x, and since $g^t g^{-t} = E$, the mapping $g^t : M \to M$ is a diffeomorphism. We remark that $\left.\frac{d}{dt}\right|_{t=0} g^t x = \boldsymbol{v}(x)$.

3. The Construction of g^t for any t

We represent t in the form $n\varepsilon_0/2 + r$, where n is an integer and $0 \le r < \varepsilon_0/2$. Such a representation exists and is unique. The diffeomorphisms $g^{\varepsilon_0/2}$ and g^r are already defined.

We set $g^t = (g^{\varepsilon_0/2})^n g^r$. This is a diffeomorphism of M onto M. For $|t| < \varepsilon_0/2$ the new definition agrees with the previous one (cf. Sect. 2). Therefore $\left.\frac{d}{dt}\right|_{t=0} g^t x = \boldsymbol{v}(x)$.

It is easy to see that for any s and t

$$g^{s+t} = g^s g^t. \tag{4}$$

Indeed, let

$$s = (m\varepsilon_0/2) + p, \quad t = (n\varepsilon_0/2) + q, \quad s + t = (k\varepsilon_0/2) + r.$$

Then the left and right-hand sides of Eq. (4) assume the form

$$(g^{\varepsilon_0/2})^k g^r \text{ and } (g^{\varepsilon_0/2})^m g^p (g^{\varepsilon_0/2})^n g^q.$$

Two cases are possible:

$$1)\ m + n = k,\ \ p + q = r, \quad 2) m + n = k - 1,\ \ p + q = r + (\varepsilon_0/2).$$

We remark that since $|p| < \varepsilon_0/2$ and $|q| < \varepsilon_0/2$, it follows that the diffeomorphisms $g^{\varepsilon/2}$, g^p, and g^q commute. Formula (4) follows from this in both the first and second cases ($g^{\varepsilon_0/2} g^r = g^p g^q$, since $|p|$, $|q|$, $|r| < \varepsilon_0/2$ and $p + q = \varepsilon_0/2 + r$).

It remains to verify that the point $g^t x$ depends differentiably on t and x. This follows, for example, from the fact that $g^t = (g^{t/N})^N$ for sufficiently large N and $g^{t/N} x$ depends differentiably on t and x (cf. Sect. 2).

---

may fail (example: the equation $\dot{x} = 1$, $\dot{y} = 1$ on the manifold obtained from two lines $\{x\}$ and $\{y\}$ by identifying points with equal negative coordinates). But if the manifold M satisfies the Hausdorff condition, then the proof of uniqueness given in § 7, Sect. 6 goes through. (The Hausdorff condition is used in proving that the values of the solutions $\varphi_1(T)$ and $\varphi_2(T)$ coincide at some first point T beyond which they do not coincide).

Thus g^t is a one-parameter group of diffeomorphisms of the manifold M; the corresponding phase velocity vector field is $\boldsymbol{v}$, and the theorem is proved. □

4. A Remark

It is easy to deduce from the theorem just proved that *every solution of a nonautonomous equation*

$$\dot{x} = \boldsymbol{v}(t, x), \quad x \in M, \quad t \in \boldsymbol{R},$$

defined by a vector field $\boldsymbol{v}$ depending on time t on a compact manifold M can be extended indefinitely.

This explains in particular the possibility of extending indefinitely the solutions of a linear equation

$$\dot{\boldsymbol{x}} = \boldsymbol{v}(t, \boldsymbol{x}), \quad \boldsymbol{v}(t, \boldsymbol{x}) = A(t)\boldsymbol{x}, \quad t \in \boldsymbol{R}, \quad \boldsymbol{x} \in \boldsymbol{R}^n. \tag{5}$$

In fact let us regard $\boldsymbol{R}^n$ as the affine part of the projective space $\boldsymbol{RP}^n$. The space $\boldsymbol{RP}^n$ is obtained from its affine part by adding an infinitely distant plane $\boldsymbol{RP}^n = \boldsymbol{R}^n \cup \boldsymbol{RP}^{n-1}$.

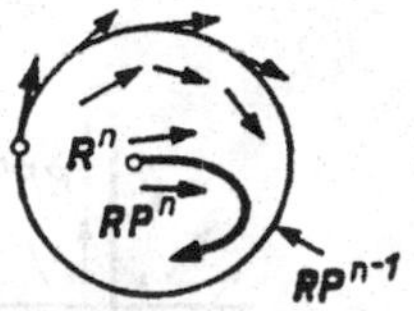

Fig. 260. Extension of a linear vector field to projective space

Let $\boldsymbol{v}$ be a linear vector field in $\boldsymbol{R}^n$ ($\boldsymbol{v}(\boldsymbol{x}) = A\boldsymbol{x}$). The following result is easily verified.

Lemma. *A vector field $\boldsymbol{v}$ on $\boldsymbol{R}^n$ has a unique extension to a smooth field $\boldsymbol{v}'$ on $\boldsymbol{RP}^n$. The field $\boldsymbol{v}'$ on the infinitely distant plane $\boldsymbol{RP}^{n-1}$ is tangent to $\boldsymbol{RP}^{n-1}$.*

In particular, let us extend the vector field $\boldsymbol{v}(t)$ (for each t) determining Eq. (5) to a field $\boldsymbol{v}'(t)$ on $\boldsymbol{RP}^n$. Consider the equation

$$\dot{\boldsymbol{x}} = \boldsymbol{v}'(t, \boldsymbol{x}), \quad \boldsymbol{x} \in \boldsymbol{RP}^n, \quad t \in \boldsymbol{R}. \tag{6}$$

Projective space is compact. Consequently each solution of Eq. (6) can be extended indefinitely (Fig. 260).

A solution with the initial condition in $\boldsymbol{RP}^{n-1}$ always remains in $\boldsymbol{RP}^{n-1}$, since the field $\boldsymbol{v}'$ is tangent to $\boldsymbol{RP}^{n-1}$.

By the uniqueness theorem the solutions of an equation with initial conditions in $\boldsymbol{R}^n$ remain inside $\boldsymbol{R}^n$ for all t. But inside $\boldsymbol{R}^n$ Eq. (6) has the form (5). Thus each solution of Eq. (5) can be extended infinitely far.

Problem 1. Prove the lemma.

Solution 1. Let $(x_1, \dots, x_n)$ be affine coordinates in $\boldsymbol{RP}^n$, and let $(y_1, \dots, y_n)$ be other affine coordinates given by:

$$y_1 = x_1^{-1}, \quad y_k = x_k x_1^{-1} \quad (k = 2, \dots, n).$$

The equation of $\boldsymbol{RP}^{n-1}$ in the new coordinates is $y_1 = 0$.

The differential equation (5)

$$\frac{dx}{dt} = \sum_{j=1}^{n} a_{i,j} x_j, \quad i = 1, \dots, n,$$

can be written in the new coordinates as (Fig. 261)

$$\frac{dy_1}{dt} = -y_1 \left(a_{i,1} + \sum a_{1,k} y_k \right), \quad k > 1;$$

$$\frac{dy_k}{dt} = a_{k,1} + \sum a_{k,l} y_l - y_k \left(a_{1,1} + \sum a_{1,i} y_i \right), \quad k > 1, \quad l > 1.$$

It is clear from these formulas, which hold for $y_1 \neq 0$, how the field should be defined for $y_1 = 0$. For $y_1 = 0$ we find $dy_1/dt = 0$, which proves the lemma. □

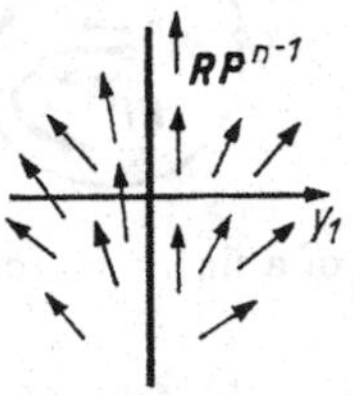

Fig. 261. The behavior of the extended field near the infinitely distant plane

Solution 2. An affine transformation can be regarded as a projective transformation that leaves the plane at infinity fixed, but not its individual points. In particular the linear transformations e^{At} can be continued to diffeomorphisms of projective space leaving the plane at infinity fixed. These diffeomorphisms form a one-parameter group; its phase velocity vector field is $\boldsymbol{v}'$.

§ 36. The Indices of the Singular Points of a Vector Field

In this section we study simple applications of topology to the study of differential equations.

1. The Index of a Curve

We begin with some intuitive considerations. They will be confirmed below by definitions and proofs (cf. Sect. 7).

Consider a vector field defined on an oriented Euclidean plane. Suppose a closed oriented curve is defined on the plane, not passing through the singular points of the field (Fig. 262). Suppose a point traverses the curve in the positive direction. The vector of the field at the point in question will rotate continuously during the motion[10]. When a point returns to a place, having traversed the curve, the vector also returns to its original position. But in doing so it may complete several revolutions in one direction or the other.

The number of revolutions of a vector field in traversing a curve is called the *index* of the curve. In computing the index a revolution is counted positive if the vector rotates in the direction given by the orientation of the plane (from the first unit vector toward the second) and negative in the opposite case.

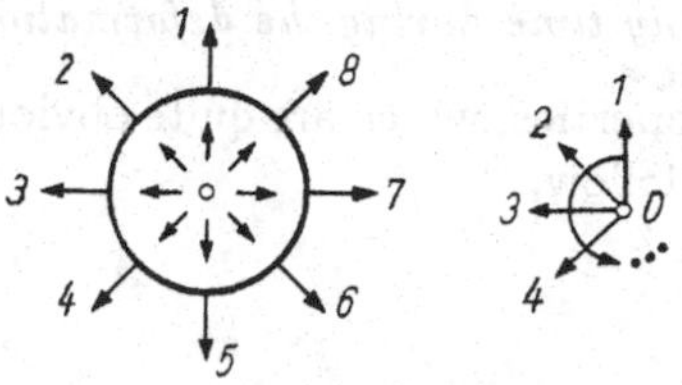

Fig. 262. A curve of index 1

Example 1. The indices of the curves α, β, γ, and δ in Fig. 263 are 1, 0, 2, and -1 respectively.

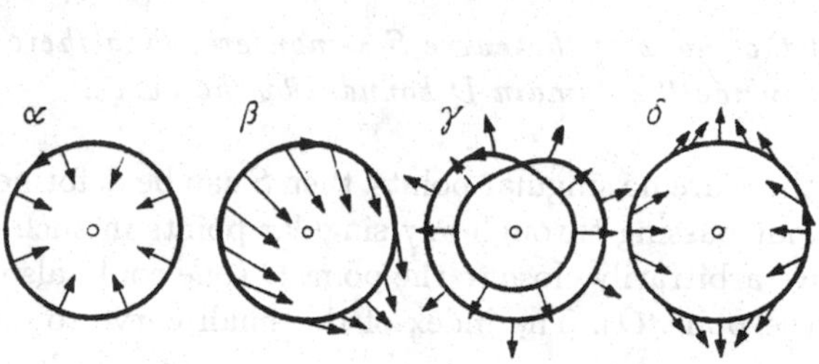

Fig. 263. Curves with different indices

Example 2. Let O be a singular point of the field. Then the index of every curve lying in a sufficiently small neighborhood of the point O is 0.

[10] To trace the rotation of a vector it is convenient to translate all vectors to the point **o** in accordance with the natural parallelization of the plane.

Indeed, the direction of the field at the point O varies continuously in a sufficiently small neighborhood of the point by less than, say $\pi/2$.

Problem 1. We define a vector field on the plane $\mathbf{R}^2 = {}^{\mathbf{R}}\mathbf{C}$ with the point O removed by the formula $\mathbf{v}(z) = z^n$ (where n is a number not necessarily positive). Calculate the index of the circle $z = e^{i\varphi}$ oriented in the direction of increasing φ (the plane is oriented by frame 1, i).

Answer. n.

2. Properties of the Index

Property 1. *Under a continuous deformation of a closed curve the index does not change as long as the curve does not pass through a singular point.*

Indeed, the direction of a vector of the field varies continuously outside the singular points; therefore the number of revolutions also varies continuously with the curve. Being an integer, it must be constant.

Property 2. *The index of a curve does not change under a continuous deformation of a vector field, provided there are no singular points of the vector field on the curve at any time during the deformation.*

From these two properties, which are quite obvious intuitively[11], a number of profound theorems follow.

3. Examples

Example 1. Consider a vector field on the plane. Let D be a disk and S its boundary circle[12].

Theorem. *If the index of the curve S is nonzero, then there is at least one singular point inside the domain D bounded by the curve.*

In fact, if there are no singular points, then S can be deformed continuously inside D without passing through any singular points in such a way that the result is a curve arbitrarily close to the point O (one could also simply deform the curve to the point O). The index of the small curve so obtained is zero.

[11] A precise formulation and proof of the propositions just stated requires certain technical apparatus from topology: homotopy, homology, or something of the sort (we shall use Green's formula for this purpose below). Cf., for example, the book of Steenrod and Chinn, *First Concepts of Topology*, Random House, New York, 1966.

[12] One can also consider the more general case where D is any plane domain bounded by a simple closed curve S.

But the index does not change under the deformation, so that it must have been zero to begin with. □

Problem 1. Prove that the system of differential equations

$$\dot{x} = x + P(x,y), \quad \dot{y} = y + Q(x,y),$$

where P and Q are bounded functions on the entire plane, has at least one equilibrium position.

Example 2. Let us prove the fundamental theorem of algebra:

Every equation $z^n + a_1 z^{n-1} + \cdots + a_n = 0$ *has at least one complex root.*

Consider the vector field $\boldsymbol{v}$ on the plane of the complex variable z defined by the formula $\boldsymbol{v}(z) = z^n + a_1 z^{n-1} + \cdots + a_n$. The singular points of the field $\boldsymbol{v}$ are the roots of this equation.

Lemma. *The index of a circle of sufficiently large radius in the field just constructed is* n (with the orientation as in the problem of Sect. 1).

In fact the formula

$$\boldsymbol{v}_t(z) = z^n + t(a_1 z^{n-1} + \cdots + a_n), \quad 0 \le t \le 1,$$

defines a continuous deformation of the original field to the field z^n. Let $r > 1 + |a_1| + \cdots + |a_n|$. Then $r^n > |a_1| r^{n-1} + \cdots + |a_n|$. Therefore on a circle of radius r there are no singular points at any time during the deformation. By Property 2 the index of this circle in the original field is the same as in the field z^n. But in the field z^n the index is n.

The lemma is now proved. □

By the preceding theorem there are singular points of the vector field, i.e., roots of the equation, inside the circle of radius r. The theorem is now proved. □

Example 3. Let us prove the following fixed-point theorem:

Theorem. *Every smooth*[13] *mapping* $f : D \to D$ *of the closed disk into itself has a fixed point.*

We shall assume that a vector space structure has been introduced into the plane of the disk D and that the origin of the vector space is at the center of the disk (Fig. 264). The fixed points of the mapping f are the singular points of the vector field $\boldsymbol{v}(\boldsymbol{x}) = f(\boldsymbol{x}) - \boldsymbol{x}$.

[13] This theorem holds for any continuous mapping, but we consider all mappings to be smooth here and prove the theorem (cf. Sect. 7) only under this assumption.

Assume that there are no fixed points in D. Then there are none on the boundary circle either.

Lemma. *The index of the disk D in the field $\boldsymbol{v}$ is 1.*

Indeed, there exists a continuous deformation of the field $\boldsymbol{v}$ into the field $-\boldsymbol{x}$, such that there are no singular points on the circle at any time during the deformation (for example, it suffices to set $\boldsymbol{v}_t(\boldsymbol{x}) = tf(\boldsymbol{x}) - \boldsymbol{x}$, $0 \leq t \leq 1$. Therefore the indices of the circle in the fields $\boldsymbol{v}_0 = -\boldsymbol{x}$ and $\boldsymbol{v}_1 = \boldsymbol{v}$ are the same. But the index of the circle $|\boldsymbol{x}| = r$ in the field $-\boldsymbol{x}$ is easy to compute directly: it equals 1.

The lemma is now proved. □

By the theorem of Example 1 there is a singular point of the field $\boldsymbol{v}$, i.e., a fixed point of the mapping f, inside the disk.

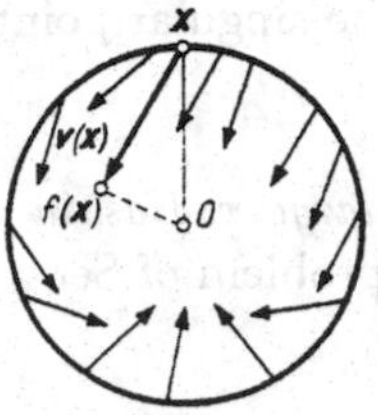

Fig. 264. A mapping of a disk into itself

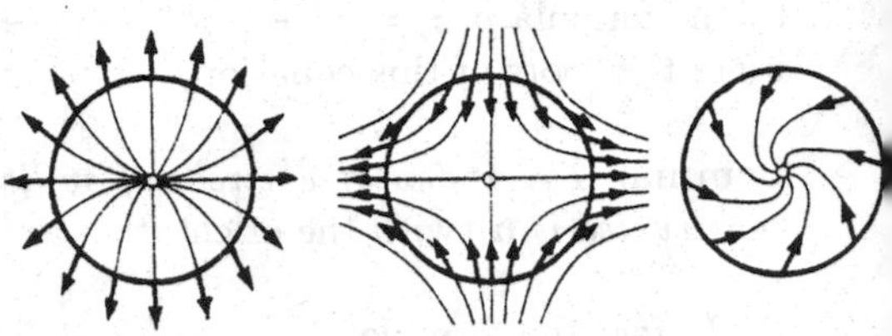

Fig. 265. The indices of the simple singular points are ±1

4. The Index of a Singular Point of a Vector Field

Let O be an isolated singular point of a vector field on the plane, i.e., suppose in some neighborhood of the point O there are no other singular points. Consider a circle of sufficiently small radius with center at this point; assume that the plane is oriented and that the orientation on the circle is chosen to be positive (as in Sect. 1).

Theorem. *The index of a circle of sufficiently small radius with center at an isolated singular point O is independent of the radius of the circle, provided it is sufficiently small.*

Indeed two such circles can be continuously deformed into each other without passing through any singular points.

We remark also that any other curve that winds around the point O once in the positive direction could have been chosen instead of a circle.

Definition. The index of some (and hence any) sufficiently small positively oriented circle with center at an isolated singular point of a vector field is called the *index of the singular point.*

Examples. The indices of singular points of the node, saddle, and focus (or center) types are respectively $+1$, -1, and $+1$ (Fig. 265).

A singular point of a vector field is called *simple* if the operator of the linear part of the field at this point is nondegenerate. The simple singular points on the plane are nodes, saddle points, foci, and centers. Thus the index of a simple singular point is always ± 1.

Problem 1. Construct a vector field with a singular point of index n.

Hint. Cf., for example, the problem of Sect. 1.

Problem 2. Prove that the index of a singular point is independent of the choice of the orientation of the plane.

Hint. When the orientation reverses, the direction for positive traversal of a circle and the direction for counting a positive revolution also reverse simultaneously.

5. The Theorem on the Sum of the Indices

Let D be a compact domain on an oriented plane bounded by a simple curve S. We give the curve S the orientation induced from D (so that the domain D stays on the left as the curve is traversed). This means that the frame formed by the velocity vector and the normal vector directed into D must define the positive orientation of the plane.

Suppose a vector field is defined on the plane having no singular points on the curve S and having only a finite number of singular points in the domain D.

Theorem. *The index of the curve S equals the sum of the indices of the singular points lying inside D.*

To prove this we remark that the index of a curve possesses the following additive property.

Consider two oriented curves γ_1 and γ_2 passing through the same point. We can form a new oriented curve $\gamma_1 + \gamma_2$ by traversing first γ_1 and then γ_2.

Lemma. *The index of the curve $\gamma_1 + \gamma_2$ is the sum of the indices of the curves γ_1 and γ_2.*

Indeed, a vector of the field will make n_1 revolutions in traversing the curve γ_1 and another n_2 in traversing the curve γ_2 for a total of $n_1 + n_2$ revolutions. The lemma is now proved. □

We now partition D into parts D_i each containing at most one singular point of the field (Fig. 266) and such that there are no singular points on the boundary of any part. We orient the curves γ_i bounding these parts as a boundary should be oriented (Fig. 266); then by the lemma

$$\operatorname{ind} \sum_i \gamma_i = \operatorname{ind} \Big(S + \sum_j \operatorname{ind} \delta_j \Big),$$

where δ_j is a closed curve representing the part of the boundary of the domain D_i located inside D and traversed twice, once in each direction.

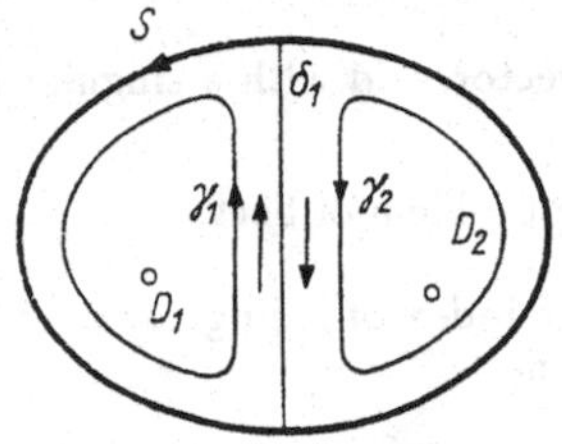

Fig. 266. The index of the curve S equals the sum of the indices of the curves γ_1 and γ_2

The index of each curve δ_j is 0, since this curve can be contracted to a point without passing through any singular points of the field (cf. Sect. 8). The index of the curve γ_i equals the index of the singular point enclosed by the curve (or 0, if there are no singular points in the domain D_i). The theorem is now proved. □

Problem 1. Let p be a polynomial of degree n in the complex variable z and D a domain in the plane of the variable z bounded by the curve S. Assume that there are no roots of the polynomial on the curve S. Prove that the number of roots of the polynomial inside D (counting multiplicities) equals the index of the curve S in the field $\boldsymbol{v} = p(z)$, i.e., the number of revolutions of the curve $p(S)$ about 0.

Remark. We thereby obtain a method of solving the Routh-Hurwitz problem (cf. § 23):

Find the number n_- of roots of a given polynomial in the left half-plane.

For this purpose we consider a semicircle of very large radius in the left half-plane with center at the point $z = 0$ and with diameter on the imaginary axis. The number of roots in the left half-plane is the index of the boundary of this semicircle (provided its radius is sufficiently large and there are no purely imaginary roots). To calculate the index of the curve S it suffices to compute the number ν of revolutions of the image of the imaginary axis oriented from $-i$ to $+i$ about the origin. Indeed, it is easy to verify that

$$n_- = \operatorname{ind} S = \nu + n/2,$$

since the image of a semicircle of sufficiently large radius under the mapping p makes approximately $n/2$ revolutions about the origin (the larger the radius, the closer to $n/2$).

In particular *all the roots of a polynomial of degree n lie in the left half-plane if and only if the point $p(it)$ revolves about the origin $n/2$ times (in the direction from 1 to i) as t varies from $-\infty$ to $+\infty$.*

6. The Sum of the Indices of the Singular Points on a Sphere

Problem *1. Prove that the index of a singular point of a vector field in the plane is preserved under a diffeomorphism.

Thus the index is a geometric concept independent of the coordinate system. This circumstance enables us to define the index of a singular point not only on the plane, but on any two-dimensional manifold. Indeed, it suffices to consider the index of the singular point on some chart: on other charts it will be the same.

Example 1. Consider the sphere $x^2+y^2+z^2 = 1$ in three-dimensional Euclidean space. The angular velocity vector field for rotation about the z-axis ($\dot{x} = y$, $\dot{y} = -x$, $\dot{z} = 0$) has two singular points: the north and south poles (Fig. 267). The index of each is $+1$.

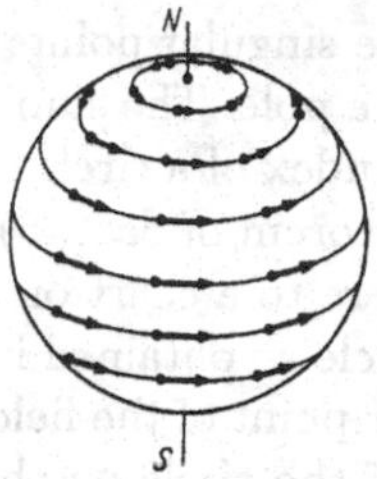

Fig. 267. A vector field on the sphere having two singular points of index 1

Assume that a vector field is given on the sphere having only isolated singular points. Then there must be only a finite number of singular points, since the sphere is compact.

Theorem *. *The sum of the indices of all the singular points of a field on the sphere is independent of the choice of field.*

It is clear from the preceding example that *this sum equals 2.*

The idea of the proof. Consider a chart of the sphere covering the whole sphere except for one point, which we shall call the pole. In the Euclidean plane of this chart consider the coordinate vector field e_1. We transfer this field to the sphere. We then obtain a field on the sphere (defined everywhere except at the pole), which we shall again denote e_1.

Now consider a chart on a neighborhood of the pole. On the plane of this chart we can also draw the vector field e_1 on the sphere, defined everywhere except at the one point O. Its form is shown in Fig. 268.

Lemma. *The index of a closed curve that revolves once about the point O in the field just constructed on the plane is equal to 2.*

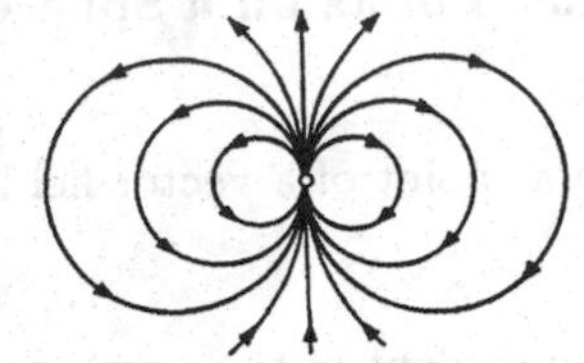

Fig. 268. A vector field that is parallel in one chart of the sphere, but is drawn in another chart

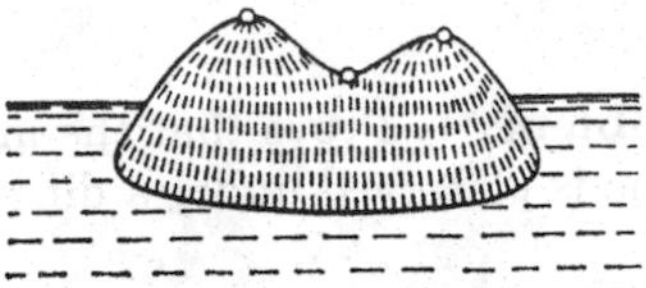

Fig. 269. On each island the sum of the number of peaks and pits is one larger than the number of passes

To prove the lemma it suffices to carry out explicitly the operations described above, taking as charts, for example, the charts of the sphere in stereographic projection (Fig. 235). The parallel straight lines of the first chart become the circles of Fig. 268 in the second, from which it is clear that the index is 2.

Now consider a vector field $\boldsymbol{v}$ on the sphere. Choose as pole a nonsingular point of the field. Then all the singular points of the field have an image on a chart of the complement of the pole. The sum of the indices of all the singular points of the field equals the index of a circle of sufficiently large radius in the plane of this chart (by the theorem of Sect. 5). We now transfer this circle to the sphere, and from the sphere to a chart on a neighborhood of the pole. On this chart the index of the circle so obtained in the field being studied is zero, since the pole is a nonsingular point of the field. Remaining on this new chart, we can interpret the index of the circle on the first chart as the "number of revolutions of the field $\boldsymbol{v}$ *with respect to the field* $\mathbf{e}_1$" during a traversal of the circle.

This number equals +2, since on the new chart the field $\boldsymbol{e}_1$ completes 2 revolutions during a traversal of a circle about the point O in the positive direction for the first chart represented on the second chart, while the field $\boldsymbol{v}$ makes zero revolutions.

Problem * 2. Let $f : S^2 \to \boldsymbol{R}^1$ be a smooth function on the sphere all of whose critical points are simple (i.e., the second differential at each critical point is nondegenerate). Prove that $m_0 - m_1 + m_2 = 2$, where m_i is the number of critical points whose negative index of inertia of the second differential is i.

In other words, *if the number of saddle points is subtracted from the number of minima and the number of maxima is added, the result is always* 2. For example, the number of mountain peaks on the Earth plus the number of pits exceeds the number of passes by 2. If we restrict ourselves to a single island or continent, i.e., consider functions on a disk without critical points on the boundary, then $m_0 - m_1 + m_2 = 1$ (Fig. 269).

Hint. Consider the gradient of the function f.

Problem *3. Prove Euler's theorem on polyhedra:

For any convex polyhedron with α_0 *vertices,* α_1 *edges, and* α_2 *faces,* $\alpha_0 - \alpha_1 + \alpha_2 = 2$.

Hint. This problem can be reduced to the preceding one.

Problem *4. Prove that *the sum of the indices χ of the singular points of a vector field on any two-dimensional compact manifold is independent of the field.*

The number χ is called the *Euler characteristic* of the manifold. For example, we saw above that the Euler characteristic of the sphere $\chi(S^2)$ is 2.

Problem 5. Find the Euler characteristic of the torus, the pretzel, and the sphere with n handles (Fig. 245).

Answer. 0, -2, $2-2n$.

Problem *6. Transfer the results of Problems 2 and 3 from the sphere to any compact two-dimensional manifold M:

$$m_0 - m_1 + m_2 = \alpha_0 - \alpha_1 + \alpha_2 = \chi(M).$$

7. Justification

We now give a precise definition of the *winding number* of a vector field.

Let $\boldsymbol{v}$ be a smooth vector field defined in a domain U of the (x_1, x_2)-plane by its components $v_1(x_1, x_2)$ and $v_2(x_1, x_2)$. The coordinate system (x_1, x_2) defines an orientation and a Euclidean structure on the plane.

Remove the singular points of the field from the domain U and denote the domain that is left by U'. We define a mapping of the domain U' into the circle by the formula $f: U' \to S^1$, $f(x) = \dfrac{\boldsymbol{v}(x)}{|\boldsymbol{v}(x)|}$.

This mapping is smooth (since we have excluded the singular points of the field). Consider some point x of the domain U'. On the circle in a neighborhood of the image $f(x)$ of the point x we can introduce an angular coordinate φ. We then obtain a smooth real-valued function $\varphi(x_1, x_2)$ defined in a neighborhood of the point x.

Let us calculate its total differential. We have for $v_1 \neq 0$

$$d\varphi = d\arctan\frac{v_2}{v_1} = \frac{v_2\,dv_1 - v_1\,dv_1}{v_1^2 + v_2^2}. \tag{1}$$

The left-hand side equals the right-hand side for $v_1 = 0$ and $v_2 \neq 0$. Thus although the function φ itself is defined only locally and only up to a multiple of 2π, its differential is a well defined smooth differential form in the whole domain U'. We shall denote this form by $d\varphi$.

Definition. The *index of the oriented closed curve* $\gamma: S^1 \to U'$ is the integral of the form (1) over the curve γ, divided by 2π:

$$\operatorname{ind}\gamma = \frac{1}{2\pi}\oint_\gamma d\varphi. \tag{2}$$

We can now give precise proofs of the theorems given above. Let us prove, for example, the index sum theorem (cf. Sect. 5).

Proof. Let D be a domain with boundary S inside which the given field v has a finite number of singular points. Denote by D' the domain obtained from D by removing small disk-shaped neighborhoods of the singular points. Then the boundary of D', counting orientation, is $\partial D' = S - \sum S_i$, where S_i is a circle enclosing the ith singular point in the positive direction (Fig. 270). We apply Green's formula to the domain D' and the integral (2). We obtain

$$\iint_{D'} 0 = \oint_S d\varphi - \sum_i \oint_{S_i} d\varphi.$$

The left-hand side is zero, since the form (1) is locally an exact differential. In view of definition (2) we obtain $\operatorname{ind} S = \sum \operatorname{ind} S_i$, which was to be proved. □

Problem *1. Prove that the index of a closed curve is an integer.

Problem *2. Carry out a complete proof of the assertions of Sects. 1, 2, 3, and 4.

8. The Multidimensional Case

The multidimensional generalization of the concept of *winding number* is known as the *degree* of a mapping.

The degree of a mapping is the number of preimages of a point counted according to signs determined by orientations. For example, the degree of the mapping of an oriented circle onto an oriented circle depicted in Fig. 271 is 2, since the number of preimages of the point y, counting signs, is $1+1-1+1 = 2$.

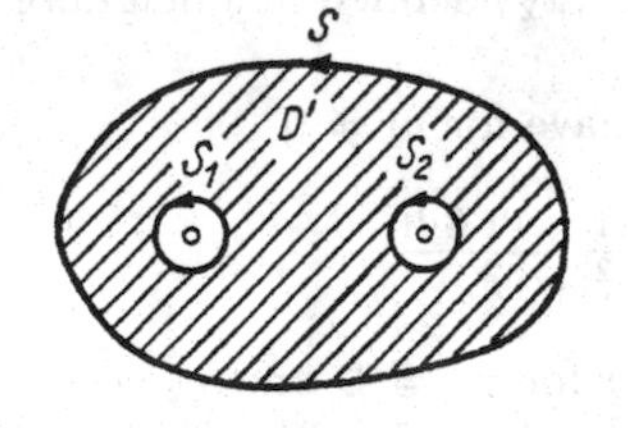

Fig. 270. A domain to which Green's formula applies

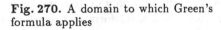

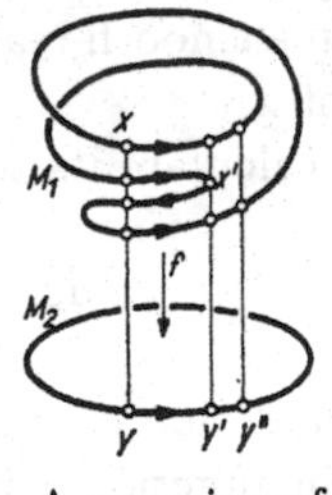

Fig. 271. A mapping of degree 2

To give a general definition we proceed as follows. Let $f : M_1^n \to M_2^n$ be a smooth mapping of one n-dimensional oriented manifold onto another. A point $x \in M_1^n$ of the domain manifold is called a *regular point* if the derivative of the mapping at the point x is a nondegenerate linear transformation $f_{*x} : T_x M_1^n \to T_{f(x)} M_2^n$.

For example, the point x in Fig. 271 is regular, but the point x' is not.

Definition. The *degree of the mapping f at the regular point x* is the number $\deg_x f$ equal to $+1$ or -1 according as f_{*x} takes the orientation of the space $T_x M_1^N$ into the orientation of the space $T_{f(x)} M_2^n$ or into the opposite orientation.

Problem 1. Prove that the degree of a linear automorphism $A : \mathbf{R}^n \to \mathbf{R}^n$ is the same at all points and equals $\deg_x A = \operatorname{sgn} \det A = (-1)^{m_-}$, where m_- is the number of eigenvalues of the operator A with negative real part.

Problem 2. Let $A : \mathbf{R}^n \to \mathbf{R}^n$ be a linear automorphism in a Euclidean space. Define a mapping of the unit sphere by the formula $f(x) = A(x)/|Ax|$. Find the degree of the mapping f at a point x.

Answer. $\deg_x f = \deg A$.

Problem 3. Let $f : S^{n-1} \to S^{n-1}$ be the mapping taking each point of the sphere into the point diametrically opposite it. What is the degree of this mapping?

Answer. $\deg_x f = (-1)^n$.

Problem 4. Let $A : \mathbf{C}^n \to \mathbf{C}^n$ be a $\mathbf{C}$-linear automorphism. Find the degree of its realification ${}^{\mathbf{R}}A$.

Answer. $+1$.

Now consider any point y of the target manifold M_2^n. A point $y \in M_2^n$ is called a *regular value of the mapping f* if all the points of its complete preimage $f^{-1}y$ are regular. For example, in Fig. 271 the point y is a regular value, but the point y' is not.

Now suppose in addition that our manifolds M_1^n and M_2^n are compact and connected. We then have the following theorem.

Theorem.

1. *There exist regular values.*
2. *The number of points in the preimage of a regular value is finite.*
3. *The sum of the degrees of the mapping at all points of the preimage of a regular value is independent of the particular regular value considered.*

The proof of this theorem is rather complicated and will not be given; it can be found in textbooks of topology[14].

Remark 1. In fact almost all the points of the manifold M_2^n are regular values: the nonregular values constitute a set of measure 0.

Remark 2. The condition of compactness is essential, not only for the second assertion of the theorem, but also for the third. (Consider, for example, the imbedding of the negative semiaxis into the number line.)

[14] Cf., for example, H. I. Levine, *Singularities of Differentiable Mappings*, Math. Inst. Univ. Bonn, 1955, Sec. 6.3.

Remark 3. The number of points of the preimage (not counting signs) may be different for different regular values (for example, in Fig. 271 the value y has four preimages and the value y'' only 2).

Definition. The sum of the degrees of the mapping f at all points of the preimage of a regular value is called the *degree of the mapping*:

$$\deg f = \sum_{x \in f^{-1}y} \deg_x f.$$

Problem 5. Find the degree of the mapping of the circle $|z| = 1$ onto itself defined by the formula $f(z) = z^n$, $n = 0, \pm 1, \pm 2, \ldots$.

Answer. n.

Problem 6. Find the degree of the mapping of the unit sphere in Euclidean space $\boldsymbol{R}^n$ onto itself given by the formula $f(z) = Az/|Az|$, where $A : \boldsymbol{R}^n \to \boldsymbol{R}^n$ is a nondegenerate linear operator.

Answer. $\deg f = \operatorname{sgn} \det A$.

Problem 7. Find the degree of the mapping of the complex projective line $\boldsymbol{CP}^1$ onto itself given by the formula a) $f(z) = z^n$, b) $f(z) = \bar{z}^n$.

Answer. a) n; b) $-n$.

Problem 8. Find the degree of the mapping of the complex line $\boldsymbol{CP}^1$ onto itself given by a polynomial of degree n.

Answer. n.

Problem *9. Prove that the index of the closed curve $\gamma : S^1 \to U'$ defined in Sect. 7 coincides with the degree of the following mapping h of a circle onto a circle.

Let $f : U' \to S^1$ be the mapping constructed in Sect. 7 using the vector field $\boldsymbol{v}$ in the domain U'. Set $h = f \circ \gamma : S^1 \to S^1$. Then

$$\operatorname{ind} \gamma = \deg h.$$

Definition. The *index of an isolated singular point* $\mathbf{o}$ of the vector field $\boldsymbol{v}$ defined in a domain of Euclidean space $\boldsymbol{R}^n$ containing $\mathbf{o}$, is the degree of the following mapping h of the sphere of small radius r with center at $\mathbf{o}$ onto itself corresponding to the field. The mapping $\boldsymbol{h} : S^{n-1} \to S^{n-1}$, $S^{n-1} = \{\boldsymbol{x} \in \boldsymbol{R}^n : |\boldsymbol{x}| = r\}$, is given by $\boldsymbol{h}(\boldsymbol{x}) = \dfrac{r\boldsymbol{v}(\boldsymbol{x})}{|\boldsymbol{v}(\boldsymbol{x})|}$.

Problem 10. Let the operator $\boldsymbol{v}_{*\mathbf{o}}$, the linear part of the field v at the point $\mathbf{o}$, is nondegenerate. Then the index of the singular point $\mathbf{o}$ equals the degree of this operator.

Problem 11. Find the index of the singular point 0 of the field in $\boldsymbol{R}^n$ corresponding to the equation $\dot{\boldsymbol{x}} = -\boldsymbol{x}$.

Answer. $(-1)^n$.

The concept of degree makes it possible to state multidimensional analogues of the two-dimensional theorems studied above. The proofs can be found in textbooks of topology.

In particular, *the sum of the indices of the singular points of a vector field on a compact manifold of any dimension is independent of the choice of the field and is determined by properties of the manifold itself.* This number is called the *Euler characteristic* of the manifold.

To compute the Euler characteristic of a manifold, it suffices to study the singular points of any differential equation defined on it.

Problem 12. Find the Euler characteristic of the sphere S^n, the projective space $\boldsymbol{RP}^n$, and the torus T^n.

Answer. $\chi(S^n) = 2\chi(\boldsymbol{RP}^n) = 1 + (-1)^n$, $\chi(T^n) = 0$.

Solution. On a torus of any dimension there is a differential equation with no singular points (cf., for example, § 24, Sect. 5), so that $\chi(T^n) = 0$.

It is clear that $\chi(S^n) = 2\chi(\boldsymbol{RP}^n)$. Indeed, consider the mapping $p : S^n \to \boldsymbol{RP}^n$ taking each point of the sphere S^n into the line joining it to the origin. The mapping p is a local diffeomorphism; here the preimage of each point of the projective space consists of two diametrically opposite points of the sphere. Consequently each vector field on $\boldsymbol{RP}^n$ defines a field on S^n with twice as many singular points, and the indices of any two diametrically opposite points on the sphere will be the same as the index of the point of projective space corresponding to them.

To compute $\chi(S^n)$ we define the sphere by the equation $x_0^2 + \cdots + x_n^2 = 1$ in the Euclidean space $\boldsymbol{R}^{n+1}$ and consider the function $x_0 : S^n \to \boldsymbol{R}$.

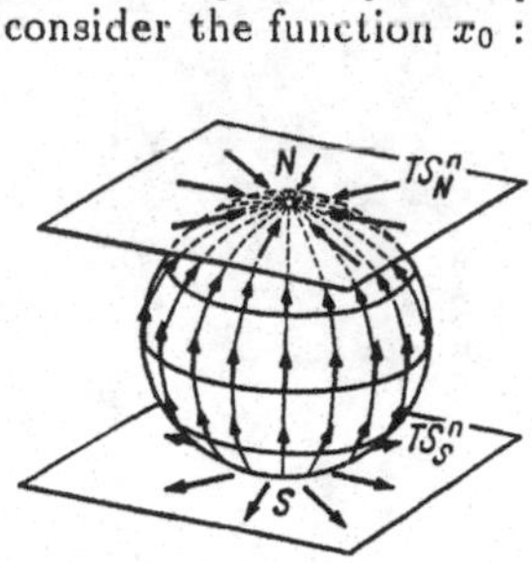

Fig. 272. Linearization of a differential equation on a sphere near its singular points

We form the differential equation on the sphere

$$\dot{\boldsymbol{x}} = \boldsymbol{grad}\, x_0$$

and study its singular points (Fig. 272). The vector field grad x_0 vanishes at two points: the north pole N ($x_0 = 1$) and the south pole S ($x_0 = -1$). Linearizing the differential equation in a neighborhood of the north and south poles respectively, we obtain the equations

$$\dot{\boldsymbol{\xi}} = -\boldsymbol{\xi}, \quad \boldsymbol{\xi} \in \boldsymbol{R}^n = T_N S^n; \quad \dot{\boldsymbol{\eta}} = \boldsymbol{\eta}, \quad \boldsymbol{\eta} \in \boldsymbol{R}^n = T_S S^n.$$

Consequently the index of the north pole is $(-1)^n$ and that of the south pole is $(+1)^n$, so that $\chi(S^n) = 1 + (-1)^n$.

In particular it follows from this that *every vector field on an even-dimensional sphere has at least one singular point.*

Problem 13. Construct a vector field on an odd-dimensional sphere S^{2n-1} having no singular points.

Hint. Consider the second-order differential equation $\ddot{\boldsymbol{x}} = -\boldsymbol{x}$, $\boldsymbol{x} \in \boldsymbol{R}^n$.

Examination Topics

1. The rectification theorem (§ 7, Sects. 1, 7) and its proof (§ 32, Sect. 5).

2. The existence, uniqueness, and differentiability theorems (§ 7, Sects. 2–5 and § 31, Sects. 1–8; § 32, Sects. 1–4). Contraction mappings (§ 30).

3. The extension theorem (§ 7, Sect. 6) and the theorem that a vector field on a compact manifold defines a phase flow (§ 35, Sects. 1–3).

4. The phase curves of an autonomous system. The theorem on closed phase curves (§ 9).

5. The derivative in the direction of a vector field and first integrals (§10, § 12).

6. The exponential of a linear operator. The exponential of a complex number and the exponential of a Jordan block (§ 14; § 15, Sects. 4, 5; § 25, Sect. 1).

7. Theorems on the connections between phase flows, linear equations, one-parameter transformation groups, and exponentials (§ 4, Sects. 2–4; § 13, Sects. 1–3; § 15, Sects. 1–3).

8. The connection between the determinant, the exponential, and the trace. Liouville's theorem on the Wronskian determinant (§ 16; § 18, Sect. 4; § 27, Sect. 6).

9. The classification of singular points of linear systems in the plane (§ 2, Sects. 4, 5; § 17, Sect. 2; § 19, Sect. 4; § 20, Sects. 3–5).

10. The solution of homogeneous linear autonomous systems in the complex and real domains in the case when the characteristic equation has only simple roots (§ 17, Sect. 1; § 18, Sect. 5; § 19; § 20).

11. The solution of homogeneous autonomous linear equations and systems in the case when the characteristic equation has multiple roots (§ 25).

12. The solution of inhomogeneous autonomous linear equations with the right-hand side in the form of a sum of quasi-polynomials (§ 26).

13. Homogeneous nonautonomous linear equations and systems. The Wronskian. The case of periodic coefficients (§ 27 and § 28, Sect. 1).

14. The solution of inhomogeneous linear equations using variation of constants (§ 29).

15. The theorem on stability in linear approximation (§ 22, Sects. 3–5; § 23).

16. The phase curves of a linear equation whose characteristic equation has purely imaginary roots. Small oscillations of conservative systems (§ 24 and § 25, Sect. 6).

Sample Examination Problems

1. [1] To stop river boats at a dock ropes are thrown from the boat and wound onto a post on the dock. How much force is needed to stop a boat if three coils of rope are wound around the post, the coefficient of friction of the rope against the post is 1/3 and the worker on the dock holds the free end of the rope with a force of 10 ks?

2. On the surface of a cylinder sketch the phase curves of a pendulum on which a constant torque acts:

$$\ddot{x} = 1 + 2\sin x.$$

What motions of the pendulum correspond to the different types of curves?

3. Compute the matrix e^{At}, where A is a given matrix of order 2 or 3.

4. Sketch the image of the square $|x_i| < 1$ under the phase flow transformation of the system

$$\dot{x}_1 = 2x_2, \quad \dot{x}_2 = x_1 + x_2.$$

over the time $t = 1$.

5. How many decimal digits are needed to write the hundredth term of the sequence $1, 1, 6, 12, 29, 59, \ldots$ ($x_n = x_{n-1} + 2x_{n-2} + n$, $x_1 = x_2 = 1$)?

6. Sketch the phase curve of the system

$$\dot{x} = x - y - z, \quad \dot{y} = x + y, \quad \dot{z} = 3x + z,$$

passing through the point $(1, 0, 0)$.

7. Find all α, β, and γ for which the three functions $\sin\alpha t$, $\sin\beta t$, and $\sin\gamma t$ are linearly independent.

8. On the (x_1, x_2)-plane sketch the trajectory of a point executing the small oscillations

$$\ddot{x}_i = -\partial U/\partial x_i, \quad U = (5x_1^2 - 8x_1x_2 + 5x_2^2)/2.$$

9. A horizontal force of 100 g acts for 1 second on a mathematical pendulum of length 1 m and weight 1 ks originally at rest. Find the amplitude (in cm) of the oscillations that will be established after the force ceases to act.

10. Investigate whether the null solution of the system

$$\begin{cases} \dot{x}_1 = x_2, \\ \dot{x}_2 = -\omega^2 x_1, \end{cases} \quad \omega(t) = \begin{cases} 0.4 & \text{for } 2k\pi \le t < (2k+1)\pi, \\ 0.6 & \text{for } (2k-1)\pi \le t < 2k\pi, \end{cases}$$

$$k = 0, \pm 1, \pm 2, \ldots$$

is Lyapunov stable.

---

[1] In all numerical problems an error of 10–20% in the answer is admissible.

11. Find all the singular points of the system

$$\dot{x} = xy + 12, \quad \dot{y} = x^2 + y^2 - 25.$$

Study their stability, determine the types of singular points, and sketch the phase curves.

12. Find all the singular points of the system on the torus $\{(x, y) \bmod 2\pi\}$:

$$\dot{x} = -\sin y, \quad \dot{y} = \sin x + \sin y.$$

Study their stability, determine the types of singular points and sketch the phase curves.

13. It is known from experiment that when light is refracted at the interface of two media, the sines of the angles formed with the normal to the interface by the incident ray and the refracted ray are inversely proportional to the indices of refraction of the media: $n_1 \sin \alpha_1 = n_2 \sin \alpha_2$.

Find the form of light rays in the (x, y)-plane with index of refraction $n(y)$. Consider the case $n(y) = 1/y$ (Lobachevsky geometry is realized in the half-plane $y > 0$ with this index of refraction).

14. Sketch the rays emanating from the origin in various directions in a plane with index of refraction $y^4 - y^2 + 1$.

The solution of this problem explains the phenomenon of the mirage. The index of refraction of the air over a desert has a maximum at a certain height, since the air is more rarefied in the layers above and in the hot layers below, and the index of refraction is inversely proportional to the velocity. The oscillation of a ray near the layer at which the index of refraction is maximal is perceived as a mirage.

Another phenomenon that is explained by the same oscillations of a ray is that of the acoustic channel in the ocean, along which sound carries for hundreds of kilometers. The cause of this phenomenon is the interplay of temperature and pressure, leading to the formation of a layer with maximal index of refraction (i.e., minimal speed of sound) at a depth of 500–1000 m. The acoustic channel can be used, for example, to warn of tidal waves.

15. Sketch the geodesics on the torus, using Clairaut's theorem: the product of the distance to the axis of rotation and the sine of the angle a geodesic makes with a meridian is constant along each geodesic of a surface of revolution.

16. Rectify the phase curves of the equation $\ddot{x} = x - x^2$ in a neighborhood of the point $x = 0$, $\dot{x} = 1$.

17. Rectify the integral curves of the equation $\dot{x} = x + \cos t$.

18. Rectify the direction field of the equation $\dot{x} = x + te^t$.

19. Rectify the phase velocity field of the equation $\dot{x} = x$ near the point $x = 1$.

20. In what coordinates are the variables separable in the equation

$$dy/dx = x^2 + y^{2/3}?$$

21. Solve the equation $\dot{x} = x + \delta(t - 2)$.

22. Find the derivative of the solution of the equation $\ddot{x} = \dot{x}^2 + Ax^3$ with initial condition $x(0) = 1$, $\dot{x}(0) = 0$, with respect to A at $A = 0$.

23. Find the eigenvalues and eigenvectors of the monodromy operator of the 2π-periodic solution of the equation $\ddot{x} - x = \sin t$.

24. Solve the equation $\dot{x} = Atx + x$, where $A : \boldsymbol{R}^n \to \boldsymbol{R}^n$ is a linear operator.

25. Can the operators A and B fail to commute if

$$e^A = e^B = e^{A+B} = E?$$

26. Find all the time-independent first integrals of the system $\dot{x} = y$, $\dot{y} = x + y$ that are continuous in the entire phase plane.

27. The numbers 1 and i are eigenvalues of $A : \boldsymbol{R}^3 \to \boldsymbol{R}^3$. Does the equation $\dot{x} = Ax$ have any nonconstant first integrals that are continuous in $\boldsymbol{R}^3$?

28. The numbers 1 and -1 are eigenvalues of $A : \boldsymbol{R}^3 \to \boldsymbol{R}^3$. Does the equation $\dot{x} = Ax$ have any nonconstant first integrals that are continuous in $\boldsymbol{R}^3$?

29. Solve the Cauchy problem

$$xu_x + u_y = 0, \quad u|_{y=0} = \sin x.$$

30. The equation $x^{(n)} = F(t, x, \ldots, x^{(n-1)})$ has solutions t and $\sin t$. Determine n.

31. Can the solutions of the equation $\dot{x} = x^3 \sin x$ be continued to the entire time axis?

32. Can all solutions of Newton's equation $\ddot{x} = -\operatorname{grad} U$, $U = x_1^4 + x_1 x_2 + x_2^6$ be extended indefinitely?

33. Can all the solutions of the equation

$$\ddot{x} = 1 + 2 \sin x$$

be extended indefinitely?

34. Can the equilibrium position of Newton's equation be Lyapunov stable without being a local minimum of the potential energy?

35. Can a periodic solution of an autonomous system represented by a limit cycle in the phase plane be asymptotically stable?

36. Can a periodic solution of an autonomous system be Lyapunov unstable if it is represented in the phase plane by a limit cycle onto which phase curves approximating the cycle wind spirally from the interior and from the exterior under a motion in the direction of increasing time?

37. Can a Lyapunov-unstable equilibrium position become stable after linearization? Can it become asymptotically stable?

38. Can an asymptotically stable equilibrium position become Lyapunov-unstable after linearization?

Supplementary Problems

1. Does the equation of variations for the equation $\ddot{x} = -\sin x$ along the solution with initial condition $x_0 = 0$, $\dot{x}_0 = 2$ have nonzero solutions that are bounded on the entire time axis?

2. Does the equation of variations of this same equation along the solution with initial condition $x_0 = 0$, $\dot{x}_0 = 1$ have unbounded solutions?

3. Solve the equation of variations in Prob. 1.

4. FInd the eigenvalues and eigenvectors of the monodromy operator for the equation in variations of Prob. 2.

5. Find the derivative with respect to ε at $\varepsilon = 0$ of the 2π-periodic solution of the equation $\ddot{x} + \sin x = \varepsilon \cos t$ that becomes $x \equiv \pi$ when $\varepsilon = 0$.

6. Find the largest value of t for which the solution of the Cauchy problem

$$u_t + uu_x = -\sin x, \quad u|_{t=0} = 0$$

can be continued to $[0, t)$.

7. Find all the finite-dimensional subspaces of the space of infinitely-differentiable functions on the line that are invariant with respect to all translations of the line.

8. Suppose the function v has a double root at zero. Prove that the equation $\dot{x} = v(x)$ can be reduced to $\dot{y} = y^2 + Cy^3$ by a diffeomorphism of a neighborhood of zero (the constant C is determined by the field).

9. Prove that the zeros of a linear combination of the first n eigenfunctions of the Sturm-Liouville problem

$$u_{xx} + q(x)u = \lambda u, \quad u(0) = u(l) = 0, \quad q > 0$$

divide the interval $[0, l]$ into at most n parts.

Hint. (I. M. Gel'fand). Convert to fermions, i.e., to skew-symmetric solutions of the equation $\sum u_{x_i x_j} + \sum q(x_i)u = \lambda u$ and use the fact that the first eigenfunction of this equation has no zeros inside the fundamental simplex $0 < x_1 < \cdots < x_n < l$.

10. (N. N. Bautin). Prove that the generalized Lotka-Volterra system

$$\dot{x} = x(a + kx + ly), \quad \dot{y} = y(b + mx + ny)$$

has no limit cycles: its closed non-point phase curves, when they exist, fill up an annulus.

11. Consider the motion of matter along a circle under the translating action of a velocity field and a small diffusion. Prove that if the velocity field has stationary points and is in general position, then almost all the mass eventually accumulates in a neighborhood of one of the points of attraction.

[The evolution equation for the density is: $\dot{u} = \varepsilon u_{xx} - (uv)_x$, where $v\partial/\partial x$ is the velocity field. On the covering line of the circle the field is potential: $v = -U_x$. If the velocity field is potential, then the stationary solution is given by a Gibbs distribution

$$u(x) = Ce^{-U(x)/\varepsilon}.$$

For small ε this distribution is concentrated near the minimum of the potential.

If the function U tends to $-\infty$ at $-\infty$, the stationary solution has the form

$$u(x) = C \int_{-\infty}^{x} e^{[U(\xi) - U(x)]/\varepsilon}\, d\xi.$$

It is concentrated near the local minimum of the potential for which the excess of the maximum value of the potential over this minimum value is maximal on the axis to the left of this minimum.]

12. (A. A. Davydov). An *involution* is a diffeomorphism whose square is the identity mapping. An involution of the plane is called *admissible* with respect to a vector field if the fixed points of the involution form a curve and under the action of the involution the vectors of the field change sign at the points of this curve.

Prove that in a neighborhood of a nonsingular point of the field all the admissible involutions in general position are equivalent (can be mapped into one another by diffeomorphisms that preserve each phase curve of the field).

[The solution of this problem provides a normal form $p^2 = x$ of an equation that is not solved for the derivative in a neighborhood of a nonregular point in general position (it was found by H. M. Cibrario in 1932). The equation $F(x, y, p) = 0$ defines a surface in three-dimensional space. At a nonregular point its tangent plane is vertical (tangent to the p-axis). In a neighborhood of a nonregular point in general position there arises an involution (it transposes the nearby points of the intersection of the surface with a vertical line). This involution is admissible with respect to the vector field tangent to the integral curves of a differential equation on the surface. Bringing the involution to normal form is equivalent to normalizing the equation by a local diffeomorphism of the (x, y)-plane.]

13. (continuation). Suppose the fixed curve of an involution that is admissible with respect to a vector field with a singular point of focus, saddle, or node type passes through a singular point, and the absolute values of the eigenvalues of the saddle or node are distinct.

Prove that any two such involutions are equivalent in a neighborhood of a singular point if the tangents to their fixed curves at the singular point are not separated by characteristic directions.

[This theorem of Davydov provides the normal form $(p-kx)^2 = y$ for an equation in general position that is not solved with respect to the derivative in a neighborhood of a nonregular point at which the plane $dy = p\,dx$ is tangent to the surface $F = 0$; k is the unique modulus (invariant under diffeomorphisms) of the "folded focus, saddle, or node" formed by the projections of integral curves on the (x, y)-plane.]

The solutions of Probs. 12 and 13 also provide normal forms of families of trajectories of slow motion in the theory of relaxation oscillations in general position with two slow variables. In this theory in three-dimensional space fibered into vertical lines over the "plane of the slow variables" two vector fields are defined: one (the "fast" one) is vertical and the other (the "perturbing" one) is arbitrary. The zeros of the fast field form the "slow surface." the planes spanned by the directions of the two fields leave tracks on the tangent plane to the slow surface which form the direction field of the "slow motion" on the slow surface. We are discussing the family of projections of integral curves of this field from the slow surface to the plane of the slow variables.

The critical values of the projection of the slow surface on the plane of the slow variables form (in a generic system) a discriminant curve with isolated cusps. In a neighborhood of a general point of this curve the family of projections is diffeomorphic to a family of semicubical parabolas $(y - c)^2 = x^3$ (this follows from the normal form of Prob. 12). At isolated points of smoothness of the discriminant curve the family of projections is diffeomorphic to a folded focus, saddle, or node (Prob. 13). In addition, in a generic system, there are isolated points of smoothness of the discriminant curve in a neighborhood of which the family can be described as follows. Label the integral curves by a parameter c and regard the family of their projections on the (x, y)-plane as a surface in three-dimensional space with coordinates (x, y, c), partitioned into the lines $c =$ const. This surface is (locally) diffeomorphic to the "folded umbrella" surface $u^2 = v^3 w^2$, partitioned into the lines $u + v + w =$ const. Finally, in a neighborhood of a cusp of the discriminant curve the family of projections is described similarly using a partition of the "swallow's tail" surface $\{u, v, w : \lambda^4 + u\lambda^2 + v\lambda + w$ has a multiple root$\}$ into the curves $u =$ const. The last family of projections, in contrast to those described above, has an infinite number of moduli, even with respect to the homeomorphisms of the (x, y)-plane (in the case of the folded umbrella there are no moduli up to infinitely-differentiable diffeomorphisms, but in the analytic case infinitely many independent moduli arise).

The solutions of Probs. 12 and 13 also describe the singularities of families of asymptotic lines on a surface in three-dimensional space (a family of semicubical parabolas at the general point of a parabolic line and a folded focus, node, or saddle at isolated points of tangency of an asymptotic direction to a parabolic line).

Written Examination Problems (Moscow State University, 1991)

(Two hours for each variant of four questions)

Variant 1.

$$\begin{cases} \dot{x} = y \\ \dot{y} = -x^2 \end{cases} \text{or} \begin{cases} \dot{x} = y^3 \\ \dot{y} = -x \end{cases} \text{or} \begin{cases} \dot{x} = y \\ \dot{y} = x^4 \end{cases}$$

$$\text{or} \begin{cases} \dot{x} = -y^2 \\ \dot{y} = x^2 \end{cases} \text{or} \begin{cases} \dot{x} = y^4 \\ \dot{y} = -x \end{cases} \text{or} \begin{cases} \dot{x} = y^2 \\ \dot{y} = x^2 \end{cases}.$$

a) Find all the equilibrium points and study their stability.
b) Are all the solutions indefinitely continuable?
c) Find the number of nonzero solutions satisfying $y(0) = x(1) = 0$.
d) Calculate the derivative of the solution with initial conditions $x(0) = y(0) = \varepsilon$ with respect to ε at $\varepsilon = 0$.

Variant 2.

$$\begin{cases} \dot{x} = x^2 y \\ \dot{y} = -xy^2 \end{cases} \text{or} \begin{cases} \dot{x} = -x^2 y^2 \\ \dot{y} = -xy^3 \end{cases} \text{or} \begin{cases} \dot{x} = -x^2 y^3 \\ \dot{y} = xy^4 \end{cases}$$

$$\text{or} \begin{cases} \dot{x} = -xy^2 \\ \dot{y} = x^2 y \end{cases} \text{or} \begin{cases} \dot{x} = xy^3 \\ \dot{y} = -x^2 y^2 \end{cases} \text{or} \begin{cases} \dot{x} = -xy^4 \\ \dot{y} = x^2 y^3 \end{cases}.$$

a) Find all the equilibrium points and study their stability.
b) Are all the solutions indefinitely continuable?
c) Find a diffeomorphism rectifying the direction field of phase curves in a neighborhood of $(1,1)$.
d) Find all the first integrals that are continuous on the whole phase plane and coincide with y for $x = 0$.

Variant 3.

$$\dot{z} = iz^2 \text{ or } \dot{z} = \bar{z}^2 \text{ or } \dot{z} = iz^2\bar{z} \text{ or } \dot{z} = z\bar{z}^2 \text{ or } \dot{z} = iz\bar{z}^2 \text{ or } \dot{z} = i\bar{z}^2.$$

a) Find all equilibrium points and study their stability.
b) Find all the initial conditions for which the solution can be continued forward indefinitely.
c) Find the image of the vector $(1,0)$ at the phase point 0 under the time 1 mapping of the phase flow.
d) Find all the first integrals that are continuous in a neighborhood of the point $z = 1$ and coincide with 1 along the real axis.

Variant 4.

$$\left.\begin{aligned} x\frac{\partial u}{\partial x} - (1 + x^4 + y^2)\frac{\partial u}{\partial y} = 2u \\ u\big|_{x=0} = 0 \end{aligned}\right\} \quad \text{or} \quad \left.\begin{aligned} (2 + x^2 + y^2)\frac{\partial u}{\partial x} + y\frac{\partial u}{\partial y} = u \\ u\big|_{y=0} = 0 \end{aligned}\right\}$$

$$\text{or} \quad \left.\begin{aligned} x\frac{\partial u}{\partial x} + (1 + 2u^4 + y^2)\frac{\partial u}{\partial y} &= 3u \\ u\big|_{x=0} &= 0 \end{aligned}\right\}.$$

a) Does there exist an unbounded solution defined on the whole plane?
b) Is the value of u bounded along the characteristics?
c) Does each characteristic intersect the surface $y = x + u^2$?
d) Does there exist a first integral of the equation of characteristics whose derivative with respect to y at the origin is equal to 1? Find the derivative of this derivative along the characteristic vector.

Variant 5.

$$\dot{x} = x^2 - \sin^2 t \text{ or } \dot{x} = x^2 - \cos^2 t \text{ or } \dot{x} = \sinh^2 x - \cos^2 t \text{ or } \dot{x} = \sin^2 t - \sinh^2 x.$$

a) Find the third derivative at the origin of the solution satisfying the initial condition $x(0) = 0$.
b) Is this solution extendable over the whole t-axis?
c) Do there exist unbounded solutions of the equation?
d) Find the number of asymptotically stable periodic solutions of the equation.

Subject Index

Universitext

Aguilar, M.; Gitler, S.; Prieto, C.: Algebraic Topology from a Homotopical Viewpoint

Aksoy, A.; Khamsi, M. A.: Methods in Fixed Point Theory

Alevras, D.; Padberg M. W.: Linear Optimization and Extensions

Andersson, M.: Topics in Complex Analysis

Aoki, M.: State Space Modeling of Time Series

Arnold, V. I.: Lectures on Partial Differential Equations

Arnold, V. I.: Ordinary Differential Equations

Audin, M.: Geometry

Aupetit, B.: A Primer on Spectral Theory

Bachem, A.; Kern, W.: Linear Programming Duality

Bachmann, G.; Narici, L.; Beckenstein, E.: Fourier and Wavelet Analysis

Badescu, L.: Algebraic Surfaces

Balakrishnan, R.; Ranganathan, K.: A Textbook of Graph Theory

Balser, W.: Formal Power Series and Linear Systems of Meromorphic Ordinary Differential Equations

Bapat, R.B.: Linear Algebra and Linear Models

Benedetti, R.; Petronio, C.: Lectures on Hyperbolic Geometry

Benth, F. E.: Option Theory with Stochastic Analysis

Berberian, S. K.: Fundamentals of Real Analysis

Berger, M.: Geometry I, and II

Bliedtner, J.; Hansen, W.: Potential Theory

Blowey, J. F.; Coleman, J. P.; Craig, A. W. (Eds.): Theory and Numerics of Differential Equations

Blowey, J.; Craig, A. (Eds.): Frontiers in Numerical Analysis. Durham 2004

Blyth, T. S.: Lattices and Ordered Algebraic Structures

Börger, E.; Grädel, E.; Gurevich, Y.: The Classical Decision Problem

Böttcher, A; Silbermann, B.: Introduction to Large Truncated Toeplitz Matrices

Boltyanski, V.; Martini, H.; Soltan, P. S.: Excursions into Combinatorial Geometry

Boltyanskii, V. G.; Efremovich, V. A.: Intuitive Combinatorial Topology

Bonnans, J. F.; Gilbert, J. C.; Lemaréchal, C.; Sagastizábal, C. A.: Numerical Optimization

Booss, B.; Bleecker, D. D.: Topology and Analysis

Borkar, V. S.: Probability Theory

Brunt B. van: The Calculus of Variations

Bühlmann, H.; Gisler, A.: A Course in Credibility Theory and Its Applications

Carleson, L.; Gamelin, T. W.: Complex Dynamics

Cecil, T. E.: Lie Sphere Geometry: With Applications of Submanifolds

Chae, S. B.: Lebesgue Integration

Chandrasekharan, K.: Classical Fourier Transform

Charlap, L. S.: Bieberbach Groups and Flat Manifolds

Chern, S.: Complex Manifolds without Potential Theory

Chorin, A. J.; Marsden, J. E.: Mathematical Introduction to Fluid Mechanics

Cohn, H.: A Classical Invitation to Algebraic Numbers and Class Fields

Curtis, M. L.: Abstract Linear Algebra

Curtis, M. L.: Matrix Groups

Cyganowski, S.; Kloeden, P.; Ombach, J.: From Elementary Probability to Stochastic Differential Equations with MAPLE

Dalen, D. van: Logic and Structure

Da Prato, G.: An Introduction to Infinite-Dimensional Analysis

Das, A.: The Special Theory of Relativity: A Mathematical Exposition

Debarre, O.: Higher-Dimensional Algebraic Geometry

Deitmar, A.: A First Course in Harmonic Analysis

Demazure, M.: Bifurcations and Catastrophes

Devlin, K. J.: Fundamentals of Contemporary Set Theory

DiBenedetto, E.: Degenerate Parabolic Equations

Diener, F.; Diener, M.(Eds.): Nonstandard Analysis in Practice

Dimca, A.: Sheaves in Topology

Dimca, A.: Singularities and Topology of Hypersurfaces

DoCarmo, M. P.: Differential Forms and Applications

Duistermaat, J. J.; Kolk, J. A. C.: Lie Groups

Dumortier.: Qualitative Theory of Planar Differential Systems

Edwards, R. E.: A Formal Background to Higher Mathematics Ia, and Ib

Edwards, R. E.: A Formal Background to Higher Mathematics IIa, and IIb

Emery, M.: Stochastic Calculus in Manifolds

Emmanouil, I.: Idempotent Matrices over Complex Group Algebras

Endler, O.: Valuation Theory

Erez, B.: Galois Modules in Arithmetic

Everest, G.; Ward, T.: Heights of Polynomials and Entropy in Algebraic Dynamics

Farenick, D. R.: Algebras of Linear Transformations

Foulds, L. R.: Graph Theory Applications

Franke, J.; Härdle, W.; Hafner, C. M.: Statistics of Financial Markets: An Introduction

Frauenthal, J. C.: Mathematical Modeling in Epidemiology

Freitag, E.; Busam, R.: Complex Analysis

Friedman, R.: Algebraic Surfaces and Holomorphic Vector Bundles

Fuks, D. B.; Rokhlin, V. A.: Beginner's Course in Topology

Fuhrmann, P. A.: A Polynomial Approach to Linear Algebra

Gallot, S.; Hulin, D.; Lafontaine, J.: Riemannian Geometry

Gardiner, C. F.: A First Course in Group Theory

Gårding, L.; Tambour, T.: Algebra for Computer Science

Godbillon, C.: Dynamical Systems on Surfaces

Godement, R.: Analysis I, and II

Goldblatt, R.: Orthogonality and Spacetime Geometry

Gouvêa, F. Q.: p-Adic Numbers

Gross, M. et al.: Calabi-Yau Manifolds and Related Geometries

Gustafson, K. E.; Rao, D. K. M.: Numerical Range. The Field of Values of Linear Operators and Matrices

Gustafson, S. J.; Sigal, I. M.: Mathematical Concepts of Quantum Mechanics

Hahn, A. J.: Quadratic Algebras, Clifford Algebras, and Arithmetic Witt Groups

Hájek, P.; Havránek, T.: Mechanizing Hypothesis Formation

Heinonen, J.: Lectures on Analysis on Metric Spaces

Hlawka, E.; Schoißengeier, J.; Taschner, R.: Geometric and Analytic Number Theory

Holmgren, R. A.: A First Course in Discrete Dynamical Systems

Howe, R., Tan, E. Ch.: Non-Abelian Harmonic Analysis

Howes, N. R.: Modern Analysis and Topology

Hsieh, P.-F.; Sibuya, Y. (Eds.): Basic Theory of Ordinary Differential Equations

Humi, M., Miller, W.: Second Course in Ordinary Differential Equations for Scientists and Engineers

Hurwitz, A.; Kritikos, N.: Lectures on Number Theory

Huybrechts, D.: Complex Geometry: An Introduction

Isaev, A.: Introduction to Mathematical Methods in Bioinformatics

Istas, J.: Mathematical Modeling for the Life Sciences

Iversen, B.: Cohomology of Sheaves

Jacod, J.; Protter, P.: Probability Essentials

Jennings, G. A.: Modern Geometry with Applications

Jones, A.; Morris, S. A.; Pearson, K. R.: Abstract Algebra and Famous Inpossibilities

Jost, J.: Compact Riemann Surfaces

Jost, J.: Dynamical Systems. Examples of Complex Behaviour

Jost, J.: Postmodern Analysis

Jost, J.: Riemannian Geometry and Geometric Analysis

Kac, V.; Cheung, P.: Quantum Calculus

Kannan, R.; Krueger, C. K.: Advanced Analysis on the Real Line

Kelly, P.; Matthews, G.: The Non-Euclidean Hyperbolic Plane

Kempf, G.: Complex Abelian Varieties and Theta Functions

Kitchens, B. P.: Symbolic Dynamics

Kloeden, P.; Ombach, J.; Cyganowski, S.: From Elementary Probability to Stochastic Differential Equations with MAPLE

Kloeden, P. E.; Platen; E.; Schurz, H.: Numerical Solution of SDE Through Computer Experiments

Kostrikin, A. I.: Introduction to Algebra

Krasnoselskii, M. A.; Pokrovskii, A. V.: Systems with Hysteresis

Kurzweil, H.; Stellmacher, B.: The Theory of Finite Groups. An Introduction

Kyprianou, A. : Introductory Lectures on Fluctuations of Lévy Processes with Applications

Lang, S.: Introduction to Differentiable Manifolds

Luecking, D. H., Rubel, L. A.: Complex Analysis. A Functional Analysis Approach

Ma, Zhi-Ming; Roeckner, M.: Introduction to the Theory of (non-symmetric) Dirichlet Forms

Mac Lane, S.; Moerdijk, I.: Sheaves in Geometry and Logic

Marcus, D. A.: Number Fields

Martinez, A.: An Introduction to Semiclassical and Microlocal Analysis

Matoušek, J.: Using the Borsuk-Ulam Theorem

Matsuki, K.: Introduction to the Mori Program

Mazzola, G.; Milmeister G.; Weissman J.: Comprehensive Mathematics for Computer Scientists 1

Mazzola, G.; Milmeister G.; Weissman J.: Comprehensive Mathematics for Computer Scientists 2

Mc Carthy, P. J.: Introduction to Arithmetical Functions

McCrimmon, K.: A Taste of Jordan Algebras

Meyer, R. M.: Essential Mathematics for Applied Field

Meyer-Nieberg, P.: Banach Lattices

Mikosch, T.: Non-Life Insurance Mathematics

Mines, R.; Richman, F.; Ruitenburg, W.: A Course in Constructive Algebra

Moise, E. E.: Introductory Problem Courses in Analysis and Topology

Montesinos-Amilibia, J. M.: Classical Tessellations and Three Manifolds

Morris, P.: Introduction to Game Theory

Nikulin, V. V.; Shafarevich, I. R.: Geometries and Groups

Oden, J. J.; Reddy, J. N.: Variational Methods in Theoretical Mechanics

Øksendal, B.: Stochastic Differential Equations

Øksendal, B.; Sulem, A.: Applied Stochastic Control of Jump Diffusions

Poizat, B.: A Course in Model Theory

Polster, B.: A Geometrical Picture Book

Porter, J. R.; Woods, R. G.: Extensions and Absolutes of Hausdorff Spaces

Radjavi, H.; Rosenthal, P.: Simultaneous Triangularization

Ramsay, A.; Richtmeyer, R. D.: Introduction to Hyperbolic Geometry

Rautenberg, W.; A Concise Introduction to Mathematical Logic

Rees, E. G.: Notes on Geometry

Reisel, R. B.: Elementary Theory of Metric Spaces

Rey, W. J. J.: Introduction to Robust and Quasi-Robust Statistical Methods

Ribenboim, P.: Classical Theory of Algebraic Numbers

Rickart, C. E.: Natural Function Algebras

Roger G.: Analysis II

Rotman, J. J.: Galois Theory

Rubel, L. A.: Entire and Meromorphic Functions

Ruiz-Tolosa, J. R.; Castillo E.: From Vectors to Tensors

Runde, V.: A Taste of Topology

Rybakowski, K. P.: The Homotopy Index and Partial Differential Equations

Sagan, H.: Space-Filling Curves

Samelson, H.: Notes on Lie Algebras

Schiff, J. L.: Normal Families

Sengupta, J. K.: Optimal Decisions under Uncertainty

Séroul, R.: Programming for Mathematicians

Seydel, R.: Tools for Computational Finance

Shafarevich, I. R.: Discourses on Algebra

Shapiro, J. H.: Composition Operators and Classical Function Theory

Simonnet, M.: Measures and Probabilities

Smith, K. E.; Kahanpää, L.; Kekäläinen, P.; Traves, W.: An Invitation to Algebraic Geometry

Smith, K. T.: Power Series from a Computational Point of View

Smoryński, C.: Logical Number Theory I. An Introduction

Stichtenoth, H.: Algebraic Function Fields and Codes

Stillwell, J.: Geometry of Surfaces

Stroock, D. W.: An Introduction to the Theory of Large Deviations

Sunder, V. S.: An Invitation to von Neumann Algebras

Tamme, G.: Introduction to Étale Cohomology

Tondeur, P.: Foliations on Riemannian Manifolds

Toth, G.: Finite Möbius Groups, Minimal Immersions of Spheres, and Moduli

Verhulst, F.: Nonlinear Differential Equations and Dynamical Systems

Weintraub, S.H.: Galois Theory

Wong, M. W.: Weyl Transforms

Xambó-Descamps, S.: Block Error-Correcting Codes

Zaanen, A.C.: Continuity, Integration and Fourier Theory

Zhang, F.: Matrix Theory

Zong, C.: Sphere Packings

Zong, C.: Strange Phenomena in Convex and Discrete Geometry

Zorich, V. A.: Mathematical Analysis I

Zorich, V. A.: Mathematical Analysis II